A series of student texts in

CONTEMPORARY BIOLOGY

General Editors:

Professor E. J. W. Barrington, F.R.S.
Professor Arthur J. Willis

Plant Anatomy

Part I

CELLS AND TISSUES

Second Edition

Elizabeth G. Cutter
B.Sc., Ph.D., D.Sc., F.R.S.E., F.L.S.

Reader in Cryptogamic Botany,
University of Manchester

ADDISON-WESLEY PUBLISHING COMPANY

READING, MASSACHUSETTS · MENLO PARK, CALIFORNIA
LONDON · AMSTERDAM · DON MILLS, ONTARIO · SYDNEY

First published 1969
by Edward Arnold (Publishers) Ltd.,

Reprinted 1970
Reprinted 1973
Second edition 1978

This edition first published in
the United States of America in 1978
by Addison-Wesley Publishing Company

ISBN 0 201 01236 7

Printed in Great Britain by
William Clowes & Sons, Limited, London, Beccles and Colchester

Preface to Second Edition

One of the aims of this book is to emphasize the central position of plant anatomy in contemporary botanical science, both by pointing out the close relationships between structure and the dynamic processes of growth, and by indicating the many practical applications of plant anatomy. In recent years, the classical descriptive study of plant anatomy has been greatly enhanced by the development of the transmission and scanning electron microscopes, each of which in its own way has opened up new structural vistas to the student of anatomy. New techniques of this kind have increased our knowledge of cell structure in numerous ways, many of which are described and illustrated in this book. It is important, however, to consider the cell as part of a tissue and, indeed, as part of the whole plant; we should not lose sight of the important inter-relationships and interactions between the various components of the whole plant. In this book organization is considered at the tissue level; its companion volume, Part 2, considers the arrangements of these tissues that characterize the various organs of the plant, and the meristems by which they are formed.

Throughout this book an attempt is made to integrate the results of the more recent experimental approach to plant anatomy with the findings of comparative anatomists. In their quest for a scientific explanation of the observed developments, experimentalists have succeeded, in particular, in implicating natural plant growth substances in the control of the differentiation of various types of cells. Thus there is a clear and important relationship between the growth and metabolism, including the production of growth substances, of a particular organ and the differentiation of its component tissues. These tissues, too, show basic relationships between structure and function.

Contrary to the belief of many students, we have not yet discovered all there is to know about plant anatomy, indeed quite the reverse. That it is a rapidly advancing subject is attested by the need for considerable amendment of and addition to the first edition of this book, published only some eight years ago. This, the second edition, has an augmented text and additional illustrations, and includes a completely new chapter. Yet the need for still further work in the field of plant anatomy is evident, and is frequently pointed out in the chapters which follow. In particular, many fundamental problems, such as the control of differentiation, and the central mystery of organization, still remain to be solved.

I am grateful to many authors, including Drs. R. Aloni, C. A. Beasley, D. W. Bierhorst, A. J. Browning, Prof. D. J. Carr, Drs. J. Cronshaw, R. E. Dengler, K. Esau, D. A. Fisher, Prof. B. E. S. Gunning, Drs. Y. Heslop-Harrison, W. A. Jensen, M. G. K. Jones, N. J. Lang, J. N. A. Lott, K. Mühlethaler, R. L. Peterson, D. J. Royle, W. A. Sakai, Prof. E. Schnepf, Drs. A. M. Siebers, M. Y. Stant, K. D. Stewart, B. W. Thair, M. Tran Thanh Van, W. S. Walker and R. H. Wetmore, for supplying original or previously published illustrations, and to these and several others, and their publishers, for permission to reproduce illustrations from their publications. These sources are acknowledged in the figure captions. I am also indebted to Miss Sonia Cook, of the University of California, Davis, for her skill in making several of the sections from which original illustrations were prepared, and to Mr. G. Grange for photographic assistance. Drs. L. J. Feldman and R. L. Peterson have provided constructive criticism of some chapters, and their help is duly acknowledged; but the sole responsibility for all errors and omissions is mine.

I am particularly grateful to Professor A. J. Willis, general editor of this series, for his wise and thoughtful handling of editorial matters, and for his continued patience and that of my publishers.

Manchester, E.G.C.
1977.

Acknowledgements

I am indebted to the following publishers and scientific journals for permission to reproduce the figures indicated:

Academic Press, Inc. (Figs. 2.2, 3.8, 4.4, 4.5, 4.6, 4.11, 4.15, 7.8, 7.12, 7.23, 8.13, 9.2, 9.5, Table 9.1); *Acta botanica Neerlandica* (Figs. 12.2, 12.3); Agricultural Publications, University of California (Fig. 9.4); American Association for the Advancement of Science (Fig. 11.7); *American Journal of Botany* (Figs. 2.1, 3.12, 5.1, 5.5, 6.1g, 6.2, 6.6, 6.7, 6.10, 6.11, 7.3a, 7.6, 7.7, 7.16b and d, 7.17, 7.28, 7.29, 8.4, 8.8, 8.11, 8.16, 11.1, 11.18); Annual Reviews Inc. (Figs. 3.3, 10.6); Arnold Arboretum, Harvard University (Figs. 6.1b, 8.14, 8.15); Edward Arnold (Publishers) Ltd. (Figs. 10.1, 11.5); Blackwell Scientific Publications (Fig. 3.9); Cambridge University Press (Figs. 3.7, 7.3b); Chapman and Hall, Ltd. (Fig. 4.1); Commonwealth Scientific and Industrial Research Organization, Australia (Figs. 11.12, 11.15); Faculty Press (Fig. 10.4); Her Majesty's Stationery Office (Figs. 7.10, 7.18); Director, Royal Botanic Gardens, Kew (Figs. 7.10, 7.18); Linnean Society of London (Fig. 11.4); Longmans, Green and Co. Ltd. (Fig. 4.9); McGraw-Hill Book Company, Ltd. (Figs. 4.7, 10.3, 10.6); C. V. Mosby Co. (Figs. 7.11a and 11.2); National Research Council of Canada (Figs. 7.1, 7.19, 10.10, 10.11, 10.12, 10.13, 12.4); *New Phytologist* (Fig. 12.8); North Holland Publishing Company (Fig. 11.8); Oxford University Press (Fig. 3.4); *Plant Physiology* (Fig. 8.12); *Phytomorphology* (Figs. 7.25, 7.27); Rockefeller University Press (Figs. 9.3, 9.6, 9.7, 9.8); Ronald Press (Fig. 4.8); Scientific American (Fig. 9.10); Springer-Verlag (Figs. 2.3, 4.12, 7.30, 9.9, 11.9, 11.10, 11.13); Syracuse University Press (Fig. 4.14); University of Chicago Press (Figs. 7.26, 7.30, 8.9).

Table of Contents

1

Introduction: Plant Anatomy and the Growing Plant

The majority of vascular plants consist of a number of different organs—usually root, stem, leaf and flower—and each of these in turn is made up of a number of different tissues. This complex structure, the whole growing plant, is derived during development from a single cell, the fertilized egg or zygote. Many complex processes of differentiation (see Chapter 2) take place during the ontogeny of the plant.

The study of plant anatomy

The study of the internal structure of these various parts of the plant is known as plant anatomy, and began a little over 300 years ago with the work of Grew and Malpighi.[88] Their work involved careful, well illustrated descriptions of plant material and, as Carlquist[88] has pointed out, Grew advocated even then that plant anatomy should incorporate comparative, developmental and seasonal studies of the whole plant. This ideal is all too seldom followed today.

Regrettably, it is commonly believed that the structure of plants is already well understood and that there is little need for further investigation. In the ensuing pages some attempt will be made to dispel this view. It has recently been pointed out that the anatomy of relatively few species is known in any detail,[88] that there is a need for greater international communication and co-operation in this field,[450] and that in *any* intensive investigation of plants one encounters aspects of their anatomy.[693] This in itself should provide an adequate reason for a sound basic knowledge of plant structure. However, comparative plant anatomy also has a number of applications outside the field of orthodox plant science, for example in pharmacognosy, forensic medicine, and archaeology or palaeoethnobotany. An interesting example is the identification of some 60 plant species from the stomach contents of Tollund Man and Grauballe Man, belonging to the Danish Ice Age.[312] The location of these bodies in peat bogs resulted in sufficiently good preservation of 1700-year-old tissues to enable identification to be made from fragments of seed coat, pericarp and even epidermal hairs. Such work clearly requires an extensive mastery of plant anatomy. Knowledge of plant structure was used also in solving the Lindbergh kidnapping case; other instances of its uses are given in later chapters.

While an accurate description of the various tissues and organs of the plant is undoubtedly essential, in order to gain a deeper understanding of the reasons underlying the observed structure it is often necessary to carry out experiments and interpret the results. The experimental approach to anatomy is, for the most part, relatively recent but its application is increasing. An attempt is made here to emphasize the relationships between structure and function, and also to integrate the results of experiments, where available, with the results already obtained by careful observation with both the light and electron microscopes. Where appropriate, new approaches to various problems, using experimental methods, have been suggested, and the paucity of our knowledge of various aspects of plant anatomy has been pointed out.

The development of the plant body

As already mentioned, the fertilized egg or zygote gives rise to the young plant. Usually the zygote divides in such a manner that a filamentous structure is formed. The first, unequal division of the zygote in higher plants results in the formation of a larger cell, which gives rise to the suspensor, and a smaller one which undergoes further division to become the embryo proper. Processes of cellular differentiation ensue, with the result that some cells of the small embryo develop differently from others. The cotyledons (two in dicotyledons, one in monocotyledons and several in gymnosperms, apart from occasional exceptions) may become evident at an early stage (Fig. 1.1a). Eventually meristematic regions, where the cells continue to divide actively for a long period of time, become delineated at each end of the embryo by the progressive vacuolation of the intervening cells. These meristematic areas are the ***apical meristems*** of the root and shoot, and during growth these important regions of the plant become

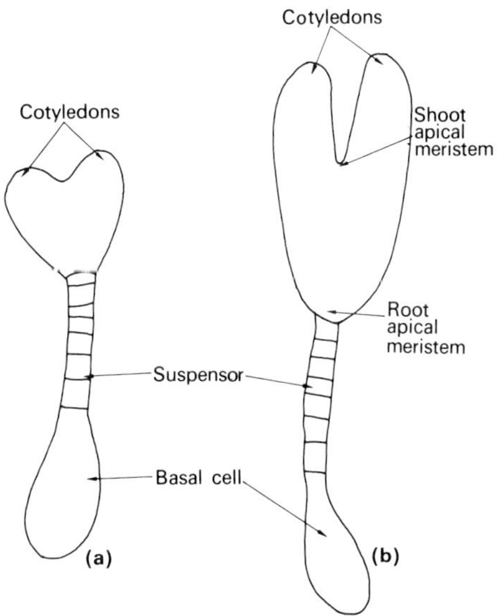

Fig. 1.1 Developing embryos of *Capsella*. (**a**) Two cotyledons are formed. (**b**) The apical meristems of the shoot and root are readily distinguishable. ×150.

progressively more distant from one another as a result of their own activity, the rest of the plant being differentiated between them (Figs. 1.1b, 1.2). Polarity is established in the very early stages of embryonic development, probably in the zygote, and this soon becomes manifest by the demarcation of a root end and a shoot end in the young embryo. This establishment of polarity, presumably as a result of physiological changes, is a critical aspect of differentiation and indeed of development in general, and will be discussed further in the next chapter.

The apical meristems of the root and shoot, established at each end of the young embryo, are extremely important regions in the plant, since they remain 'perennially embryonic' and by their activity give rise to all the tissues of the root and the shoot respectively. This type of 'open growth' from an apical meristem is peculiar to plants, and nothing like it is known in animals. Meristematic cells are usually considered to be thin-walled, often relatively rich in cytoplasm and with only small vacuoles, and capable of continued active division. Usually they stain more densely than adjacent more extensively vacuolated cells.

The organs and tissues formed by the activity of these apical meristems constitute the primary plant body (see Esau[187]), and this is

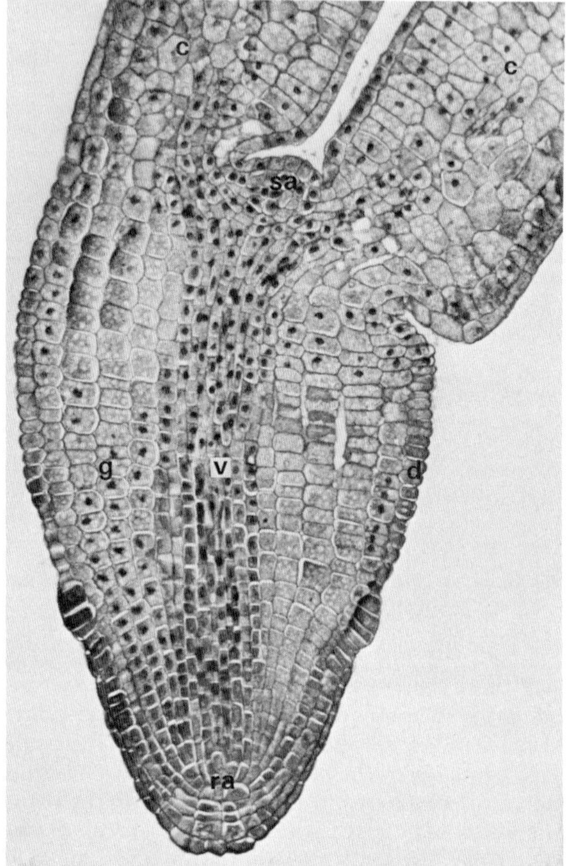

Fig. 1.2 Longitudinal section of fully developed embryo of *Clarkia rubicunda* subsp. *rubicunda*, showing the two cotyledons (c) and the axis between the shoot apex (sa) and radicle apex (ra). The forerunners of the dermal (d), vascular (v) and ground (g) tissue systems are indicated. (By courtesy of Dr. R. E. Dengler.) ×180.

composed of primary ***tissues***. These are made up of cells, the units of the plant body, each tissue being composed of a restricted range of cell types; in many tissues only one type of cell is present. The three main systems of tissues in the plant may be called the dermal, vascular and ground systems.[187] The dermal system comprises the outer covering layer, the epidermis, during primary growth. The vascular system is made up of the phloem and the xylem, the conducting elements of the plant, which during primary growth develop from the procambium, and the ground tissue includes those tissues distinct from the dermal and vascular components. The ground tissue is often composed of thin-walled parenchyma cells, but more thick-walled strengthening elements, collenchyma and sclerenchyma, may also be present. These various tissues are described in more detail in subsequent chapters.

The tissues are arranged differently in the various ***organs*** of which the plant is composed. The structure of organs is discussed in detail in Part 2 of this work.[136] The vascular elements often form a rod or cylinder, with ground tissue peripheral, and sometimes also central, to them, and dermal tissue on the periphery. The ground tissue between the dermal and vascular components is commonly called cortex; that within the vascular cylinder, pith. In the root there is often no pith. In the stem the vascular tissue most frequently consists of numerous separate strands, the vascular bundles, but this is not always so; it may also form a cylinder or be arranged in some other way. In the leaf, the vascular system usually forms a network of strands embedded in the ground tissue. In leaves the ground tissue is called mesophyll, and is usually composed of thin-walled cells which contain chloroplasts and function in photosynthesis.

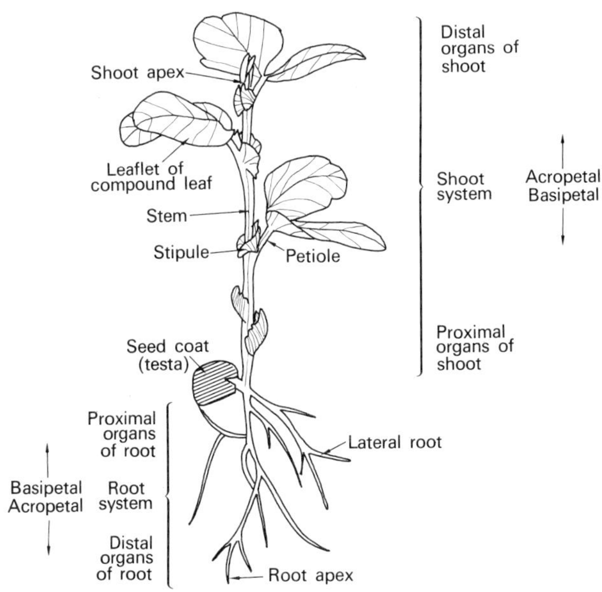

Fig. 1.3 Young plant of *Vicia faba* showing the various organs and illustrating the meaning of certain terms. ×1/3.

During the ontogeny of most dicotyledons and gymnosperms, and a few monocotyledons, secondary growth occurs. By the activity of special meristems, secondary tissues are formed, which add to the girth of the primary plant body. These meristems are the vascular cambium, which gives rise to secondary xylem and phloem (the secondary conducting tissues), and the phellogen or cork cambium, which gives rise to the periderm, an outer covering which may replace the epidermis when this is ruptured due to the expansion in girth which occurs during secondary growth. The vascular cambium and phellogen are sometimes called ***lateral meristems***, on account of their position, to distinguish them from apical meristems. Detailed discussion of apical meristems is deferred to Part 2,[136] when they are considered along with the organs to which they give rise.

The lateral organs of the shoot are formed by the apical meristem. This gives rise to the leaf primordia, meristematic mounds of tissue which eventually develop into leaves. Bud primordia, which usually occur in the axils of leaves, are generally also formed by the shoot apex, but may sometimes dedifferentiate from already more or less mature tissues.

Lateral roots are not formed at the apex of the root, but at some distance away from it. Usually they develop from primordia formed in the pericycle, the outermost layer of the vascular cylinder.

The flower or inflorescence is formed either from the terminal shoot apex or from lateral shoot apices, as a consequence of changes that take place in these regions as a result of certain factors such as daylength.

TERMINOLOGY

Certain terms referring to planes of cell division, the direction in which tissues differentiate, and components of the plant body, will be used in the following chapters. These terms, some of which are illustrated in Fig. 1.3, are as follows:

Proximal —situated near or towards the point of attachment of an organ.

Distal —situated away from the point of attachment.

Basipetal —from the apex towards the base; for example, differentiation may occur basipetally.

Acropetal —from the base towards the apex.

Anticlinal —used to describe a cell wall formed at right angles to the surface of the organ. This is an anticlinal wall.

Periclinal —used to describe a cell wall formed parallel to the surface of the organ.

Apoplast —the non-living component of the plant, comprising cell walls and intercellular spaces.

Symplast —the living component, or protoplasm, rendered a continuous system, bounded by the plasmalemma, by intercellular cytoplasmic connections (the plasmodesmata) which pass through channels in the cell wall. Both apoplastic and symplastic transport of substances through the plant occur.

2

Differentiation

During the development of a whole plant from a single cell, the fertilized egg or zygote, various processes of differentiation must take place. By repeated division of the zygote and its immediate products, a group of initially fairly homogeneous meristematic cells is formed. By processes of differentiation cells become recognizably different both from the cells which have given rise to them and from neighbouring cells having the same origin as themselves. All this involves complex biochemical and biophysical processes that we are only beginning to understand. It is sometimes said, indeed, that differentiation is one of the most complex unsolved problems of biology. While it is relatively easy to describe the visible manifestation of differentiation, it is much more difficult to explain the underlying causes in terms of the biochemical changes taking place in the cells. It is important to remember that invisible biochemical changes always precede the visible morphological ones; anatomical features may be said to mirror in visible form the less obvious physiological events that took place some time before.

The differentiation of the organs and tissues of the plant is not random, but in fact takes place in a very orderly fashion, according to the species. The control of organization, or orderly development, in plants is as yet little understood, and poses many fascinating and difficult problems. Planes of cell division are usually considered to be very important in the control both of the position and of the shape of the developing organ, but recent work with irradiated wheat seedlings suggests that leaves, for example, may develop their characteristic shape despite the prevention of nuclear division by irradiation of the seeds.[291,292] Irradiation of wheat seeds prevented the synthesis of deoxyribonucleic acid (DNA) and stopped mitosis, but did not prevent germination and growth of the seedlings. Such seedlings, designated 'gamma plantlets', thus afford material in which some development takes place in the absence of cell division. However, this growth, interesting as it is, merely involves an expansion of the primordia already present in the embryo, and it seems likely that mitoses which have taken place during embryogenesis may already have determined the shape and polarization of the organs present. This is not to deny, however, that cell enlargement may play a greater role in the determination or development of organ shape than has been attributed to it hitherto. These same workers have recently shown that within the roots of irradiated seedlings cells of the pericycle enlarged but did not divide, giving rise to structures the initial form and position of which resembled those of lateral root primordia.[244] Foard[242] also claims that potential photosynthetic cells in leaves of *Camellia* can differentiate as sclereids without prior division, and hence that there is only an indirect relationship between cell division and differentiation.

In the irradiated wheat seedlings tissue differentiation and maturation took place. Thus the later stages of differentiation, at least, can ensue in the absence of mitosis, but as the authors themselves point out, early stages of differentiation had already occurred in the embryo prior to irradiation.[243] This work seems to lay emphasis on the importance of very early phases of development and differentiation in the determination and control of subsequent stages. All cells which differentiate have originated from cells which have previously undergone division.[666] However, the precise relationship between cytokinesis and differentiation does seem to require further investigation.

In many cases some of the somatic cells of the plant body are polyploid, and it has sometimes been suggested that the level of ploidy of the cells might control differentiation. However, it seems that in general, at least, polyploidy is merely one manifestation of cell differentiation, rather than a necessary precursor or controlling factor.[493] Measurement of the relative amount of DNA in the nuclei of various tissues of three species revealed that *Helianthus* had no polyploid cells in any tissue. It appears, therefore, that polyploidy is not an essential condition for differentiation of tissues or organs.[210] Thus we must look not to the number of chromosomes, but to the genes themselves, or to factors in the cytoplasm, for the ultimate arbiters of cell and tissue differentiation, at the same time heeding a salutary reminder that the genome is itself answerable to external control.[275]

Genetic control of differentiation

At this stage we may consider whether differentiation is genetically controlled. There are, of course, many examples of the occurrence in higher plants of structures, e.g. particular types of hairs or sclereids, that are characteristic of the species in which they occur. The best evidence, however, comes from a relatively simple organism, the somewhat toadstool-shaped unicellular green alga *Acetabularia*. Different species of this organism, which has a single, large nucleus, situated in the basal rhizoid, have different shapes of cap. If the alga is cut in pieces, only that portion containing the nucleus will regenerate a whole cell, though the tip will form a cap. If excision of the middle portion is delayed, it will acquire from the nucleus the ability to form a cap prior to enucleation. Differentiation of a cap can occur in the absence of a nucleus but depends on substances released by the nucleus.[302] By transplanting nuclei from one species to another it can be shown that the shape of the cap is genetically controlled. If differentiation is under genetic control, evidently the basic problem is that from a population of cells that are initially derived from one cell—the zygote—and are presumably genetically identical, many different tissues become differentiated. This suggests that differentiation cannot be under direct genetic control. Several possibilities which might modify this view, however, may be considered. Firstly, all the cells may not, in fact, be genetically identical. Differentiation might result, for example, from the selective loss of certain genes; this cannot be generally true, however,[650] since whole plants can regenerate from single somatic cells (see below). Alternatively, increases in the genome might occur. If the genome replicates completely, values of nuclear DNA will form a doubling series and polyploidy will occur. We have already seen that, although polyploidy may accompany differentiation, it is not a necessary prerequisite. If a non-doubling series of DNA values is observed, the genome may be replicated differentially, certain genes perhaps being amplified in particular cells; that is, there might be an increase in one gene independently of others. This is known to happen in certain animal oocytes. There is as yet no report of any comparable massive gene amplification in plants, though there is some evidence for differential replication of the genome in some plant cells.[339,604] It is clear, at all events, that gene amplification cannot provide a general explanation for cell differentiation,[288] the results of nucleic acid hybridization experiments in particular providing evidence against this.[245]

If all the cells of the organism are indeed genetically identical, differences between them must be due to cytoplasmic differences, or to the selective activation and repression of particular genes. Support for the latter view would come from demonstrations of the totipotency of individual cells or nuclei; such evidence is discussed below.

Totipotency

Successful growth of whole plants has now been reported from isolated cells or small groups of cells, obtained from various tissues of the embryo or adult plant. This possibility has now been demonstrated in a number of species, using cells derived from various tissues and grown under a variety of conditions. Some examples will be briefly discussed.

Steward and his co-workers[643,646] devised a method for growing carrot tissue by excising small discs from the secondary phloem region of carrot roots and placing these in a liquid medium, under aseptic conditions. In the presence of coconut milk—the liquid endosperm which nourishes the embryo of the coconut—the phloem tissue began to grow actively, in a manner which it would not have done if left in its normal environment. When explants of carrot tissue were grown in liquid medium with coconut milk in special culture flasks, which were shaken and rotated, some single cells and small groups of cells were loosened from the surface. When these detached cells were grown separately, it was found that only a fairly small proportion of single cells would continue development. Some, however, did grow and develop in a number of ways. Some divided giving rise to filaments of cells which in certain respects resembled normal carrot embryos. These groups of cells eventually formed small roots, and if transferred on to a medium solidified with agar would also form shoots opposite to the root. These 'test-tube' plants could be grown to maturity; indeed, they flowered and produced viable seeds.

If, instead of pieces of phloem, embryos of the wild carrot were caused to proliferate and to separate into free cells, these cells could very readily be induced to develop as embryo-like structures, or ***embryoids***, if grown on a medium containing coconut milk.[645] Thus apparently every cell from a young embryo was capable of developing in a manner comparable to that of a normal zygote, if provided with appropriate nutritional conditions. Halperin[299] has also clearly demonstrated the importance of the chemical constitution of the medium in controlling the development of pieces of callus from the petiole of wild carrot as either root-bearing structures or as embryoids. The resemblance of the latter to stages in normal embryogenesis is very striking.

These experiments indicate that, not only can cells from an embryo behave like zygotes, but so also can mature, fully differentiated cells or their derivatives. The necessary conditions were thought to be (i) removal of the cell from its normal environment, and (ii) supplying it with an appropriate chemical environment. Physical isolation of the cell is apparently not necessary,[300] however, since cells which are about to give rise to embryoids may remain in continuity with adjacent cells through plasmodesmata.[649,685] In later stages the young embryoids do become isolated. Coconut milk is also not essential, and indeed the nutritional requirements of the embryoids may vary at different times during their ontogeny. Such sequential changes in the medium may in fact be important in many tissue cultures.[666] Steward[644] has pointed out that since differentiation normally involves a set sequence of events, a particular sequence may be necessary for the reverse process of dedifferentiation leading to induced embryo development from adult cells. For example, the auxin level necessary for embryoid formation must be reduced in order to promote their further development.[649]

Embryoids can develop also from haploid cells, namely pollen grains at certain stages of development (see Part 2,[136] Chapter 6), and also from isolated protoplasts. Whole plants have been formed from protoplasts of at least 6 species, including carrot,[112,260] and from somatic hybrid cells formed by the fusion of protoplasts.[446]

A fully differentiated cell thus retains the potentialities for growth and differentiation normally manifested only by the zygote, but in its normal environment within the plant exhibits only a few of these potentialities. The limitations on its capacity for development must be imposed by its environment, i.e. by its position within the organism.[643] Such cells are sometimes said to be ***totipotent***, a concept which is by no means new.[685] It has been pointed out, however, that the cells which develop as embryoids and show totipotency, are usually parenchyma, and that it cannot be demonstrated that all the cells of the plant are and remain totipotent.[271,290,339] In particular, many unsuccessful attempts to induce embryoid development in various species have been made,[649] and the causes of these failures remain unknown.

Evidence that genes are not lost or the genome irreversibly altered during differentiation comes also from the results of experiments on transplantation of animal nuclei, notably in amphibians. Enucleate unfertilized egg cells are not capable of further development, but if the nucleus of a differentiated cell, e.g. from skin, is transplanted into such an egg, normal adult frogs can develop.[288,289] It is apparent that in such cases, nucleocytoplasmic collaboration is involved and the nucleus receives instructions or signals of some kind from the cytoplasm.[288, 635] Labelling experiments show a correlation, and thus a possible causal relationship, between the induction of nuclear activity by the cytoplasm and movement of proteins from the cytoplasm to the nucleus.[289] The current belief is that differentiation is a consequence of selective gene activation, which depends in turn upon interactions between nucleus and cytoplasm, and between the cell and its immediate environment.

As long ago as 1878 Vöchting[687] was emphasizing the importance of the position of a cell in the organism, and later Driesch[167] summed this up for animal development by stating that the fate of a cell is a function of its position. The results of many experiments now convincingly demonstrate that the normal fate of a cell can sometimes be altered by changing environmental factors. This is illustrated, for example, in many every-day horticultural practices. For example, in many species removal of a leaf from the plant will often lead to the formation of roots and a bud from tissues at the base of the leaf that would never have shown any sign of these potentialities if left in their normal environment.

The potentiality of a cell to differentiate in different ways, and its competence to react to stimuli, may be retained for long periods of time without being manifested. The parenchymatous cells of the pith of the sahuaro cactus (*Cereus giganteus*) may remain in an active state during the 100 or 150 years of the plant's life;[428] furthermore, if an incision is made into the pith, the cells can divide actively and form a cambium-like tissue which gives rise to cork,[521] thus exhibiting a potentiality for differentiation that would not normally have been expressed. As has been aptly pointed out,[645] there must be a system of restraints normally operating within the plant that allows cells to manifest only a small fraction of their actual potentialities for differentiation and development while they remain undisturbed *in situ*.

Polarity

One of the most important factors in early differentiation is the establishment of polarity. This is the setting-up of a difference, structural or physiological, between one end of a cell or organ and the other. Establishment of polarity in the developing zygote, for example, is very important; it may be controlled to some extent by the environment in which the zygote develops. Recent work on embryos with the electron microscope[356] demonstrates very clearly the polarized distribution of cytoplasm at the time of early divisions of the zygote (Fig. 2.1).

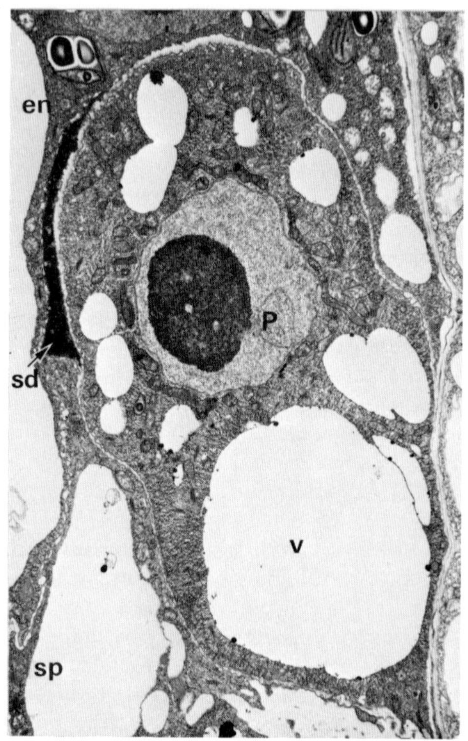

Fig. 2.1 Zygote of *Capsella bursa-pastoris* soon after fertilization. The cell is already polarized, with a large vacuole (v) at one end and dense cytoplasm at the other. en, endosperm; p, pocket in nucleus, a temporary result of fusion of the egg and sperm nuclei; sd, part of degenerating synergid; sp, persistent synergid. ×4000. (By courtesy of Sister Richardis Schulz and Dr. W. A. Jensen. From *Am. J. Bot.*, **55**, Fig. 11, p. 813, 1968.)

The importance of environment in establishing polarity has been studied in experiments with fertilized eggs of the brown seaweed *Fucus*. These zygotes develop freely in sea water, i.e. entirely free from a cellular environment, and it is thus possible to impose various single environmental factors upon them. The first visible sign of differentiation in the fertilized egg of *Fucus* is the outgrowth of a rhizoid at one position in the cell wall. The first new cell wall is formed in a plane at right angles to the emerging rhizoid. The factors which determine the position of the rhizoid thus determine the polarity of the developing zygote. Whitaker[714, 715] and others have carried out many experiments with *Fucus*. For example, if fertilized eggs of *Fucus* are aggregated together in groups, the rhizoids are formed pointing towards the centre of the group, even if the eggs belong to different species. It is thought that diffusion gradients of metabolic products are set up. Gradients of temperature and of pH are also effective in determining polarity, rhizoids developing on the more acid side. If *Fucus* zygotes are centrifuged in sea water, stratification of the cytoplasm takes place, and the rhizoid develops at the centrifugal pole.[714] However, if the visible contents of the zygote become redistributed again uniformly after centrifugation but before the formation of rhizoids, these develop at random in relation to the previous stratification. Distribution of the cytoplasm within the cell is thus apparently important. Clearly polarity may be determined, at least to a considerable extent, by external environmental factors.

Such effects can be enhanced by feedback mechanisms. If *Fucus* eggs are placed in series, it can be shown that each drives an electric current through itself from rhizoidal to thallus end. It is thought that this current may feed back, amplifying the differences between the two ends and thus helping to establish polarity by a kind of self-electrophoresis.[349] Electrophoresis depends upon the fact that proteins carry a net electric charge, which may be positive or negative. A solution of different proteins subjected to an electric current will fractionate themselves, e.g. within the cell, by moving towards the pole of opposite charge. In recent ingenious experiments with the related brown alga *Pelvetia*, it was shown that a substantial calcium current passed through eggs while they were being polarized by unilateral light.[549] This current could create an intracellular gradient of calcium ions.

Polarity occurs also in multicellular organisms, in which gradients of various kinds are evidently of some importance. The simplest examples are perhaps uniseriate filamentous algae. It is evident from the results of various experiments that each cell of such a filament is polarized. When the filamentous red alga *Griffithsia* was maintained under unfavourable conditions, the axis tended to separate into individual cells, each of which could form a new filament.[105] In more recent experiments, single intercalary cells were excised and maintained in favourable conditions in culture. Within one or two days such isolated cells formed a shoot cell at the original apical end, and a rhizoidal cell at the base.[170] Similar results had much earlier been obtained with single cells of the green alga *Cladophora*, isolated either surgically or physiologically, by plasmolysis.[140] Polarity could be reversed by centrifuging. In these experiments on algae of relatively simple form, it was thus shown some 60 years ago that single isolated cells could regenerate whole plants. If polarity can occur in all the cells of a filament, it is likely that it can do so also in the cells of tissues of higher plants.

If polarity is imposed by a gradient of some sort, clearly the cells must have the capacity to recognize it;[127] and if the topographic position of a cell is important in affecting its differentiation, this, too, must somehow be detected. Wolpert and his co-workers[727] have pointed out that cells must have their position specified in relation to certain boundary regions, and that a gradient between a sink at one boundary and a source at the other could provide such positional information. Grafting experiments on the invertebrate animal *Hydra* have shown that the time required for the transmission of inhibitory effects is increased with distance,[727] and it is agreed that a mechanism based on diffusion would fit the known facts, including the probable speed of transmission.[127,128] Gradients, then, may be set up by diffusion; the 'messenger' substance need not be highly specific, and may indeed be not one but several substances moving at different velocities.[402]

It has been proposed that 'cellular differentiation may result from the segregation of specific biochemical systems *within* the single parent cell, and that this separation becomes finalized by the laying down of the wall between the two sister cells'.[115] It is often difficult to distinguish between the establishment of polarity and its consequences. Moreover, it is perhaps arguable whether polarity is a factor in the control of differentiation or an early manifestation of it.

Unequal cell divisions

Establishment of polarity within a cell may lead to its subsequent unequal division, and to a different fate for the two daughter cells formed by that division. Such unequal or asymmetric divisions are important in the differentiation of various structures.[75] For example, in the formation of many stomata and root hair initials and of some other structures, a cell divides unequally to give two derivatives, a small one with dense cytoplasm, and a larger one with less activity (Fig. 2.2). The small cell differentiates as the guard cell mother cell or root hair initial, the larger one as an ordinary epidermal cell. The visible differentiation of these cells is preceded or accompanied by various differences in enzyme distribution, nucleolar size, etc. (see Chapter 7). It is noteworthy that those structures whose differentiation is preceded by unequal cell divisions of this kind are lacking from gamma plantlets developing after irradiation.[243] Thus asymmetric mitosis does indeed seem to be a factor in their differentiation.

On the other hand, under some circumstances roots of the grass *Phleum* can form root hairs from long cells as well as short, and it is argued that the polarized unequal division cannot be the factor which determines the hair-bearing capacity of the cells.[242] Presumably in such cases conditions are somehow rendered similar in both cells. In general, however, unequal divisions seem to be important in cell differentiation throughout the plant kingdom.[698] It is thought that all daughter cells would differentiate in the same way as the parent cell unless the cytoplasm was altered in some way, and that prior to an

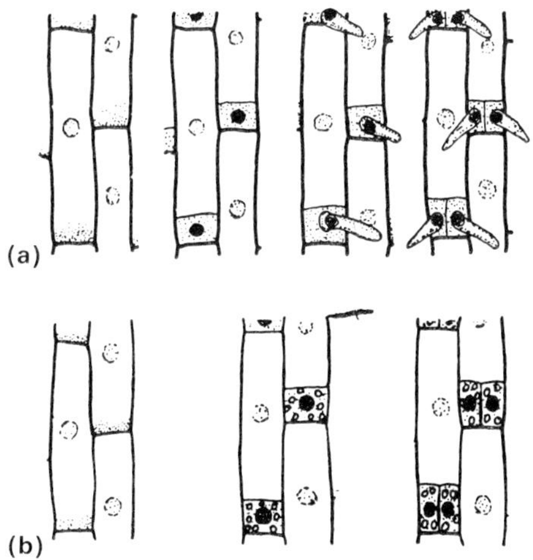

Fig. 2.2 Differentiation of (**a**) root hairs and (**b**) stomata in monocotyledons. Successive stages of development are shown from left to right in each figure. Unequal, polarized distribution of cytoplasm is followed by nuclear and cell division resulting in the formation of a large cell and a small one, the latter developing as the root hair or guard cell mother cell. (From Bünning,[74] Fig. 2 (II and III), p. 111.)

unequal division there is a polarized distribution of determinants in the cell, which then become distributed differentially between the two unequal daughter cells.[271,288] Stebbins[635] considers that the only reasonable mechanism leading to polarization of the cytoplasm is the self-electrophoresis hypothesis of Jaffe[349] already mentioned. The divergent differentiation of the two daughter cells following an unequal division could be attributable to feedback into the nucleus of different proteins synthesized in the cytoplasm.[289,635] It is considered that unequal distribution of the cytoplasm may initiate the processes of cell differentiation, but cannot fully explain them.[289]

Pattern formation

Differentiated regions or structures, such as root hairs, stomata and procambial strands, are often maintained at some distance from one another, forming a more or less regular pattern. Bünning[74] has applied to these active loci of growth the term ***meristemoid***, and has pointed out that they show a degree of mutual incompatibility by means of which the pattern is established and maintained. Meristemoids of a particular kind, e.g. stomata, can inhibit not only other stomata, but also meristemoids with a different destiny, such as hairs. When these regions of active growth become inactive or separated from each other by some distance, new ones can form between them. For example, in the ground tissue of the developing root of *Pandanus* a pattern of cells which will contain raphide (needle-shaped) crystals is found. These cells are initially recognizable by the density of their cytoplasm. After these cells have formed raphides and died, a new pattern of meristemoids is able to form in positions between the raphide cells. These new meristemoids develop as bundles of fibres.[76] In other instances, for example in the differentiation of stomata, the distance between the meristemoids may simply increase as a result of expansion of the organ concerned, this enabling new ones to differentiate in the enlarged spaces between. The differentiation of new stomata in close proximity to existing ones in some species poses a difficulty for this theory.[489] Pattern formation, like differentiation itself, is a general biological phenomenon, and experimental analysis of epidermal hairs on the milkweed bug, *Oncopeltus*, also showed that the first hairs originated in the centre of the largest spaces, the pattern being built up in a precise order.[405] Analysis of stomatal patterns by computer methods is discussed in Chapter 7.

The nature of the incompatibility which apparently exists between some meristemoids is not yet fully understood. Competition for substances necessary for growth and differentiation could be involved as well as, or instead of, the production of inhibitory substances by the regions of active growth. Whatever its nature, the inhibitory field can apparently extend over an area much greater than that occupied by the meristemoid itself, at least in some instances. In view of the previous discussion on gradients, it is interesting to note that analysis of the pattern of hairs on the milkweed bug is consistent with the view that morphogenetic substances moving by diffusion could lead to pattern formation.[405]

In assessing the importance of the factors involved in differentiation, the effects of one tissue on another should be considered. For example, not only can cells that give rise to hairs affect the differen-

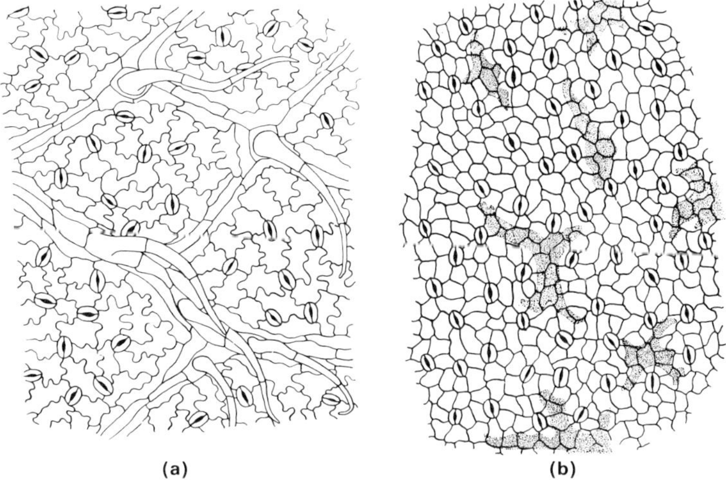

Fig. 2.3 Patterns of differentiation. (**a**) Lower epidermis of the leaf of *Helianthus rigidus*. Hairs are situated over the veins. (**b**) Leaf of *Nymphaea alba*. The sclereids (stippled) in the mesophyll never occur below stomata. (From Bünning[76], Figs. 10 and 11, p. 391.)

tiation of stomata within a single tissue, the epidermis, by mutual inhibition, but the distribution of the hairs themselves may depend upon the underlying tissues. In the leaf of *Helianthus*, for example, the hair-bearing epidermal cells follow the pattern of the underlying vascular tissue forming the veins (Fig. 2.3a). Again, in the leaf of *Nymphaea alba* the growth and development of the sclereids in the mesophyll and the stomata in the epidermis are apparently interrelated, since the sclereids never occur below stomata (Fig. 2.3b).[76] These visible morphological relationships can be readily observed, but the underlying physiological relationships between the various cells and tissues are not yet well understood and some interesting experiments could be designed to study them. For example, incisions could be made in young leaves to separate those tissues that normally differentiate in proximity to each other. Also, the effects of auxin, which has been shown in other species to affect the formation of both stomata and sclereids, might profitably be studied in developing leaves of *Nymphaea*.

The importance of the position of a cell in relation to other tissues is shown also by certain aspects of differentiation in *Monstera deliciosa*, in which unequal cell divisions occur in the cells of various tissue layers.

In the root epidermis, the small cells give rise to root hairs; in the hypodermis, they differentiate merely as short cells alternating with longer ones, and in the cortex the short cells develop as tricho-sclereids.[60]

Control of cell differentiation: nuclear–cytoplasmic interactions

In his experiments on *Acetabularia*, Hämmerling[302] found that if excision of the middle portion of the cell was delayed, it could acquire from the nucleus the ability to form a cap, though that part itself was enucleate. He therefore postulated that morphogenetic substances produced by the nucleus were stored in the cytoplasm. It is now believed that these substances are stable mRNAs. Further work with *Acetabularia* has shown that these messages from the nucleus are not all translated simultaneously, some being stored for quite long periods. The cytoplasm, probably by means of proteins, regulates the sequential reading of messages from the nucleus.[529]

From the discussion in this chapter we can now draw certain conclusions: (i) differentiation is genetically controlled; (ii) some, though probably not all, living, nucleate plant cells are totipotent; there have therefore been no irreversible cytoplasmic changes, no loss of genes, and no irreversible changes in the genome during maturation; (iii) genetic expression of the still totipotent nuclei of differentiated cells is greatly influenced by the surrounding cytoplasm, as shown by nuclear transplant experiments; (iv) normally not all the potentialities of a cell are expressed, these being evoked or repressed by factors in its immediate environment; (v) such factors include those leading to the establishment of polarity, often with subsequent unequal cell division which results in the establishment of cytoplasmic differences between adjacent cells.

Thus, although all the cells of the plant may have the same genetic complement not all the genes may be able to express their activity in a cell at any one time. For example, those genes which control floral development presumably do not come into play until the plant has been induced to flower. We may say that during vegetative growth the genes controlling flowering are repressed, becoming active, or derepressed, as the result of appropriate external stimuli, for example exposure to a certain number of hours of alternating darkness and light. Bonner[62] has pointed out that a whole sequence of genetic switching mechanisms may be set in train by an initial induction or trigger. The nature of the agents which can trigger this sequential activation of genes, and the mechanism of their action, are consequently the keys to the control of tissue and organ differentiation. The cytoplasm must contain components that are capable of regulating gene transcription, perhaps particular proteins that pass into the nucleus.[288,635,699] While selective gene activation may be thus controlled, the time of gene expression may be controlled at the translation level.[271] This is indeed evident from the work on *Acetabularia*.

In the ensuing chapters the differentiation of various individual tissues is further discussed, and the reader will be struck by the growing evidence for the control of differentiation by hormones, and particularly by the interaction of hormones with other substances, or of one hormone with another. Also, one hormone (e.g. auxin) may be important in the differentiation of several quite different tissues. How, then, does the hormone act? In the next chapter the specific effect of a hormone, gibberellic acid, on stimulating the formation of the enzyme α-amylase in a particular tissue, the aleurone layer of barley grains, is described. The differential synthesis of enzymes is considered by some[629] to be the basic process of cell differentiation, and certainly it is now known that concentrations of particular enzymes often precede the morphological differentiation of certain cells and tissues. The effect of the gibberellic acid in this case, however, must be affected or determined by the so-called 'pre-programming' of the aleurone cells, since it will produce different effects in other tissues.[294] In general, it is thought that the effect of hormones on cell differentiation in animals, at least, is to enhance cytoplasmic differences which already exist.[288] The whole history of a cell is thus of great importance in its differentiation. Hormones affect nucleic acid metabolism; several of them seem to regulate gene expression, but their precise mode of action is still not understood.[259] Further work on the fascinating problems of differentiation is clearly called for, especially the further application of experimental methods.

3

The Plant Cell

Plants, like animals, are composed of cells. Some plants consist of only one cell, but the flowering plants, with which this book is concerned, are made up of many cells which at maturity differ greatly in structure. As we have seen, these differences are the result of processes of differentiation. The cell is the unit of construction of plants, just as atoms are the units of molecules. Recognition of the cellular construction of plants goes back to the seventeenth century, and knowledge of the cell and its contents has progressed hand-in-hand with the development of the microscopes employed for its observation. The advance of biological knowledge is dependent to a considerable extent upon advances in those other disciplines on which biology relies for its tools, for example physics and electronics as well as biochemistry.

In the year 1665 Robert Hooke studied sections of a bottle cork with a microscope which we would now consider extremely primitive. He observed that the cork resembled a honeycomb, consisting of pores that were separated by walls, and he called the individual units ***cells***, because of the resemblance to the cells of a honeycomb. Thus at this time attention was largely focused on the cell wall, although botanists were aware that living cells contained liquid contents. Shortly after this, green bodies, the chloroplasts, were observed inside the cells, and in 1883 Robert Brown observed the nucleus, a larger body present in all living cells[355] (with the probable exception of mature sieve elements).

By the middle of the nineteenth century it had been realized that all organisms consist of cells, and, further, that all such units are derived from the division of existing cells. It was shown that chromosomes were present in the nucleus, and that these divided during nuclear division. Nuclear divisons were of two kinds: those that gave rise to the somatic cells of the plant, in which the chromosomes duplicated and the daughter cells had the same number of chromosomes as the original cell (mitosis); and those that gave rise to the reproductive cells of the plant, in which the daughter cells had only half the original number of chromosomes (meiosis).

The cell, then, was known to consist of a cell wall and contents, the protoplasm. This comprised a more or less spherical body, the nucleus, containing chromosomes which were the bearers of the hereditary units or genes, embedded in the granular matrix, the cytoplasm. Various other inclusions were also observed with the light microscope, and biochemical studies led to some understanding of their function. The great expansion in knowledge of cell structure, however, followed the development of the electron microscope, an instrument that employs a beam of electrons instead of a beam of light. This enabled anatomists to observe much smaller structures than are visible with the light microscope, and the study of the fine structure or ultrastructure of plant cells began. Although some valuable work on whole mounts or pieces of cell wall was done earlier, the work on sectioned material dates only from about 1950, and work on surface cells with the scanning electron microscope only from about 1965.

When Robert Hooke observed the cells of the bottle cork, he remarked that he had found 'a new invisible world'. The development of the electron microscope has revealed yet another 'invisible world', one on a much smaller scale. Whereas the upper magnification of the light microscope is about 1200 times, that of the electron microscope is 160 000 times.[355] With the light microscope—where the lower limit of visibility is determined by the wavelength of light—the smallest objects that could be observed had a diameter of about 0·3μm (micrometre); with the electron microscope objects of considerably less than 10 nm (nanometres) (equivalent to 100 Å (Ångströms)) can be studied, and the limits of resolution are of the order of 0·8 or 1·0 nm (1 mm = 1000 μm; 1 μm = 1000 nm or 10 000 Å). The scanning electron microscope (SEM) has a depth of focus of about 300 Å, compared with the 10 Å of the transmission electron microscope (TEM). Thus whole, relatively large objects can be seen and photographed in focus.[633]

Although electron microscopy has greatly extended our knowledge of plant structure, essentially it is not a new discipline, but merely comparative anatomy on a different scale. When electron microscopy—or, for that matter, classical comparative anatomy—is combined with a biochemical approach, the aim of establishing the function of the various components of the cell can be achieved, at least to some extent. Indeed, investigation of the activities of the cell components on a molecular scale is now possible.

The various components of the plant cell (Fig. 3.1) that are known at the present time will now be listed; each is discussed at more length below. The plant cell is surrounded by a **cell wall** (see Chapter 4), which may be a primary wall only or may comprise both primary and secondary walls; in the wall there may be depressions or pits, and the wall may be traversed by cytoplasmic strands, the ***plasmodesmata***. Immediately within the wall the cell contents are delimited by a membrane, the plasma membrane or ***plasmalemma***; in the cytoplasm there may be one or more ***vacuoles***, containing cell sap and each bounded by a membrane, the ***tonoplast***. There are also membrane systems, the ***endoplasmic reticulum*** and the ***dictyosomes***. In addition there are a number of organelles, including the ***nucleus***, bounded by the nuclear envelope and containing the chromatin and one or more nucleoli; the ***plastids***, of which one type, the chloroplasts, are active in photosynthesis; the ***mitochondria***, active in respiration; and the ***ribosomes***, active in protein synthesis. Also present are ***microbodies***, the site of various enzymes, ***spherosomes***, and ***microtubules***, rod-shaped structures apparently involved in wall formation. All these small structures are known as ***organelles***, the essential characteristic of which is to set apart a group of particular functions within the cell.[415] In some cells substances such as fats and oils, starch, protein bodies and crystals may also be present; these are known as ergastic substances.

Bonner[62] gives an estimate of the approximate numbers of the various organelles that may be present in a plant cell, which may help in visualizing the cell and its continuous activities. He says that the cell contains 'a nucleus (generally one); some chloroplasts, fifty or so . . .; mitochondria, five hundred or so; ribosomes, five hundred thousand or so; and enzyme molecules, five hundred thousand thousand or so.' This emphasizes not only the necessarily extremely small size of these bodies, but also the great physiological activities of various kinds that are going on within each cell.

Over the past several years it has been conclusively demonstrated that both mitochondria and plastids contain DNA, DNA and RNA polymerases, and ribosomes. Both organelles thus possess all that is required for protein synthesis, and for potential autonomy.[114] It is thought that a unit of chloroplast DNA could code for approximately 3000 small proteins.[671] These and other observations have led to the hypothesis (to some extent a resurrection of an old hypothesis) that mitochondria and chloroplasts originated (in an evolutionary sense)

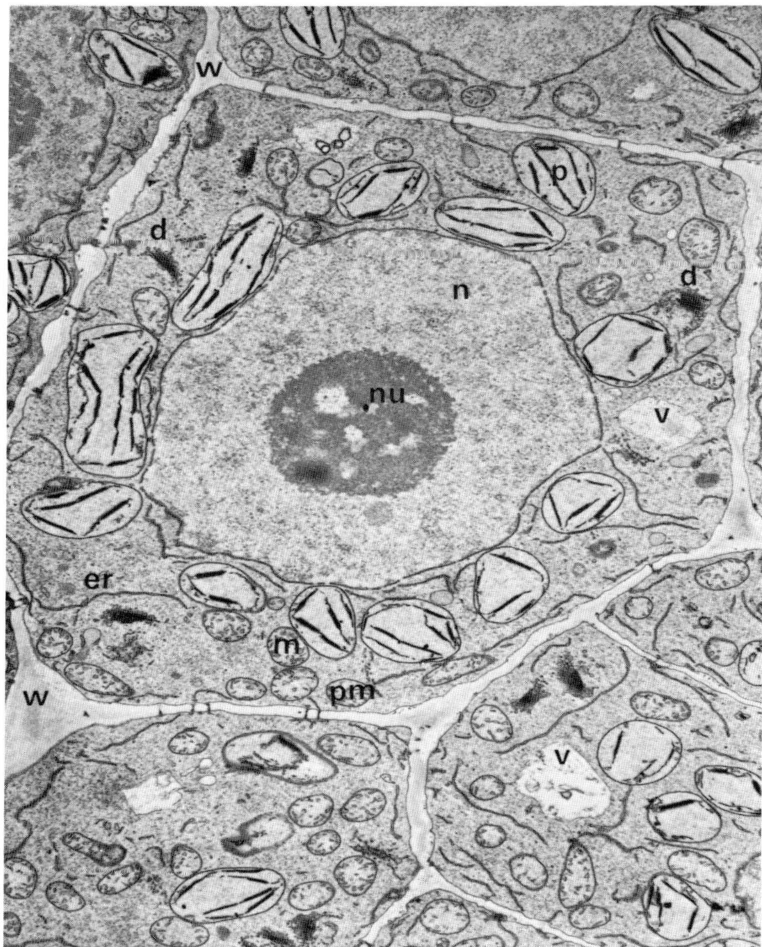

Fig. 3.1 Cell from the ground meristem of the root of *Hydrocharis morsus-ranae* showing the nucleus and various organelles in the cytoplasm. d, dictyosomes; er, endoplasmic reticulum; m, mitochondria; n, nucleus; nu, nucleolus; p, plastids; pm, plasmodesma; v, vacuole; w, cell wall. × c. 14 000. (Photo by Dr. K. D. Stewart.)

from prokaryotes which were able to live as endosymbionts in the cells of higher plants.[439] Evidence which supports this view includes the resemblance between the DNA of these organelles and that of prokaryotic bacteria and blue-green algae (the molecules are often circular, the DNA lacks histone, and plastid DNA is capable of hybridizing with that of blue-green algae).[114,286] The ribosomal RNA of the chloroplast of the eukaryotic alga *Euglena* has also recently been shown to be typically prokaryotic.[73] In mitochondria of higher plants, however, DNA molecules are linear.[480] There is at present no conclusive evidence which would enable the hypothesis to be accepted or rejected unequivocally. Evidence on the whole is against the idea that mitochondria and plastids are autonomous, since they are known to be under the control of nuclear genes.[51,114]

It is important to remember that cells that we see in illustrations, all those that we observe with the electron microscope, and many of those seen with the light microscope, have been killed, fixed and often also stained. But in life the cell is very different. For in the living cell there is movement of various kinds: Brownian movement; active and often rapid cytoplasmic streaming, effecting the movement of organelles such as plastids and mitochondria; movement of the chromosomes first towards and then away from the spindle; and so on. This movement can be seen in living cells observed with the light microscope, and observations of killed cells with the electron microscope indicate that movement must have occurred.[65,66]

CELL CONTENTS

Endoplasmic reticulum

Within the cytoplasm there is an elaborate system of membranes, the endoplasmic reticulum. This consists of double membranes that enclose spaces or cisternae; these may be more or less cylindrical in shape, or flat and ribbon-like.[626] The membranes of the endoplasmic reticulum, or ER, which are of lipo-protein nature, are connected with the nuclear envelope and also extend to the margins of the cell; connections have also been observed between the ER of adjacent cells.[522] Thus there exists a complicated network of membranes which connects the nucleus of one cell with the cytoplasm and nuclei of adjacent cells. Usually actively growing cells have more ER than resting ones.

The ER occurs in two forms, smooth and rough. The rough form is rendered rough or particulate by the occurrence of numerous small particles of ribonucleoprotein on its outer surface; these ribosomes are involved in protein synthesis. There is continuity between the smooth and rough forms of the ER.

The functions of the ER are not yet well established, and some of those proposed are still controversial. It has been suggested that it forms a transport system for proteins, etc., and that it has a role in wall formation. It is also thought that it may give rise to at least some of the membrane of the dictyosomes, and of microbodies.[286] Changes in the amount and type of ER in a cell can occur rapidly; for example, formation of rough ER can be stimulated by hormones such as gibberellic acid.[686]

Dictyosomes

The Golgi apparatus of plant cells consists of a system of dictyosomes occurring throughout the cytoplasm. A dictyosome consists of a small stack (from 2 to 20) of smooth double membranes enclosing cisternae, which are often dilated at the ends; they are associated at the edges with a number of small vesicles which are thought to have been constricted from them. This is clearly seen in a cross-sectional view of a dictyosome (Figs. 3.1, 3.2). Contrary to former belief, the cisternae comprising the dictyosome are not merely flattened sacs, but have a central region of this kind with, in addition, anastomosing tubular proliferations extending from it. The tubules branch and rejoin to form a complex fenestrated system,[459] the vesicles at the edge being attached to the cisternae by one or more tubules (Fig. 3.3). The amount of peripheral fenestration is variable, and may even differ during development. Individual dictyosomes often do not show all features depicted in Fig. 3.3.[710] The number of vesicles seems to vary with cell activity.[712] The number of dictyosomes per cell also varies widely, and in some plant cells may be as many as several thousand,[459] increasing during differentiation. Usually, the so-called forming face of the dictyosome is convex, the maturing face concave. Materials which are packaged by the dictyosomes are thought to enter the cisterna at the forming face, via the ER; the cisternae then migrate towards the maturing face and their contents are passed across the plasmalemma to the outside of the cell by a mechanism which is not fully understood.[117,286] One of the most spectacular functions of dictyosome cisternae is the packaging of complete complex scales in some unicellular algae, and their subsequent transport across the plasma membrane to the cell exterior.

Dictyosomes are now known to function in secretion (see Chapter 11). Their function in the synthesis and transport of polysaccharides has recently been clearly demonstrated experimentally by combining autoradiography with electron microscopy. Wheat roots were supplied with radioactive glucose and after the material was fixed and sectioned for the electron microscope it was placed in contact with a photographic emulsion sensitive to radioactivity. The point of this technique is that after developing, dark regions are present in the film; these lie

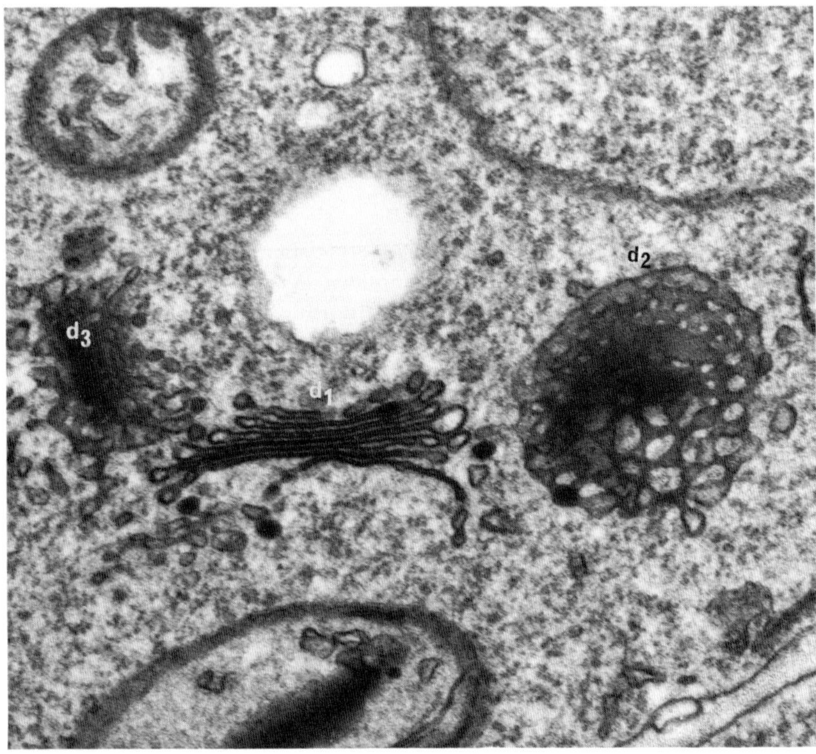

Fig. 3.2 Dictyosomes in a trichoblast of *Hydrocharis morsus-ranae*. d_1, dictyosome in section, showing the stacked cisternae; d_2, section of a dictyosome at right angles to d_1, probably showing a single cisterna (compare Fig. 3.3); d_3, obliquely sectioned dictyosome. ×45 000. (Photo by Dr. K. D. Stewart.)

over organelles which have incorporated the radioactive substance. By use of this method, labelled material was found in the dictyosomes and associated vesicles, and also in the cell wall. By supplying the radioactive glucose for various short periods of time it could be shown that there was a loss of radioactivity from the dictyosomes with time and an increase in radioactivity in the cell wall, indicating that the product of the dictyosomes was transported there.[474] In the root cap of maize, dictyosomes secreting polysaccharides were characterized by hypertrophy of the cisternae and the formation of large vesicles.[367, 463]

Dictyosomes occur generally in plant cells, and are not restricted to those actively involved in secretion. Vesicles produced by the dictyo-

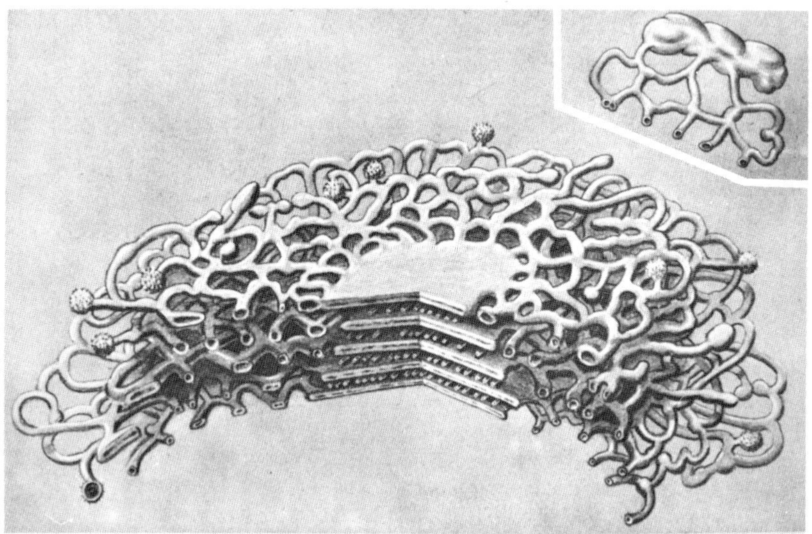

Fig. 3.3 Diagram of part of a dictyosome, composed of five cisternae, from a plant cell. The cisternae are fenestrated, and are separated by intercisternal regions. Vesicles are formed at the edges of the cisternae (see inset). (From Mollenhauer and Morré,[459] Fig. 1, p. 29.)

somes are incorporated into the cell wall; this activity of the dictyosomes is particularly evident in root hairs.

Microtubules and microfilaments

Small elongated structures, the microtubules, are present in peripheral regions of the cytoplasm, and in some other parts of the cell.[408] They are made of a protein called tubulin, which in its polymerized form is visible as microtubules.[286] The frequent similarity of alignment of the microtubules and the microfibrils in the cell wall, led to the view that the microtubules might be responsible for orientation of cellulose synthesis; this is discussed more fully in the next chapter. Microtubules are also thought to participate in the establishment and maintenance of cell shape.[316] The observation of Halperin and Jensen[301] that microtubules were rarely seen in carrot cells cultured in a medium containing auxin, but became visible parallel to the walls in every cell after removal of auxin, is interesting because of the view that auxin affects wall plasticity.

Microfilaments, believed to resemble the actin-containing filaments of muscle, occur in some plant cells, especially components of vascular tissue. They are thought to be involved in the control of cytoplasmic streaming.[316]

Nucleus

In meristematic cells the more or less spherical nucleus occupies a considerable proportion of the volume of the cell, sometimes as much as $\frac{2}{3}$ or $\frac{3}{4}$; this proportion becomes less during differentiation. The nucleus is bounded by a double membrane, the nuclear envelope, which has pores in it at rather regular intervals. The nuclear envelope is connected with the endoplasmic reticulum. Electron-dense chromatic material is present in a ground substance which resembles the cytoplasm but is often of a different density.[712] One or more nucleoli may be present; these are not bounded by membranes. The structure of the chromosomes themselves, the bearers of the genes, is now understood to some extent. The chromosomes consist of nucleoprotein, comprising deoxyribonucleic acid (DNA) and proteins, including histone. The nucleolus, on the other hand, contains ribonucleic acid (RNA) and protein.

The nucleus controls the development of the cell by means of its DNA, which is capable of controlling the synthesis of a kind of RNA known as messenger RNA. This substance is transported out of the nucleus into the cytoplasm, and there transmits a coded message conveying instructions for the synthesis by the ribosomes of particular proteins. According to which of the genes present in the nucleus are active, or derepressed (not all of them are functional at any one time in ontogeny), a different kind of messenger RNA will be synthesized, and different proteins will be made; this in turn will affect the development of the cell. The connections which, as we have seen, exist between nucleus and cytoplasm, and between nuclear envelope and ER (and hence ribosomes), are thus seen to be of great importance in the functioning of the nucleus in the control of development. Experiments involving nuclear transplants in both animals and plants have emphasized that the nucleus is not autonomous, but has an intricate relationship with the cytoplasm.[287]

Ribosomes

Ribosomes are small, more or less spherical organelles which occur free in the cytoplasm, on the outside of the endoplasmic reticulum, and in the nucleus, chloroplasts, and mitochondria. In older electron micrographs the ER often appears devoid of ribosomes, since the fixative then used did not preserve them. Ribosomes have a diameter of about 10–15 nm and consist of RNA and protein, mainly histone. They are functionally very important, being the sites of protein synthesis, but as yet no distinctive sub-structure has been ascribed to them. Clusters of ribosomes known as polyribosomes or polysomes may be the actual structures most important in protein synthesis. A striking example in plants are those observed in root hairs of radish.[63]

Spherosomes

The majority of spherosomes consist of lipid droplets which are not bounded by a membrane.[286] Their frequency of occurrence varies in different cells.

Microbodies

Apart from some which have not yet been categorized, there are two types of these small bodies, glyoxysomes and peroxisomes. Both have a single bounding membrane, and are widespread in plant cells; they differ chiefly in the enzymes present in them. ***Peroxisomes*** occur in leaves, often contiguous with the chloroplast envelope.[286] They contain most of the enzymes for the glycolate pathway from the photosynthetic carbon cycle, and occur in smaller numbers in C_4 plants (see below, p. 38).[665] ***Glyoxysomes*** occur in the storage tissues of fatty seeds and contain the β-oxidation system for breaking down fatty acids in the glyoxylate cycle for redirecting carbon into photosynthesis. The enzyme catalase, which breaks down H_2O_2, is present, apparently in the matrix rather than the membrane of the organelles.[53] In greening cotyledons of a number of plants there is a transition from glyoxysomes to peroxisomes in the same cells;[286] the question whether the one gives rise to the other remains in doubt, though recent experiments involving extraction of organelles from developing cucurbit cotyledons favour the view that the two kinds of organelle arise independently.[368]

Mitochondria

These are organelles which lie at about the limits of resolution of the light microscope, but are just visible in the form of small rods or spheres. In the living condition they are selectively stained by Janus Green B. The electron microscope reveals that they are bounded by a double membrane. Extensions of the inner membrane form tubular projections or microvilli (Fig. 3.4), or plate-like folds (true cristae). Small stalked particles project inwards from the inner membrane; these have sometimes been thought to be artefacts, but it has proved possible to isolate them. They have been observed in plant as well as animal mitochondria.[286, 480] The matrix of the mitochondrion is mainly made up of protein. Ribosomes are present, and are slightly smaller than those in the cytoplasm. DNA fibrils occur in regions of the matrix.

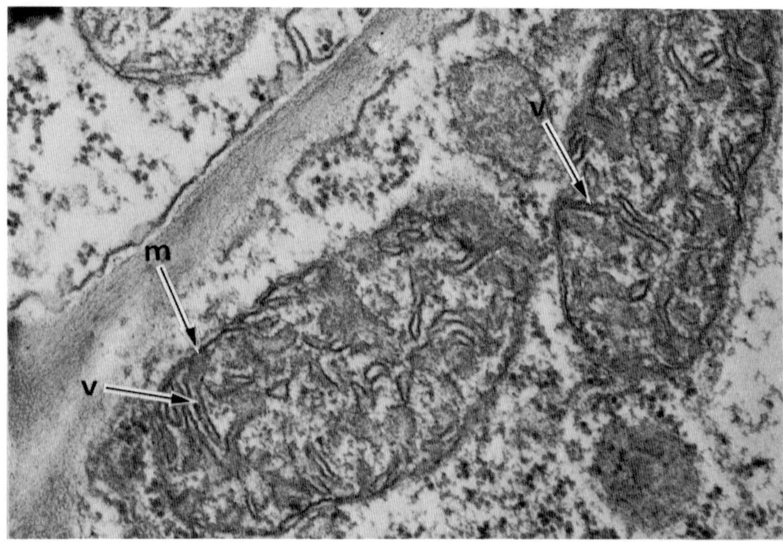

Fig. 3.4 Mitochondria in the mesophyll of the cucumber cotyledon. m, membrane; v, villi. ×48 000. (From Butler, R. D. (1967), The fine structure of senescing cotyledons of cucumber. *J. exp. Bot.*, **18**, 535–543, Plate 5A. Courtesy of Clarendon Press.)

The mitochondria are concerned with processes of energy conversion, and have been called the powerhouses of the cell. They are the locus of many enzymes in the cell, especially those of the Krebs cycle. Mitochondria are thus concerned with respiration. Numerous mitochondria occur in each plant cell, whether meristematic or differentiated; glandular cells and other highly active cells usually contain a greater number.[256] In some cells, e.g. sieve tube elements, the mitochondria undergo degeneration.

The formation of mitochondria has proved a controversial matter, but it is probable that they are formed by fission of existing mitochondria, and indeed the results of experiments involving labelling with radioactive isotopes are compatible only with this view.[480]

Plastids

Plastids may be divided into two types; pigmented and non-pigmented. Pigmented plastids consist of ***chloroplasts***, which are green in colour and in which the pigment chlorophyll predominates, and ***chromoplasts***, which are usually yellow, orange or red and contain the pigment carotene. Non-pigmented plastids or leucoplasts include ***amyloplasts***, which synthesize starch, ***elaioplasts***, which synthesize fats or oils, and some authors include plastids which synthesize storage protein. A comprehensive account of the chemistry, inheritançe and structure of plastids as understood at that time was published some years ago;[382] more recent reviews include those of Thomson[659] and Gunning and Steer.[286]

These plastids are all derived from very small bodies, the proplastids, which are present in egg cells and in cells of the apical meristems.[273] The development of these bodies into amyloplasts has been followed, for example, in the root cap of maize.[712] The proplastids may undergo fission three or four times. Leucoplasts and chromoplasts may originate by the arrest at particular phases of the normal sequence of development of a chloroplast from a proplastid.[273] If the development of a proplastid into a chloroplast is interrupted by lack of light, an ***etioplast*** is formed.[286]

Amyloplasts are usually found in storage organs, such as the potato tuber, and in other deep-seated tissues. They are capable of developing into chloroplasts on exposure to light. The amyloplasts of the root cap, starch sheath of the stem, and the coleoptiles of seedling grasses are thought to function in the perception of gravity[286, 328] (see Part 2,[136] Chapter 2 for a fuller discussion).

Elaioplasts are mainly found in certain monocotyledons. Chromoplasts occur in many flowering parts, e.g. petals and fruits (Fig. 3.5), and also in roots, e.g. carrot. They vary greatly in shape and size, and are often rather angular. They may develop from chloroplasts, e.g. in the fruits of orange or tomato, but may also develop directly from proplastids.

Chloroplasts occur principally in leaves and the cortex of young stems, in parenchymatous or collenchymatous tissues exposed to the light. They are concerned with energy reactions in the cell, lipid synthesis, nitrate reductase activity, and with photosynthesis and the build-up of temporary starch. There are usually several chloroplasts in each cell; for example, 30–500 in leaf mesophyll cells.[286] According to one estimate there are 403 000 chloroplasts in each square millimetre of the leaf of the castor bean, *Ricinus communis*.[273] As has been pointed out, movement of various kinds occurs within plant cells, and the chloroplasts possess a degree of mobility. For example, in many plants light intensity affects the orientation of the chloroplasts; at low light intensity the broad surface of the plastid faces the light, and at high intensities the edge of the plastid faces the light, indicating some movement of the organelles.

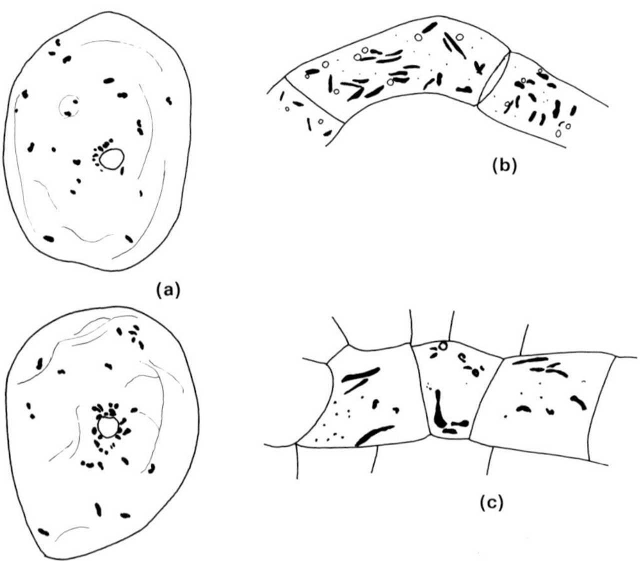

Fig. 3.5 Chromoplasts (black) in cells from (**a**) fruit of *Cyphomandra*; (**b**) fruit of *Capsicum*; (**c**) root of *Daucus* (carrot). (**a**) ×150. (**b**) and (**c**) ×240.

Chloroplasts appear more or less homogeneous or slightly granular under the light microscope, but with the electron microscope they are seen to have a complex structure. They are bounded by a double semi-permeable membrane, within which is a colourless matrix, or ***stroma***, which often contains starch grains. Stacks of lamellae, called ***grana***, constitute the third major component. The stroma is proteinaceous, one of its main constituents in most plants being the enzyme ribulose biphosphate carboxylase, which catalyzes the reaction between CO_2 and its acceptor.[659] Also often present are electron-dense droplets called plastoglobuli. These are not membrane-bounded, and are composed mainly of lipid. Phytoferritin, an iron-protein complex, is often stored in the stroma. Proteinaceous crystals may also be present.[286,659] As already mentioned, ribosomes and fibrils of DNA occur in the stroma, in clear areas called nucleoids. Based on the number of nucleoids and the amount of DNA present, it is thought that chloroplasts are often polyploid.[125] Recently a 'chromosome' active in transcription has been isolated from the chloroplast of the alga *Euglena*. Synthesis of RNA continued for 1 h or more in vitro.[298] Thus transcription can occur without the presence of nuclear DNA.

The most evident bodies present in the stroma are several ***grana***, each of which is composed of a series of double-membrane lamellae, or thylakoids, stacked one upon the other like a pile of discs or pennies (Fig. 3.6a, b). The grana are interconnected by a system of intergranal lamellae, or frets, that pass through the stroma (Fig. 3.6a).[273,702] Several grana are present in each chloroplast; the chloroplasts of tobacco, for example, each have 40–80 grana.[273] There may be from 2 to 100 thylakoids per granum.[671]

Realization of the complexity of the grana-fretwork system of the chloroplast is due to the comparatively recent discovery that the connections between thylakoids of the grana are much more frequent than was at first supposed.[319,492] Every fret ascends a granum in a right-handed helix, and is connected to every thylakoid in the granum; there are also connections with the thylakoids of other grana (Fig. 3.7). Thus the grana-fretwork system is an integrated unit, consisting of a continuous membrane-bounded, inner compartment, and a continuous system of membranes which abut upon the stroma.[286,492] The grana contain chlorophyll.[659] The light reactions of photosynthesis occur in the membrane system of chloroplasts and dark reactions in the stroma.[286]

Suspending isolated chloroplasts in an organic buffer solution, in which the membrane unfolds, indicates the continuity of the grana-fretwork system. Addition of inorganic salts results in reassembly of the grana and frets. Areas of membrane must somehow be able to adhere together to form the grana.[286]

During the formation of chloroplasts from proplastids, flattened vesicles bud off from the inner membrane of the proplastid. These increase in number and form collapsed double-membrane lamellae, which are aggregated in rows in some areas to give rise to the grana and become green. It has recently been claimed that membrane-bounded inclusions observed in proplastids and immature plastids may play a role in thylakoid formation.[126] If barley seedlings are grown in the dark, the rate of fusion of vesicles is slow and instead they accumulate to form a structure known as the prolamellar body. If left in the dark for

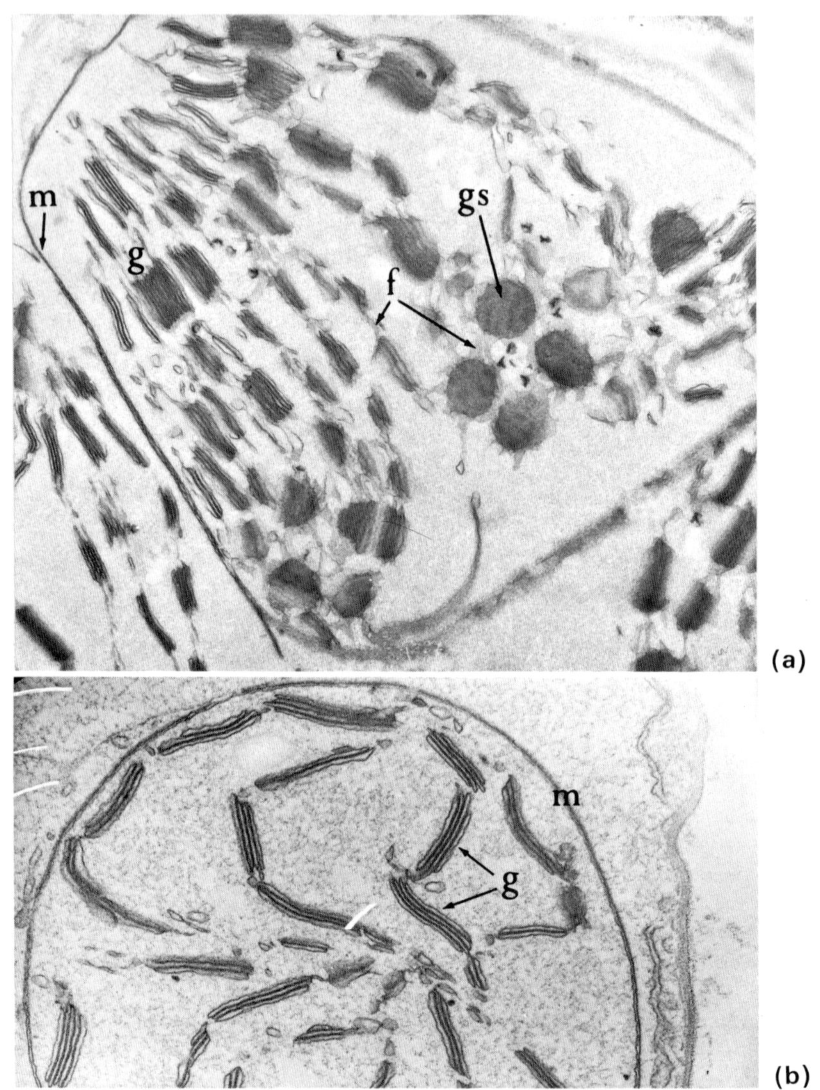

Fig. 3.6 Plastids of *Lemna minor*. (**a**) Parts of two adjacent plastids. The grana are seen both in side and surface view. ×14 200. (**b**) Part of a younger plastid showing grana composed of only a few lamellae. ×19 500. f, frets; g, grana in side view; gs, grana in surface view; m, plastid membrane. (By courtesy of Dr. Norma Lang.)

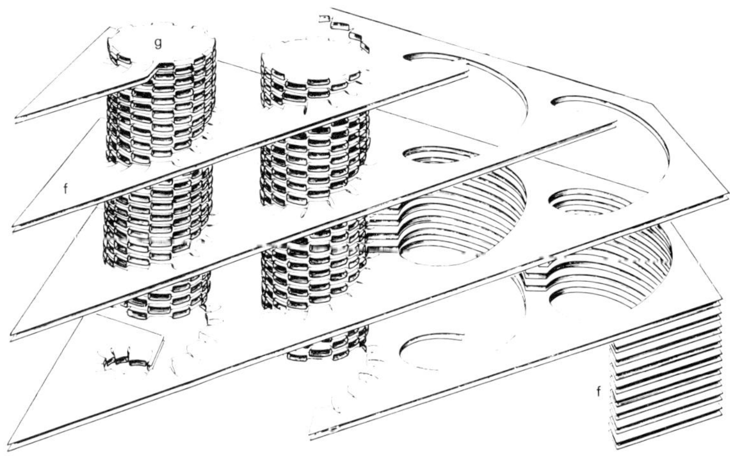

Fig. 3.7 Idealized diagram of the grana-fretwork system of part of a chloroplast in which four grana (g) occur in a line. On the left seven of the eight frets (f) are not drawn in but their connections to the thylakoids of the grana are shown. On the right all eight frets are shown but the grana are omitted. The complexity of the system even under strictly specified conditions is evident. (From Paolillo,[492] Fig. 1, p. 246.)

three to ten days a 'crystal lattice' structure is developed; this may give rise eventually to concentric lamellae if the seedlings are maintained in darkness (Fig. 3.8).[273] Such plastids, formed in the absence of light, are known as etioplasts. The form of the prolamellar body is very variable; it provides a store of membranes and contains the pigment protochlorophyllide, which can give rise to chlorophyll in the light.[286]

It is evident that many cells that do not normally form chloroplasts retain the ability to do so. For example, plants derived from single phloem cells, or small groups of cells, of the carrot root, which are normally devoid of chloroplasts, and buds derived from epidermal cells of flax stems, in which plastid development is arrested at an early stage, both produce chloroplasts. The question of why plastid development is arrested in the cells of the epidermis, but not in the underlying mesophyll or cortex, poses an interesting developmental problem.

The discovery[307, 390] that certain tropical grasses, such as sorghum, maize and sugarcane, possess an additional pathway of photosynthetic carbon fixation has led to extensive work on their biochemistry and fine structure. These grasses are said to exhibit C_4-photosynthesis because the first formed product is a four-carbon dicarboxylic acid (oxaloacetic

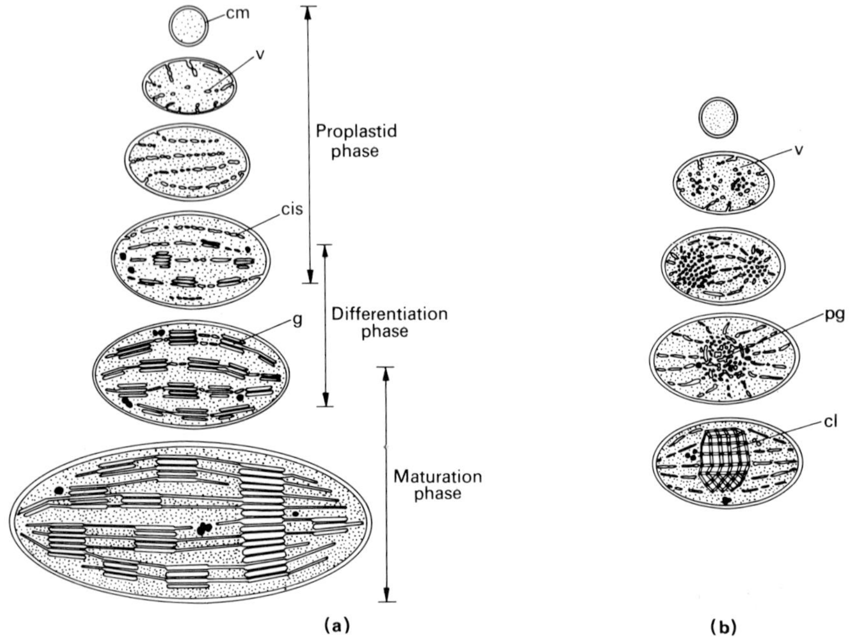

Fig. 3.8 Stages in the development of a proplastid into (**a**) a chloroplast, in the light; (**b**) an etioplast, in the dark, showing the development of the prolamellar body (pg) and crystal lattice (cl). cm, double membrane of the chloroplast; v, vesicle; cis, flattened cisterna; g, granum. (From Granick,[273] Fig. 25, p. 535. Modified after von Wettstein.)

acid) whereas in the majority of plants, in which CO_2 is principally assimilated through the reactions of the Benson–Calvin cycle (or reductive pentose phosphate pathway), the first product (3-phosphoglycerate) contains only 3 carbons. These are commonly called C_3 plants. C_4 plants also contain the enzymes of the C_3 pathway, but the two photosynthetic cycles are contained in different cells within the leaf. In most C_4 species, which include not only tropical grasses but also some dicotyledons, members of the Amaranthaceae, Chenopodiaceae and other families (primarily in the Caryophyllales),[396] one or more layers of parenchyma are arranged round the vascular bundles to form a bundle sheath (Kranz-type anatomy). In these plants CO_2 is initially incorporated into oxaloacetate by the action of phosphoenolpyruvate carboxylase in the cytoplasm of the mesophyll cells, which form the outer compartment. Following reduction or transamination, the product is then translocated to the inner (bundle-sheath) compartment where it is decarboxylated. The CO_2 which is released is then assimilated in the reaction catalyzed by ribulose biphosphate carboxylase which, in these species, is believed to be restricted to the bundle sheath chloroplasts.[172] This fixation, decarboxylation, assimilation sequence increases the CO_2 concentration within the inner compartment, thereby repressing photorespiration and favouring maximal operation of the Benson–Calvin cycle. C_4 plants thus have a high capacity for net photosynthesis and, in their natural environment, crop growth rates for C_4 plants may be double those of C_3 plants.[744] In the majority of C_4 plants grana are present in the chloroplasts of the mesophyll cells, but are absent from or greatly reduced in those of the bundle sheath cells, which contain starch grains.[394 395] The dimorphic plastids of sugarcane, *Saccharum officinarum*, are illustrated in Fig. 3.9. Structural dimorphism of this type does not always exist in C_4 plants, however, and may be replaced by pronounced size differences between the chloroplasts.[396]

Chloroplasts of C_4 plants have a peripheral system of anastomosing tubules formed by invaginations of the inner membrane.[659] However, some workers consider that this peripheral reticulum is not restricted to the chloroplasts of C_4 plants.[286]

An interesting ontogenetic study of the sugarcane leaf[398] has shown that the plastids of both bundle sheath and mesophyll cells, eventually so different in structure and function, both originate from proplastids which are not morphologically distinguishable. During development the chloroplasts in both types of cell develop grana, but those in the mesophyll have more thylakoids in the grana and also have well developed prolamellar bodies. During the terminal phase of chloroplast growth the plastids in the bundle sheath cells lose their grana; at maturity they have no grana but can accumulate large amounts of starch (Fig. 3.9). Thus in sugarcane the structure of the chloroplasts in the specialized bundle sheath cells is a result of reduction;[398] in maize the chloroplasts in the bundle sheath cells may not lose all their grana. During development the grana in bundle sheath chloroplasts of maize enlarge initially, but later most of the membrane stacking is lost.[381]

Despite the frequent correlation between type of chloroplast and photosynthetic pathway, observations on tissue cultures of a member of the Amaranthaceae indicate that the structural specialization of chloroplasts has no obligate causal relationship with the C_4 pathway. These tissue cultures utilized both the C_4 and the C_3 pathways, but had chloroplasts of only one type, with well developed grana and no peripheral reticulum.[396, 397]

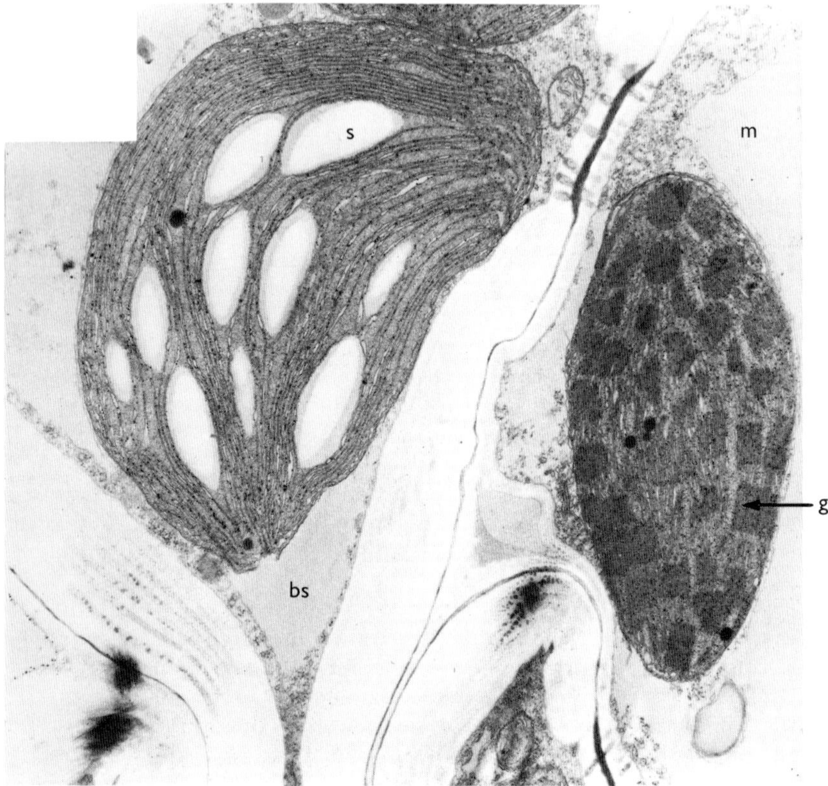

Fig. 3.9 Plastids in the bundle sheath (bs) and mesophyll (m) cells of the leaf of sugarcane, *Saccharum officinarum*. Plastids in the mesophyll cells have evident grana (g), those in the bundle sheath cells lack grana but store starch grains (s). × 12 300. (From Laetsch,[395] Fig. 5, p. 329.)

Ergastic substances

Starch

Starch grains, and the other cell inclusions to be discussed below, are formed as a result of metabolic activity in the cell, and are sometimes known as ergastic substances. Some of these are waste products, others are stored food material. Starch grains characteristically stain bluish-black with a solution of iodine in potassium iodide.

The carbohydrate starch is composed of long-chain molecules which are symmetrically spaced and consequently confer some crystalline properties on the starch grains. This can be seen by viewing the grains between crossed polarizing filters, when they will appear luminous against a black background, as many crystals do. The arms of a dark cross intersect at the hilum of the starch grain. The hilum is the centre of origin of the grain; around it layers of carbohydrate are deposited. These appear as striations, like contours, in the grain, because of a difference in diffraction between successive layers of starch. This gives the starch grain some superficial resemblance to the shell of a mussel. The hilum may be centrally situated in the grain, or it may be eccentric, as in potato starch (Fig. 3.10a). This contributes to the considerable range of form among starch grains. Compound starch grains with two or more hila are characteristic of some plants, e.g. rice, *Oryza sativa*, sweet potato, *Ipomoea batatas* (Fig. 3.10b).

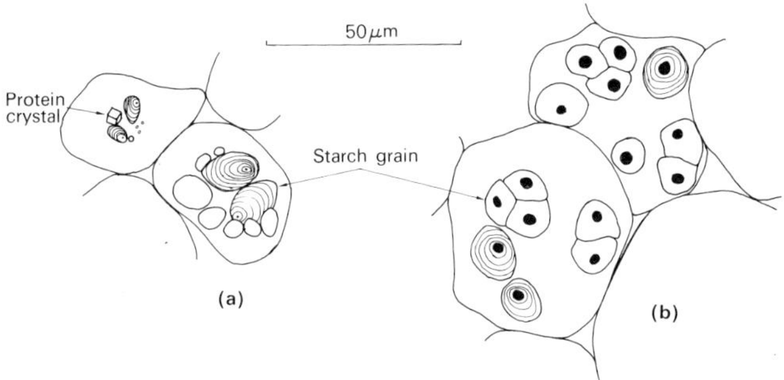

Fig. 3.10 Cells with starch grains. (**a**) *Solanum tuberosum*, potato. Starch grains are simple with an eccentric hilum. A protein crystal is also present. (**b**) *Ipomoea batatas*, sweet potato, with some compound grains. Striations are shown in some of the starch grains. ×430.

In the starch grains of cereals, the number of striations correspond to the number of days' growth, whereas in potato starch the layers are sub-divided into a thick region formed in $18\frac{1}{2}$ h and a thinner region formed in 2 h. Experiments in which plants were grown under conditions of constant temperature and continuous light showed that the cereal starches no longer exhibited stratification whereas potato starch still did. Periodicity in the latter is thus apparently independent of external conditions.[256]

Starch grains may be found in all parenchymatous tissues, but especially in storage organs such as tubers, corms, rhizomes, endosperm or cotyledons of seeds. Starch is synthesized by the chloroplasts; it is subsequently broken down and transported as sugar to other parts of the plant, where it is re-synthesized by amyloplasts.

Commercial starches are extracted from various different parts of the plant, e.g. sago starch from stems; potato starch from tubers; wheat, maize and rice starches from seeds; tapioca and cassava starches from roots.

Protein

Proteins sometimes occur as reserve material. Storage proteins may be amorphous or crystalline or may occur in the form of definite bodies, the aleurone grains. Cuboidal crystalloids of protein may be observed in the cells of the peripheral regions of the potato tuber (Fig. 3.10a). Peeling potatoes may thus remove much of the reserve protein, and is nutritionally a bad practice. These crystals consist of protein whose amino acid component is mainly lysine.[329]

Aleurone grains sometimes combine amorphous and crystalloid forms of protein. They are found in many seeds, in the cells of the endosperm and embryo, sometimes occurring in specific layers. An example of this is the aleurone layer of cereals, the outermost layer or layers of the endosperm, just below the coat of the caryopsis.

Aleurone grains are bounded by a proteinaceous membrane which in the simplest type merely encloses a mass of amorphous protein. The more complex types have various inclusions—crystalloids, globoids or crystals of calcium oxalate. For example, the aleurone grains of the castor oil plant, *Ricinus*, usually have one crystalloid and one globoid, those of the nutmeg, *Myristica fragrans*, have only a crystalloid, and those of some members of the Umbelliferae contain a rosette crystal of calcium oxalate.

In general aleurone grains are rather small, smaller than most starch grains. They stain brown with iodine in potassium iodide, and yellow with an alcoholic solution of picric acid.

Work with the electron microscope has confirmed the vacuolar origin of aleurone grains.[81,475] In wheat endosperm, storage proteins are believed to be formed on the ribosomes of the ER and subsequently secreted internally, appearing to be localized in vacuoles.[272] Study with the electron microscope reveals a peripheral bounding membrane—at least in some species—enclosing a ground substance in which, as a rule, little structure is distinguished; in some species, however, electron-dense globoids are observed.[475] The protein bodies of the cotyledons of squash, *Cucurbita maxima*, are characterized by four components: the proteinaceous matrix, a protein crystalloid, a soft globoid and a globoid crystal.[424]

Recently, a function of aleurone-containing cells of great importance to the embryo has been demonstrated. In barley it has been shown that the hormone gibberellic acid (GA) stimulated both germination and the activity of the starch-digesting enzyme amylase. When, after removal of the embryo, slices of endosperm with and without the special aleurone layer characteristic of cereals were treated with GA, it was found that amylase activity was stimulated only in the presence of the aleurone layer.[481] The cells of the aleurone layer specifically respond to GA by secreting α-amylase.[684] Under normal conditions in germinating seeds it is the embryo which produces gibberellin, which stimulates the aleurone layer to produce α-amylase; and this in turn acts upon the starch present in the cells of the endosperm to convert it to sugar, thus rendering it available to the growing embryo. In the language of molecular biology, in this system the gibberellin is acting as an effector which activates the normally repressed genes controlling the synthesis of α-amylase. After treatment with GA, the rates of synthesis of RNA and of protein by the cells of the aleurone layer are approximately doubled, and nearly half of the total protein formed is α-amylase.[62]

Changes in the aleurone grains and their membranes can be observed with the electron microscope as early as 8 h after treatment with GA, and the changes which occur in fine structure induced by GA in excised aleurone tissue are identical with those occurring during normal germination.[481]

This is a good example of a structurally specialized region which has also a distinctive physiological role in the plant. It is evident that physiological events and structural changes in the tissues are very closely associated.

Fats and oils

Fats and oils are widely distributed in plant cells; they are chemically similar, fats usually being solid and oils liquid, though this is a rather arbitrary distinction. Waxes, cutin and suberin, which occur in and upon the cell wall, are also fatty.

Fats and oils are valuable reserve food materials, and are most commonly found in the tissues of seeds. They occur in the form of solids or as liquid droplets, often dispersed in the cytoplasm of the cell sap. They may be formed by elaioplasts, or by small organelles bounded by a unit membrane, the spherosomes. Fats and oils are frequently located in the cells of the endosperm or perisperm of the seed, but essential oils may occur in special layers, e.g. in the second layer below the epidermis in seeds of cardamom (*Elettaria cardamomum*). Essential or volatile oils are usually formed by specialized secretory tissues, and are discussed in Chapter 11.

Oils and fats stain a reddish colour on warming with Sudan III or IV, and are turned black by osmic acid.

Seeds and fruits are important commercial sources of oil. For example, almond, linseed and castor oil are derived respectively from the seeds of *Prunus amygdalus*, *Linum* and *Ricinus*, and olive oil from the fruit of *Olea*.

Crystals

Crystalline deposits in various forms occur in the cells of many plants. Most consist of salts of calcium; the commonest is calcium oxalate, but calcium carbonate also occurs. Crystals are generally considered to constitute depositions of waste products, and originate within vacuoles.[253] They are visible as bright objects on a dark background when viewed by polarized light.

There are several different forms of crystals in plants, but all originate from a single crystal. Subsequently crystals may aggregate together.

Single or twin ***prisms***, either rectangular prisms or pyramids, may occur (Fig. 3.11b). Prismatic crystals are found in leaves of *Hyoscyamus niger*, *Vicia sativa*, etc., and very large prismatic crystals in the secondary phloem of *Quillaia saponaria*.

Another common form of crystal is the ***druse***. These are more or less spherical aggregates of crystals composed of many prisms or pyramids with projecting points all over the surface (Fig. 3.11a). Such crystals are present in the rhizome of rhubarb, *Rheum rhaponticum*, the leaves of *Datura stramonium*, etc.

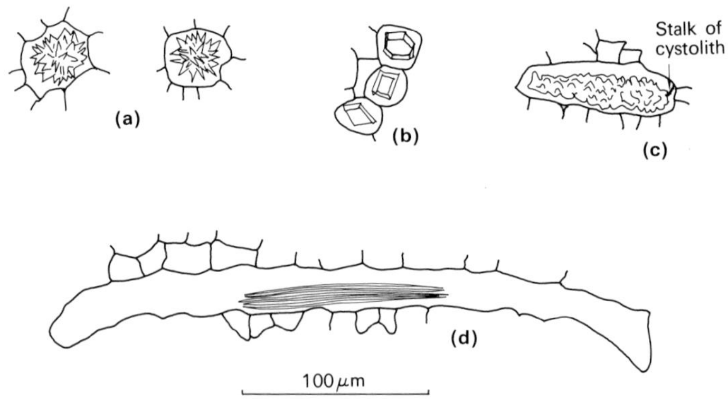

Fig. 3.11 Crystals. (**a**) Druses from the mesophyll of the leaf of *Salix*. (**b**) Prismatic crystals from the leaf of *Citrus*. (**c**) Cystolith from the leaf of *Hygrophila*; the cystolith lies horizontally in the cell (compare Fig. 3.12). (**d**) Raphides in the petal of *Impatiens*. In (**c**) and (**d**) the crystal-containing cells are considerably larger than neighbouring cells. ×245.

Rosette crystals are also aggregates having a fairly large, uniform centre. The components are all nearly equal in length, and radiate out from this centre, giving the whole crystal the appearance of a toothed edge. Rosette crystals occur in the aleurone grains of Umbelliferous seeds.

Raphides or acicular crystals are long and needle-shaped, pointed at both ends, and are usually aggregated in bundles. They occur in the cells of the bulbs of squill, *Scilla urginea*, and petals of *Impatiens* (Fig. 3.11d).

Sandy crystals, or crystal sand, are very small crystals frequently massed together. They occur in particular families, e.g. Solanaceae. The individual crystals are often wedge-shaped, or microsphenoidal, in form, as in the leaf of deadly nightshade, *Atropa belladonna*.

The types of crystal described above are usually deposits of calcium oxalate. Specialized depositions of calcium carbonate, known as ***cystoliths***, also occur in some species (Figs. 3.11c, 3.12). These consist of depositions of $CaCO_3$ around peg-like ingrowths of the cell wall, and are found in parenchyma and epidermal cells, including trichomes or hairs, e.g. those of the hop, *Humulus lupulus*. In the leaf of *Ficus elastica*, the india rubber plant, cystoliths occur in enlarged epidermal cells known as ***lithocysts***. At early stages of leaf development all the epidermal cells appear similar, but slightly later some cells, the future lithocysts, become distinguishable by their denser cytoplasm and larger nucleus. These cells do not divide in harmony with their neighbours, but enlarge while the outer wall thickens and a cellulose stalk develops and projects into the lumen of the cell (Fig. 3.12b, c). Deposition of $CaCO_3$ begins at a relatively late stage of leaf development (Fig. 3.12c, d). The nucleus apparently remains in a functional condition within the lithocyst.[4]

Crystals are most commonly found in the parenchymatous cells of the pith, cortex and secondary phloem. They may be randomly distributed in the cells, or they may occur in ***idioblasts***. These are specialized cells which differ from their neighbours in size, contents or function. For example, the bundles of raphides in squill bulb and in the petals of *Impatiens* occur in cells much larger than those around them; the raphide-containing cells may also contain mucilage. In *Canavalia*, also, crystal cells are larger than their neighbours, and show considerable protein synthesis.[253] In the raphide idioblasts of some members of the Araceae histochemical staining showed that the crystals were surrounded by a matrix of polysaccharide. This brought about the release of the raphides by swelling, thus breaking the wall.[567] The lithocysts of *Ficus* are obviously also idioblasts. There may be one or many crystals in a cell. Crystals may be restricted to or concentrated in

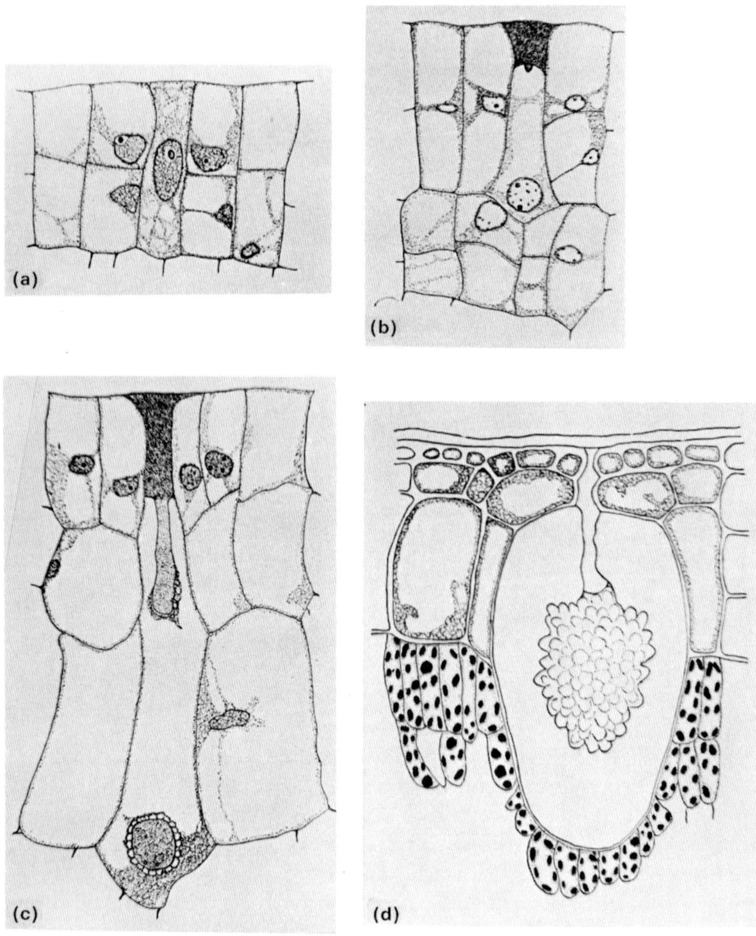

Fig. 3.12 Stages in the development of lithocysts in the leaf of *Ficus elastica*. (**a**) An incipient lithocyst with enlarged nucleus and dense cytoplasm. (**b**) The outer wall has thickened and the peg or stalk of the cystolith is just developing. The nucleus is at the bottom of the cell. (**c**) A late stage in stalk formation. The lithocyst has failed to divide, whereas neighbouring cells have divided. Cytoplasm is aggregated round the nucleus at the base of the cell. (**d**) Mature lithocyst with deposit of calcium carbonate on the stalk. (**a**), (**b**) and (**c**) $\times$ 855. (**d**) $\times$ 700. (From Ajello,[4] Figs. 4, 6, 9 and 17, pp. 590 and 592.)

the cells of a particular region, e.g. the prismatic crystals in the phloem parenchyma cells surrounding a bundle of fibres in the bark of cascara, *Rhamnus purshianus*, and in the cells adjacent to the veins in leaves of *Vicia sativa*. Such a localized distribution of crystal-containing cells is very interesting, and deserves further investigation. At present the underlying causes are not understood.

Studies with the electron microscope have shown that various kinds of crystals have a complex internal structure.[18, 19] Externally, too, they may be grooved and bear excrescences such as barbs, as in the raphides of some edible aroids, which are capable of causing irritation of the mouth and throat. This is thought to be attributable to the purely physical effects of the barbs and tapering point of the crystals.[567]

Crystals may sometimes be of taxonomic value. For example, Dormer[165] has shown that the crystals in the ovary wall of members of the Compositae vary in form in different species. In a more extensive study of the genus *Centaurea*, he showed[166] that the 112 species studied could be divided into two groups on the basis of crystal form; the crystals were either prismatic or curvilinear.

4

The Cell Wall

The prevailing view of the cell wall is that essentially it is not a living system, but in the absence of the protoplast that formed it is merely a non-living shell. However, it is by no means independent of the cytoplasm.

The wall is formed during the growth of the cell, and its properties are affected by the cell's environment, nutrition and stage of differentiation.[473]

Plant cell walls are of considerable importance to man. Cell walls themselves constitute timber and are also used directly as fibres, cotton, etc.; materials extracted from them serve as glue, food and food additives.[525] The properties of cell walls can be important in the texture of apple fruits or the cooking quality of potatoes. Proteins which are the allergens responsible for hay fever are contained in the cell walls of pollen grains, and can be recognized not only by human tissues but also by the plant stigma; if the proteins emitted by the pollen are not compatible with the genotype of the stigma, its cells form another cell wall substance, callose, which prevents entry of the pollen tube;[321] (see also Part 2,[136] Chapter 6). Sporopollenin, a constituent of the cell wall of spores and pollen grains, and of some algae, is one of the most resistant substances known. Other cell walls can be fairly readily broken down by polysaccharide-degrading enzymes; the ability to produce such enzymes accounts in part for the success of many plant pathogens, which gain entry to the host plant in this way.[6]

In summary, the cell wall consists of a crystalline polysaccharide, which in higher plants is cellulose, embedded in an amorphous matrix composed of a variety of polysaccharides and other compounds. The cellulose occurs in bundles of chains which comprise positively birefringent fibre-like structures, the microfibrils.[526] The whole structure, comprising as it does fibrils embedded in a matrix, has been compared with fibreglass or reinforced concrete.[286]

Formation of the cell wall

The cell wall is formed during the process of cell division. There is evidence that the presence of the nucleus is necessary for wall formation.[90] During nuclear division a plate is gradually produced at the position of the equator of the spindle. Work with the electron microscope indicates that vesicles formed by the dictyosomes apparently fuse to form the cell plate[466,711] and in higher plants the process continues at both ends until the cell plate reaches the existing cell walls. Elements of the endoplasmic reticulum become incorporated into the cell plate at intervals and mark the positions of the future plasmodesmata, the cytoplasmic connections which traverse the wall between adjacent cells.

The precise chemical nature of the cell plate in the early stages of its development is not known, but it gives rise to the middle lamella, which is composed of pectic substances. This layer, which is sometimes called intercellular substance, holds together the primary walls of adjacent cells. It can be dissolved by various substances, including the enzyme pectinase; the various techniques used to macerate plant tissues in order to observe dissociated cells depend upon this fact. Some fungi are able to produce pectinase, and can penetrate plant tissues by causing dissolution of the middle lamella.

The question of how the components of the cell wall are synthesized and reach the exterior of the plasmalemma is one which has not yet been satisfactorily resolved. It is considered that polysaccharides are probably secreted by dictyosome cisternae and move thence to fuse with the plasmalemma; vesicles of endoplasmic reticulum (ER) may also be involved.[473] However, O'Brien[476] has rightly warned against the dangers of deducing the progress of dynamic events from static electron microscopic images.

More recently, the view that small, ordered granules occurring on the outer face of the plasmalemma are involved in cellulose synthesis and orientation of the microfibrils has received considerable support (Fig. 4.1). The occurrence of such granules, comprising synthetases or cellulose-synthesizing enzymes, was predicted in 1964 and has since been confirmed by observation, usually in specimens treated by freeze-etching.[525,526] Many of these observations were on lower plants, but the particles have also been observed in xylem, collenchyma and

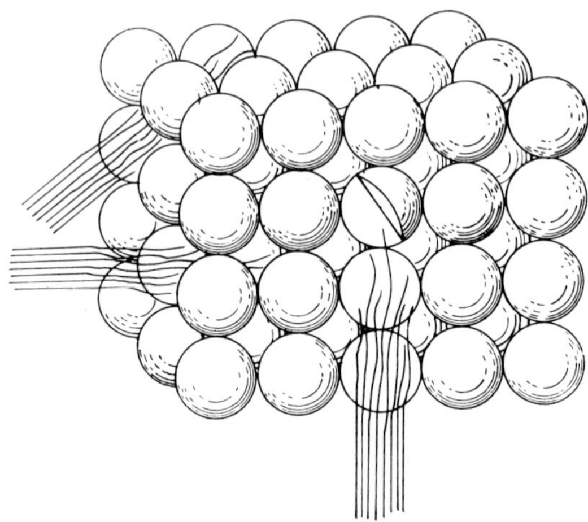

Fig. 4.1 Diagram showing the ordered array of granules containing glucan synthetases thought to be associated with the plasmalemma, which would be parallel to the plane of the page. Microfibrils are shown as bundles of threads representing individual cellulose chains, though they are not drawn to scale; their ends are in contact with the granules. (From Preston,[526] Fig. 13.8, p. 445.)

parenchyma cells of higher plants.[97, 526] When the glucan synthetases contained in the granules come into contact with the end of a microfibril, this is extended terminally by condensation of glucose residues. On this view, the orientation of the microfibrils is determined by the square packing of the granules (Fig. 4.1). Precursors of cellulose would be received from dictyosome vesicles, ER or microtubules.[526] This mechanism would have to be able to change with time,[286] because of changes which take place in the wall during ontogeny (see below). The successful isolation of plasmalemmas and the demonstration that these could polymerize glucose, a process promoted by the synthetic auxin 2, 4-D, lends support to this hypothesis.[683]

Recent techniques which have enabled isolated plant protoplasts to be grown in defined culture media (see Chapter 3) have offered a further means of studying cell wall formation. In such isolated protoplasts, initial cell wall formation is not associated with cyto-kinesis. There is no cell plate, but the wall is formed at the surface of the plasmalemma. A membranous envelope is first formed, apparently involving activity of the dictyosomes, and later cellulose microfibrils are formed between the outer envelope and the plasmalemma.[113] Cell walls can be regenerated by isolated protoplasts by the third day of culture;[469] indeed, scanning electron microscopy of tobacco protoplasts indicates substantial wall formation after 48 and even 24 h of culture.[79] Cultured protoplasts could be used to study the effects of hormones, nutrients or other environmental conditions on the process of wall formation.[1]

Primary cell wall

The primary wall is the first wall to be formed by the cell, and is deposited on either side of the middle lamella by the contiguous cells. Chemically it consists mainly of cellulose, hemicellulose and other polysaccharides. The cellulose lamellae may be separated by layers of pectic substances, for example in the walls of epidermal cells.

All meristematic cells have primary walls, and also many mature cells which still have living contents. Since the wall is formed when the cell is young it must undergo considerable growth; the mechanism by which it grows is discussed below. It is considered that one of the functions of auxin is to increase the plasticity, and thus the extensibility, of the cell wall. The wall may not only undergo surface growth, but may also increase markedly in thickness. Changes of thickness of the primary wall during growth are considered to be reversible, as opposed to the more permanent changes which occur in the secondary wall.[187] The primary walls of some plants, e.g. those of the endosperm in date, *Phoenix dactylifera* (Fig. 4.2), and persimmon, *Diospyros virginiana*, are very thick and serve as a source of reserve carbohydrate.

Secondary cell wall

Secondary walls are usually formed after a cell has completed its elongation, and therefore do not normally extend to any considerable

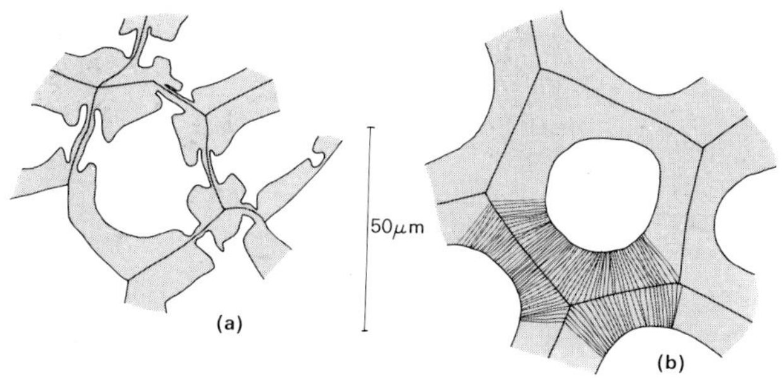

Fig. 4.2 Endosperm cells. (**a**) *Phoenix dactylifera*, date. (**b**) *Strychnos nux-vomica*, nux vomica. The thick cellulosic cell walls are stippled. In (**a**) slightly bordered pits are present in the walls; in (**b**) some of the numerous plasmodesmata are shown. ×430.

degree. Where a secondary wall is formed it is deposited on the inner side of the existing wall, next to the cell lumen. It consists of cellulose and other polysaccharides, but hemicelluloses are relatively less important than they are in the primary wall. Hemicelluloses are like cellulose, but are built up, not of glucose molecules, but of those of other sugars. Various other substances, notably lignin, may be deposited in the wall. The structure of lignin is not fully understood; it is not a carbohydrate, but a polymer made up of units of phenyl-propane derivatives. Where the secondary wall becomes lignified, the primary wall usually does so also, and indeed lignification commonly begins at the primary wall or the middle lamella. The secondary wall often consists of three layers, so that a cell wall may consist altogether of five layers: the middle lamella, the primary wall and a three-layered secondary wall (Fig. 4.3).

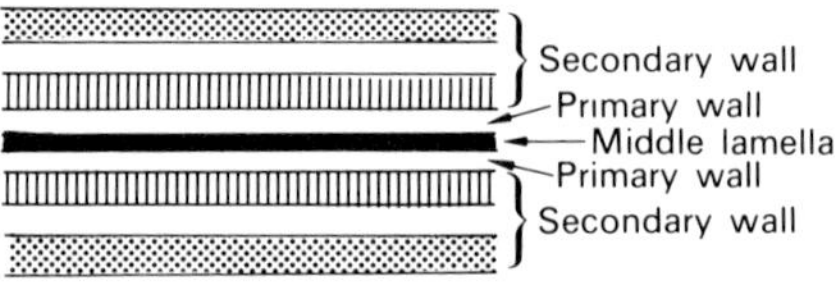

Fig. 4.3 Diagram indicating the structure of two adjacent secondary cell walls.

Secondary walls are generally present in cells which are non-living at maturity, such as sclereids, fibres and vessel elements. In the still elongating elements of the protoxylem the secondary wall is not continuous but is laid down in annular or helical bands; the primary wall between these regions continues to extend and grow harmoniously with the organ in which the elements occur. The reasons for this localized deposition of secondary wall are not fully understood, and are discussed further below. The secondary cell wall is considered to provide mechanical strength.

Primary and secondary cell walls differ both physically and chemically. One consequence of this is that most plant pathogens have much less ability to degrade secondary walls.[6]

Components and structure of the cell wall

This complex field is the topic of a recent book.[526] Cell walls may contain cellulose, hemicelluloses, pectic compounds, lignin, suberin, protein, etc. The proportions of these substances differ both in different species, and in primary and secondary walls of the same species. For example, on a weight to weight basis the proportion of cellulose varies from 1–10% in primary walls of angiosperms to about 50% in some secondary walls.[525] Water is one of the most important, and also one of the most variable, components of cell walls.[286,473] It is thought that the extensibility of collenchyma may be due to its high water content, which enables the microfibrils to slip past one another in a dilute pectin matrix.[526]

Work with the electron microscope shows that the cellulose in cell walls consists of many fine strands or microfibrils, approximately 10 nm in diameter. These may be arranged randomly or in a more or less regular fashion. Within the microfibrils themselves are smaller units, the micelles, which are small aggregations of cellulose molecules that lie parallel to one another and thus confer a crystalline structure upon the microfibrils. It has been claimed that the ultimate structural units of the cell wall are elementary fibrils about 3·5 nm in diameter, which are not aggregated into larger strands,[467] but this is now considered to be a misconception.[526] The spaces between the less regularly arranged molecules in the microfibrils are filled with water, pectic substances, hemicelluloses and, in secondary walls, lignin, cutin, etc.[626] Because of this deposition of lignin between the existing cellulose framework, there is always a swelling of the cell wall during lignification.[465] By use of walls isolated from cell suspension cultures, it has been possible to examine the architecture of the primary walls of higher plants. In sycamore and other dicotyledons, the cellulose microfibrils are coated with a single layer of hemicellulose, in this case xyloglucan. The hemicellulose is cross-linked by pectic polymers, so that the microfibrils are interconnected (Fig. 4.4).[5] A check was made that walls from cultured cells were the same as those in intact plants of the same species. Preliminary work with monocotyledons suggests that their structure may be based on a similar plan, except that arabinoxylan takes the place of xyloglucan.[5]

It has been claimed that there is a group of proteins containing hydroxyproline in the primary walls of various tissues. The amount present increases during growth, and it is thought that the proteins may serve enzymatic as well as structural functions.[380] Experiments with radioactive isotopes show that protein synthesized in the cytoplasm is regularly transported into the cell wall. The protein may be involved in orientation of the fibrils.[467] Other workers consider that the wall protein plays an important role in cell extension,[401] and accordingly it has been given the name extensin. Although the presence of protein in the wall is still not generally accepted, the balance of opinion seems to regard it as proven. Labelling experiments suggest that the protein is deposited throughout the wall matrix.[545]

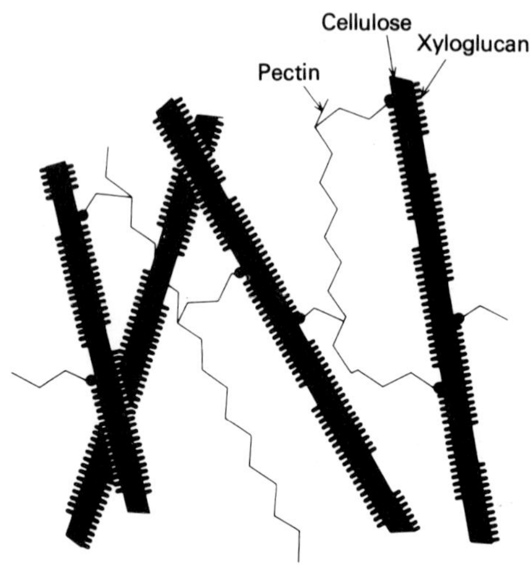

Fig. 4.4 Simplified model of primary cell walls in dicotyledons. Molecules of hemicellulose (xyloglucan) are strongly attached to the cellulose fibres by hydrogen bonds. The fibres are cross-linked by covalent attachment of some of the xyloglucan chains to a pectic polysaccharide. (From Albersheim,[5] Fig. 1, p. 382.)

The protein is firmly bound, apparently by attachment to polysaccharides.[525,526]

The microfibrils are oriented in various ways in cell walls, usually more regularly in the secondary wall. In the primary wall the microfibrils are often oriented in a direction more or less transverse to the long axis; they become arranged more longitudinally during growth of the cell. As subsequent wall layers are formed the microfibrils come to be oriented more and more longitudinally. This transition is gradual; the change in direction of the microfibrils in successive wall layers may be about 120° (Fig. 4.5). As the last stage of wall formation, a tertiary wall may be formed; this differs from the primary and secondary ones and is probably not cellulosic. In gymnosperms this layer may be covered with warts. In the cotton hair, the subject of much work on the cell wall, there is a gradual transition from an approximately axial orientation of microfibrils on the outer surface, through a central region of crossed microfibrils to the transverse orientation of the inner layer (Fig. 4.6).[550]

Recent studies of cell walls of various cell types stained with permanganate have shown that in parenchyma cell walls the microfibrils are laid down in lamellae, the direction of the fibrils alternating in

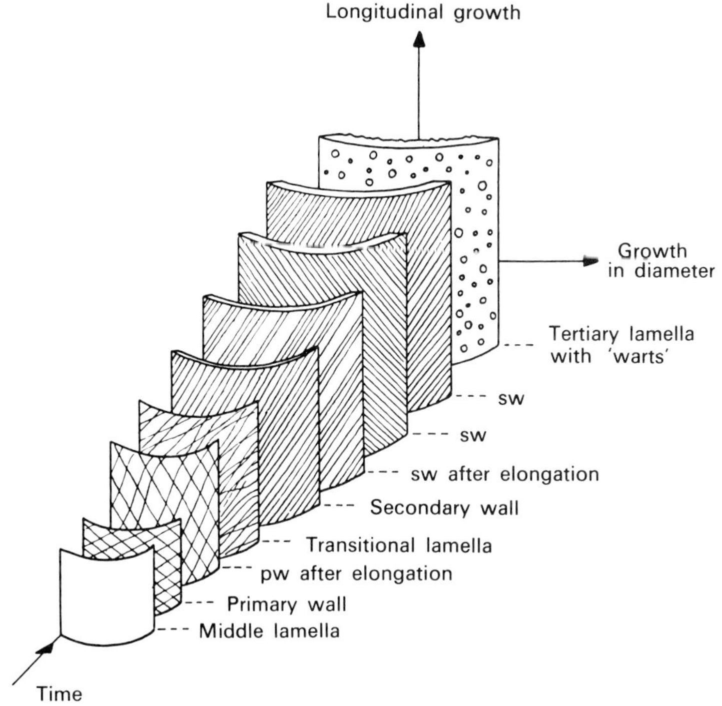

Fig. 4.5 Time sequence of the cell wall layers in a tracheid. pw, primary wall; sw, secondary wall. (From Mühlethaler,[465] Fig. 8, p. 105.)

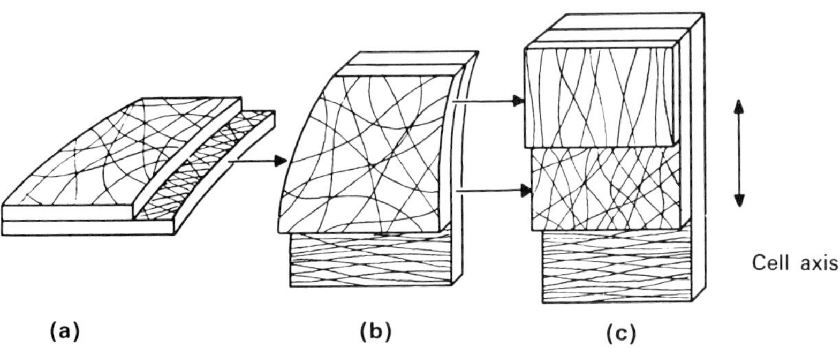

Fig. 4.6 Multi-net growth of the cell wall in a growing cotton hair. (**a**) Near the tip; (**b**) where the tip merges into the tubular part of the hair; (**c**) in the tubular part. The transition is more gradual than the diagram indicates. (From Mühlethaler,[465] Fig. 13, p. 112; after Houwink and Roelofsen (1954).)

successive lamellae; collenchyma cell walls and the nacreous wall of sieve elements are polylamellate.[157,158,159]

Not only can the orientation of microfibrils change from the primary to the secondary wall, but the secondary wall itself may consist of several layers with different orientation (Fig. 4.7). This occurs in wood fibres and tracheids. Layer S_1 is outermost, adjacent to the primary wall, and S_2 and S_3 are laid down within this. The microfibrils form a lax helix in S_1 and S_3, and a steep one in S_2 (Fig. 4.7).[525 526] The chemical composition of these layers also differs.

The question of what controls the orientation of these microfibrils, and the deposition of wall materials in localized sites, as in annular or spiral vessel elements, is a complex one. In many instances ***microtubules*** are aligned with the microfibrils, and this has led to suggestions that they control the orientation of the microfibrils.[97,316,517,526] The fact that the orientation of microtubules and

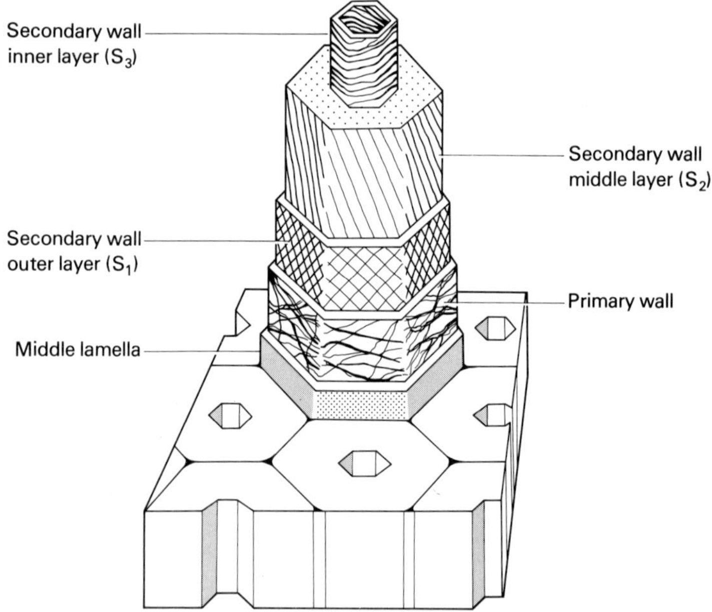

Fig. 4.7 Diagram showing the layered wall structure of a xylem fibre or tracheid. The angle of the microfibrils is shown by the oblique lines in the primary wall and layers S_1, S_2 and S_3 of the secondary wall. (From Preston,[525] Fig. 7.16, p. 280. Copyright © 1974 McGraw-Hill Book Co. (UK) Ltd. From Robards: *Dynamic Aspects of Plant Ultrastructure*. Reproduced by permission.)

microfibrils does not always coincide[97,476] could be explained by supposing that both are orientated by a third factor.[526] Treatment with colchicine, which prevents synthesis of microtubules, also disrupts the pattern of microfibrils.[316] However, synthesis of wall materials continues, as it does also in sites from which microtubules are normally absent,[476,517] so it seems unlikely that microtubules control the actual deposition of microfibrils.

Microtubules are especially clearly associated with the thickened walls of guard cells (see Chapter 7), nacreous walls of sieve elements, and bands of secondary wall in tracheary elements and the cells of the velamen of aerial roots.[316,472] While the tracheary elements of wheat coleoptiles still have primary walls, microtubules become grouped at the sites of subsequent deposition of secondary walls. These may channel wall materials into the thickened regions.[517] It thus seems clear that microtubules play an important role in cell wall formation, though its precise nature is not yet established.

In the cells of different regions of a developing organ the walls may show differential orientations of the microfibrils. For example, in the root of onion (*Allium cepa*) the cell walls of the apical initials show a loosely woven mesh of microfibrils; in slightly older cells the microfibrils are mainly aligned horizontally, and this holds true also during active elongation. But in older elongating cells the pattern changes, and an interwoven mesh of microfibrils is again present. In the root hair zone of the root successive sheets of helical microfibrils, alternately clockwise and anticlockwise, resulting in a criss-cross pattern, are deposited.[586]

Another interesting study on the onion root[354] reveals that these changes in wall structure can apparently be correlated with changes in the relative amounts of cell wall components both in cells at different stages of development, i.e. at different distances from the root tip, and in cells of the different tissues at any one level (Fig. 4.8). In particular, the transitional region between radial enlargement and rapid elongation of the root is characterized by changing relationships between wall components.

Specialized cell wall components

The fatty substance cutin is found in association with most epidermal cell walls. The cuticular membrane is composed of a cutinized layer, a layer of cellulose encrusted with cutin, and an outer cuticularized layer or cuticle consisting of cutin adcrusted on the cell wall (Fig. 4.9).[626] The boundary between the cellulose layer and the cutin is sharply delineated by a layer of pectin (Fig. 4.10).[465] The

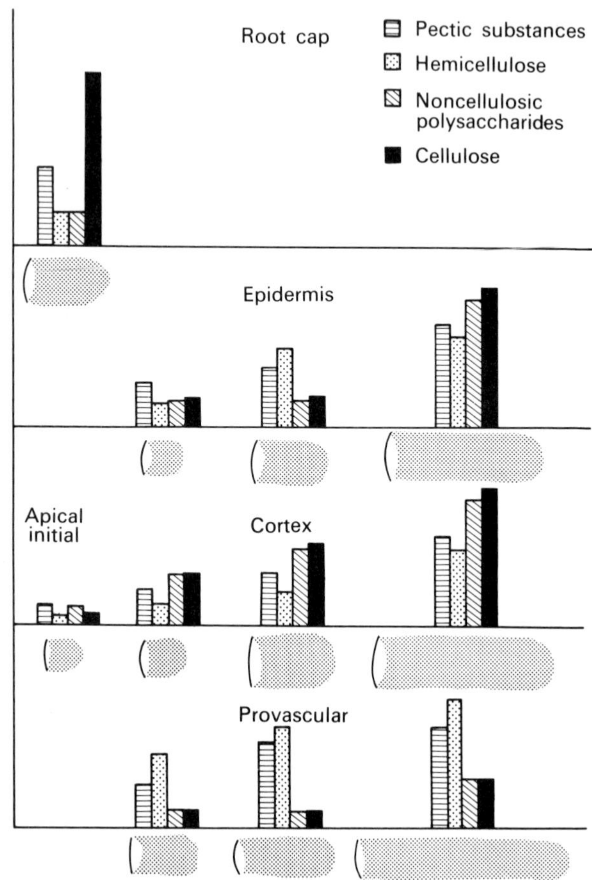

Fig. 4.8 Constituents of the cell wall in the various tissues of a developing root of onion, *Allium cepa*, based on quantitative histochemical procedures and cytochemical data. Outlines of stages of cell development in the various tissues are given on the horizontal axis. (From Jensen,[354] Fig. 4–6, p. 101.)

cuticular layer may produce cuticular pegs between the walls of adjacent epidermal cells. No internal structure has been observed in the cutin itself. Recent work has revealed pores in the cuticle, through which, for example, secretions may pass (see Chapter 7). Wax may be present on the surface of the epidermis in leaves and fruits (see Chapter 7; Fig. 7.11).

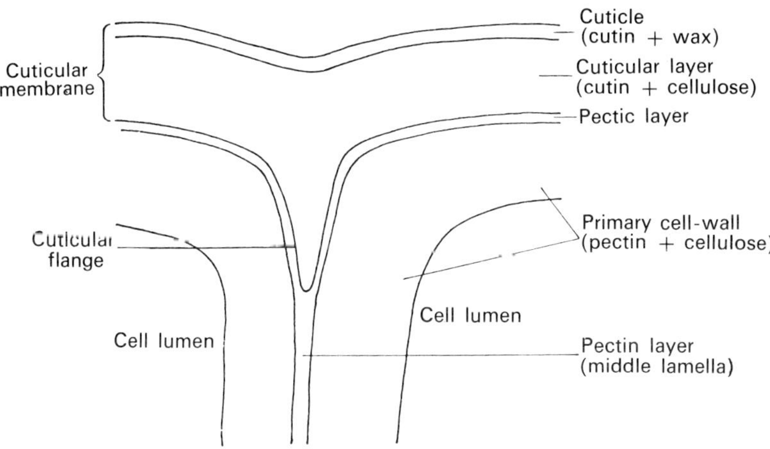

Fig. 4.9 Diagram illustrating the structure of two cuticularized epidermal cell walls at their junction. (From Stace,[626] Fig. 13, p. 63.)

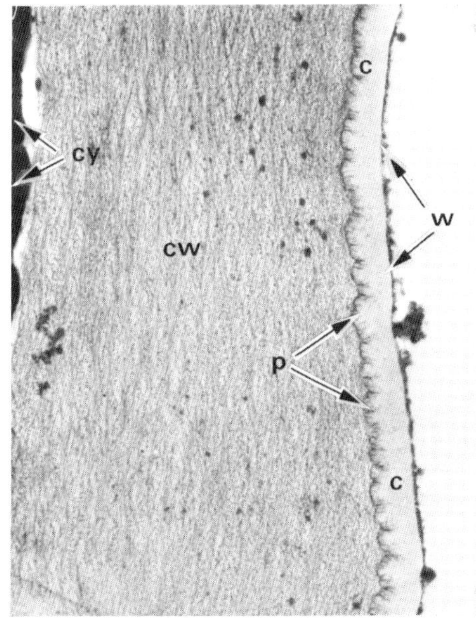

Fig. 4.10 Section through the cell wall of leaf epidermis of *Plantago major*. c, cutin; cw, cell wall; cy, cytoplasm of cell; p, dark layer of pectic material; w, dark layer of wax. ×22 000. (By courtesy of Dr. D. A. Fisher.)

In some cells, notably those of the phellem or cork (see Chapter 12), the walls are encrusted with another fatty substance, suberin. This contains no cellulose. Suberin strips also occur in the radial walls of some cells, such as the endodermis of roots and the basal cells of some secretory trichomes. This Casparian strip of impermeable material in the cell wall, to which the plasmalemma of adjacent cells becomes fused, is thought to enforce passage of materials through the protoplasts of the cells. Suberized lamellae have also been observed in the inner sheath surrounding the vascular bundles of some members of the Gramineae.[478] The carbohydrate callose (see Chapter 9) is another specialized component of some cell walls, as is lignin (see p. 52).

Growth of the cell wall

Formerly two theories were held regarding how the cell wall grows in thickness: that of growth by ***intussusception***, where new microfibrils were held to be laid down between existing microfibrils (Fig. 4.11a), and that of growth by ***apposition***, where new microfibrils were laid down on top of existing ones, forming a new layer (Fig. 4.11b). It is now considered that the formation of both primary and secondary cell walls occurs principally by the mechanism of apposition.[465] It is probable, however, that some growth by intussusception does occur. Growth by apposition is shown convincingly in the successive lamellae visible in some secondary walls, e.g.

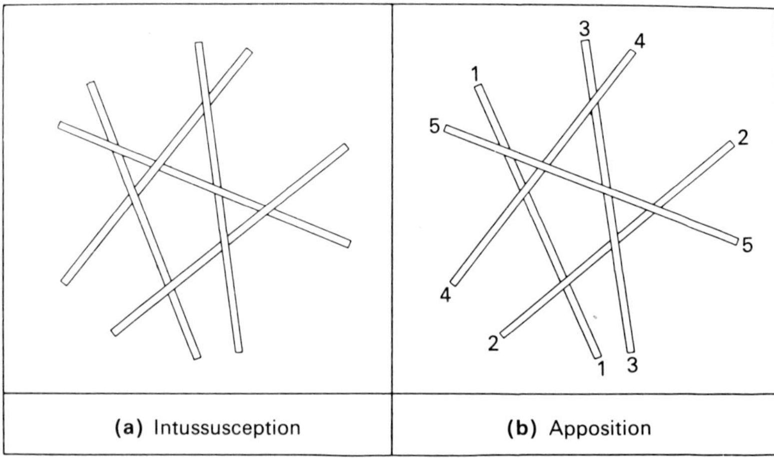

Fig. 4.11 Microfibrils in a primary cell wall. Diagrams showing growth by (**a**) intussusception and (**b**) apposition. The first formed fibril is 1, the second 2, etc. (From Mühlethaler,[465] Fig. 7, p. 103.)

those of cotton hairs and phloem fibres (Fig. 4.6).[550] Deposition of cellulose uniformly over the whole surface of the cell has been demonstrated by the use of the radioactive isotope ^{14}C; this was incorporated into the whole length of the primary cell wall.[550]

With respect to longitudinal growth, the theory now most widely held is the ***multi-net*** theory of cell wall growth; this accounts also for the observed orientation of the microfibrils in successive layers of the wall. On this view microfibrils are first deposited more or less transversely to the long axis of the cell, and this layer is later pushed outwards as a result of the formation by apposition of a layer internal to it. During cell elongation the first-formed layers of microfibrils are stretched and thus become oriented in a progressively more longitudinal plane (Fig. 4.6). Studies of cell wall formation in fibres and tracheids using the electron microscope (Fig. 4.5), and also the technique of autoradiography whereby the path of radioactive isotopes is followed, are consistent with the multi-net theory of cell wall growth.[695] As in earlier work with primary cell walls, labelled carbon was found to be deposited more or less uniformly over the secondary walls of fibres and tracheids. Recent work has tended to support the multi-net theory or some modification of it.[158,472,526]

In fibres the formation of the secondary wall may begin near the centre of the cell and progress towards the tips. The wall is thus thicker near the centre. In velamen cells of aerial roots, on the other hand, the formation of the strands of wall thickening begins near the ends of the cells.[472]

In some cells, e.g. root hairs, pollen tubes, tracheids and fibres, growth occurs only at the tip.[550] This tip growth is regarded as a localized type of multi-net growth. The role of determining whether the wall of the whole cell will grow or only a localized part of it—as, for example, in root hairs or stellate parenchyma cells—is attributed to the cytoplasm.[551]

The question of synthesis of wall materials has been considered earlier. Cell growth, however, involves the stretching of the existing wall as well as addition of new materials. Above a certain critical pressure, the rate of cell elongation is proportional to the turgor pressure.[108] The possible role of the protein extensin in cell growth is still controversial.[108,525] Cell growth requires a factor which will bring about loosening of the wall; it is considered unlikely that any single enzyme would suffice.[108] An ideal factor would have not only to break bonds existing in the wall, but also to catalyze new cross-connections. In enhancing elongation of pea stems, auxin is known both to stimulate removal of molecules of xyloglucan, and to promote the inclusion of new ones.[5]

The ability of auxin to increase wall plasticity has been known for some 40 years, though its mechanism of action is not fully understood; it seems likely that its site of action must be close to the plasmalemma.[526]

It is indeed remarkable that the cells of a tissue are apparently able to enlarge so harmoniously that the plasmodesmata connecting them are not sheared.[286] It is difficult to imagine how factors promoting cell growth can be so uniformly distributed, or how the complex process of wall growth can proceed so smoothly.

Intercellular spaces

In mature tissues, spaces are frequently present between cells. These spaces are formed by a splitting apart of the walls of contiguous cells, and are therefore said to be schizogenous. In many plants growing in aquatic habitats a complex continuous system of well developed intercellular air spaces is present (see Chapter 5). Even in terrestrial plants, intercellular spaces sometimes become enlarged and give rise to secretory glands or ducts (see Chapter 11).

In meristematic regions the cells are usually thought to be in close contact all round. Observation of the onion root tip with the electron microscope, however, revealed the presence of intercellular spaces only 20 μm from the root apex.[586] Thus very small spaces may exist between cells even where they are not readily observed with the light microscope.

Plasmodesmata

Thin strands of cytoplasm, the plasmodesmata, pass through the cell walls at intervals, thus connecting the living protoplasts of adjacent cells. The plasmodesmata sometimes occur in the future sites of primary pit fields (see below), but may be randomly distributed through the wall. They can often be observed traversing the thick cell walls of the endosperm of certain seeds (Fig. 4.2), e.g. date, *Phoenix dactylifera*, coffee, *Coffea arabica*. Plasmodesmata are evident in cell walls viewed with the electron microscope; they may be very numerous.

Although the existence and possible significance of plasmodesmata have been recognized since 1879, and the early light microscopists achieved considerable knowledge of them,[91] the advent of electron microscopy has so increased the amount of information on plasmodesmata that they are the subject of a recent book.[284]

Plasmodesmata are formed during cytokinesis, apparently at sites in the cell plate where strands of ER prevent fusion of vesicles.[359]

Secondary plasmodesmata can form between more mature cells which have not undergone division, for example in cells of organs (e.g. carpels) which have undergone postgenital fusion, or between the haustoria of certain parasites and the cells of their host plant. In this instance, wall-degrading enzymes must be present in order to perforate an already existing wall, and the two cells involved must somehow exchange information in order to align the two halves of the plasmodesma.[90, 359] However formed, plasmodesmata are narrow channels through the cell wall, bounded by the plasmalemma, containing cytoplasm and often a desmotubule (Fig. 4.12). Plasmodesmata are small, up to 60 nm in diameter, and organelles cannot normally pass through them.[276, 543, 544] The central core or desmotubule, which is not always present, is composed of protein sub-units and is a modified membranous structure which is continuous with, and sometimes thought to be a derivative of, the ER of the adjoining cells.[276, 543, 544] A plasmodesma of this general type is illustrated in Fig. 4.12. Continuity between the desmotubule and the ER is especially well seen in electron micrographs of phloem.[277]

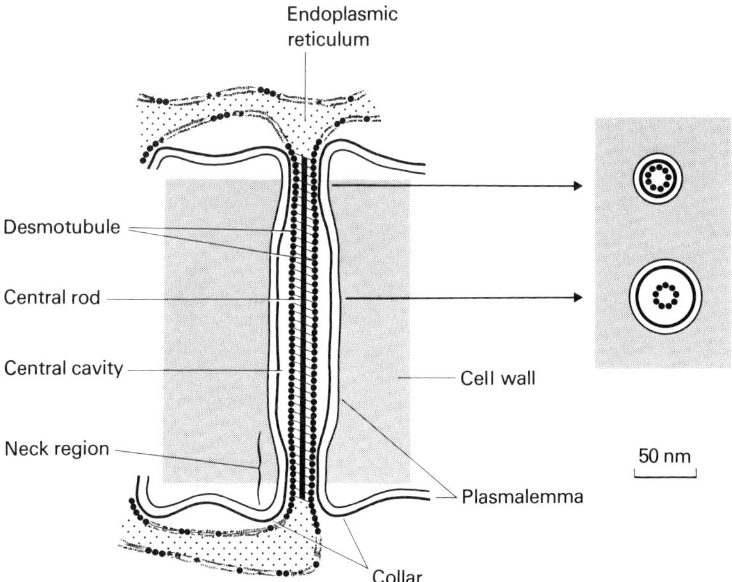

Fig. 4.12 Diagram of a plasmodesma, based on images seen with the electron microscope. The left-hand diagram is the view seen in transverse sections of adjoining cell walls; on the right sectional views of the plasmodesma itself at different levels are shown. (From Robards,[544] Fig. 2.2, p. 31.)

The distribution and number of plasmodesmata per unit area are usually determined during cell division. Plasmodesmata can be quite numerous, and meristematic cells are likely to have between 1000 and 10 000 such connections with neighbouring cells.[544] Plasmodesmata are not uniformly distributed; certain walls of a cell may have a greater number per unit area than others, and there may be a greater number of plasmodesmata between certain components of a tissue than others, for example between A-type transfer cells (see Chapter 10) and sieve elements in the phloem.[277] All living cells of a plant are believed to be connected initially by plasmodesmata, although the latter may be subsequently lost.[544]

The importance of plasmodesmata is believed to lie in their ability to act as channels for symplastic transport. The activity of various parts of the now continuously connected symplast could be regulated through these pathways, although the evidence for transport is in fact still largely circumstantial.[543,544] Calculations show that the plasmodesmata are of sufficient dimensions to be able to cope with the known rates of short-distance transport of substances.[276,286] Calculations also show that there is considerable uniformity in flux per plasmodesma in various different tissues and even species, suggesting that plants may be able in some way to regulate the number of plasmodesmata per unit area of wall according to the demand for transport capacity.[277]

It is not thought likely that most plant hormones move preferentially through the plasmodesmata, though the as yet unidentified flowering hormone may do so, and perhaps the geotropic stimulus in the root.[90]

Intercellular transport of substances may be bidirectional, and it is not known whether this occurs through a single plasmodesma or not. In most plasmodesmata there are, however, two potential conducting channels, the outer cytoplasmic channel and the desmotubule.[285] It has also been suggested that the desmotubule might act as a valve to control the direction of flow through the outer channel.[286]

The fact that cells which are undergoing, or have already undergone, very different pathways of differentiation are effectively connected by these cytoplasmic channels poses a number of interesting problems.

Pits

Primary cell walls are usually not of uniform thickness but have conspicuous depressions in them at intervals; these are called primary pit fields. Secondary walls also have cavities of various kinds, the pits. However thin the wall may be in the area of the pits or primary pit fields, it always forms a continuous membrane across these regions;

actual holes or pores in the wall large enough to be visible with the light microscope are uncommon. Plasmodesmata traverse the primary pit fields, but may occur in other regions of the wall also. The pits may correspond in position to the primary pit fields of an earlier stage in wall development, but this topographic correspondence is by no means absolute. No thickening is laid down over the primary wall in the region of the pit, i.e. the secondary wall is completely interrupted.

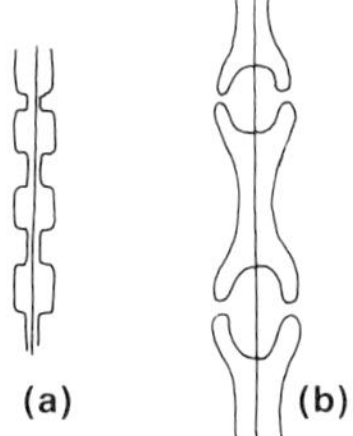

Fig. 4.13 Structure of simple and bordered pits as seen in longitudinal section. (**a**) Simple; (**b**) bordered.

A pit consists of a pit cavity and a pit-closing membrane, which comprises the middle lamella and a thin layer of primary wall. Where pits in the walls of adjacent cells correspond in position, as they frequently do, a pit-pair is formed, and the pit-closing membrane consists of the middle lamella and the primary walls of both cells. Pits may be divided into two main types: simple pits, in which the secondary wall does not arch over the pit cavity, and bordered pits, in which it does (Fig. 4.13).

Simple pits

Conspicuous areas of simple pits may be present in the walls of certain parenchymatous cells, e.g. those of the pulp of the fruit of *Citrullus colocynthis*, or the pith of the stem of elder, *Sambucus nigra*. In some simple pits, especially those found in thick-walled sclereids, the pit cavities may branch, although the opening to the interior of the cell, the pit aperture, is always simple. Such pits are called ramiform pits (Fig. 6.1).

The pit cavity may be of uniform width throughout, or it may be wider or narrower where it abuts upon the lumen of the cell. If it becomes narrower at the end towards the lumen, it is approaching the structure typical of a bordered pit.

In the cells of the onion root tip as many as six or seven primary pit fields per square micron may be present in the walls. This means that a meristematic cell 20 μm long on each side could possess of the order of 20 000 cytoplasmic connections with adjacent cells, passing through the primary pit fields. During cell growth the number of primary pit fields may remain constant, the distance between them increasing.[465] This is another indication that in most cells growth occurs over the whole of the primary wall.

These primary pit fields in the cells of the onion root tip can be observed first as minute circular depressions in the walls of the apical initials. Under the electron microscope it can be seen that in the young stages each primary pit field consists of numerous pores, some of which contain the remains of plasmodesmata, which, however, are not restricted to the primary pit fields. In the region of elongation the pits may be divided into two or more sections by strands of microfibrils. In the final stage of development of the pit it is subdivided in this way into two sections, each with several clearly defined pores between the microfibrils.[586] These pores are, of course, extremely small, and are not visible with the light microscope.

Bordered pits

These usually occur in the elements of the xylem, e.g. vessels, tracheids, fibres, and are more complex in structure than simple pits. The secondary wall arches over the pit cavity, forming the pit border. This encloses the pit chamber, which opens into the cell lumen through the pit aperture (Figs. 4.13, 4.14). The torus is a thickened area on the pit membrane which is present in gymnosperms. If the wall is very thick there may also be a pit canal leading from the cell lumen into the pit chamber. There is then an outer aperture, towards the cell wall, and an inner aperture towards the cell lumen. The latter is variable in shape; the thicker the cell wall the longer and narrower is

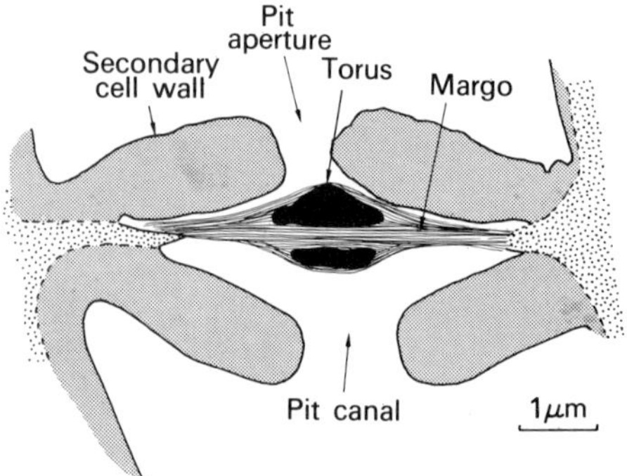

Fig. 4.14 Cross section through a bordered pit of a transfusion tracheid from a needle of *Pinus sylvestris*, showing the pit membrane. The torus (black) is covered on both sides by microfibrils originating from the margo. The secondary wall is lightly stippled, the lignified rim of the pit chamber dotted. (Drawn from an electron micrograph of Liese,[416] Fig. 17, p. 287.)

the inner aperture, and it may even be slit-like in shape. When this occurs the two apertures often cross one another, giving the 'cross pits' frequently seen in lignified fibres or tracheids.

Bordered pits in tracheids can act as valves which control the flow of water through the cell; they thus assume some physiological importance in the plant. The main path of water from the roots to the leaves of conifers is undoubtedly through the bordered pits.[525] The torus acts as a stopper in the valve. If the fluids move too rapidly through the pit, the torus is pushed into a position where it rests against the pit border, preventing further flow.[465] The structure of the pit membrane is important in understanding the mechanism for entry of fluids. The torus, which is apparently impermeable, is suspended by loosely arranged, radially oriented groups of microfibrils which form the raised border of the pit membrane, known as the margo; small perforations occur in this region (Figs. 4.14, 4.15). This structure of the pit membrane, which has been confirmed in studies with the electron microscope,[416,674] was postulated in its entirety by Bailey in 1913, on the basis of light microscope observations and experiments on the passage of an aqueous suspension of carbon particles through the woody tissues.[36] This is one good illustration of the fact that the skill,

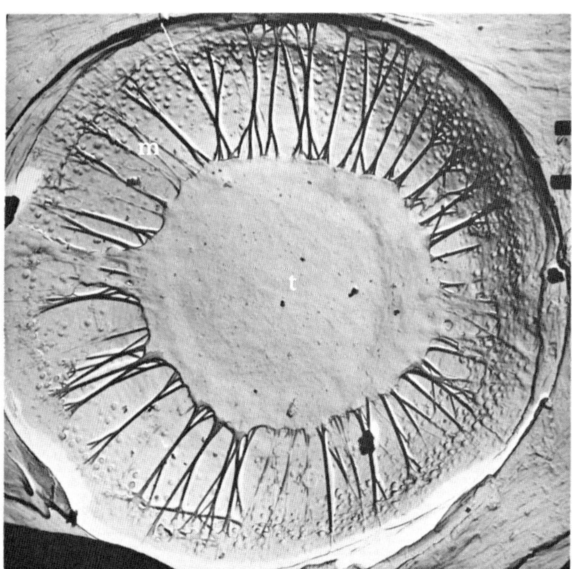

Fig. 4.15 Bordered pit of *Pinus sylvestris*, showing the torus (t) and margo (m) with radiating fibrils. ×5000. (From Mühlethaler,[465] Fig. 31, p. 122.)

ingenuity and insight of the research worker are more important than the possession of elaborate equipment. Much valuable work can be done with very simple tools, such as are readily available in any school or college laboratory. Access to more elaborate equipment, however, may make the research work easier and permit a more certain interpretation of the results.

In the bordered pits of angiospermous woods (hardwoods) both the pit canal and the pit chamber are lined by tertiary wall. Plasmodesmata are usually present in pit membranes which separate living cells. In some dicotyledons, vestured pits occur; these are characterized by small outgrowths from the pit wall which project into the pit cavity.[696] These vestures apparently consist of accumulations of cytoplasmic material at the cell wall, which are covered by the warty layer.[572]

The arrangement of pits in a cell or element may vary considerably. There are three main types of arrangement: scalariform (ladder-like), opposite and alternate. Some work suggests, however, that one arrangement may readily change to another during growth and development.[54]

5

Parenchyma and Collenchyma

PARENCHYMA

Much of the discussion of the plant cell in the preceding chapters is relevant to parenchyma, the most basic type of differentiated cell. Parenchyma cells are usually relatively unspecialized. They form the ground tissue of plants, and occur in the pith and cortex of stems and roots, the mesophyll of the leaf, the endosperm of the seed, the flesh of fruits and in the medullary rays. Parenchymatous cells are also present in association with the conducting elements of the primary and secondary xylem and phloem. Their origin may thus be diverse: from the apical meristems of stem or root, the marginal meristems of leaves, or from the vascular cambium or even the phellogen in more mature organs with secondary growth.

Parenchyma consists usually of thin-walled, vacuolated cells with living protoplasts; the cells are often, but by no means always, more or less isodiametric. The topographic situation of these cells within the plant, outlined above, gives some indication of the importance of parenchymatous cells in many functional activities. Examples of these are photosynthesis, respiration, secretion, and storage of food materials of various types. Transfer cells (see Chapter 10) are essentially specialized parenchyma cells. The contents of parenchyma cells may include crystals, tannins, oils and other secretions, starch, aleurone grains and plastids. Parenchyma which contains numerous chloroplasts and is concerned principally with photosynthesis is termed ***chlorenchyma***. It may be found not only in leaves but frequently also in the peripheral regions of young stems. Food material may be stored not only in cell inclusions such as aleurone grains or starch grains within the cell (Fig. 3.10), but also sometimes in the thick cell walls of the endosperm of certain seeds, e.g. *Strychnos nux-vomica*, *Diospyros virginiana* (persimmon) or *Phoenix dactylifera*, the date (Fig. 4.2). The hemicelluloses within the thick walls of these cells may be regarded as reserve materials. The walls of such endosperm cells, although thick, are primary walls, but parenchyma cells may sometimes have secondary, lignified walls, as in secondary xylem and, occasionally, pith parenchyma.

Although parenchyma cells are usually described as isodiametric, their shape is by no means simple. In isolation, parenchyma cells may be more or less spherical, but when they form part of a tissue various forces act upon them and affect their form. They are in fact polyhedral, having many facets along which they are in contact with neighbouring cells; ideally, they have 14 sides or facets (Fig. 5.1).[333] Recently Korn[387] has constructed a three-dimensional model of dividing parenchyma cells based on certain rules, and has obtained theoretical values for the number of facets, which agree closely with those obtained by Hulbary[333] in an actual tissue, the pith of *Ailanthus*. With a model of epidermal cells, also, the same rules were found to apply, at least in part.[389] Both pressure and surface tension have been considered to play a part in influencing the shape of a cell within a tissue.

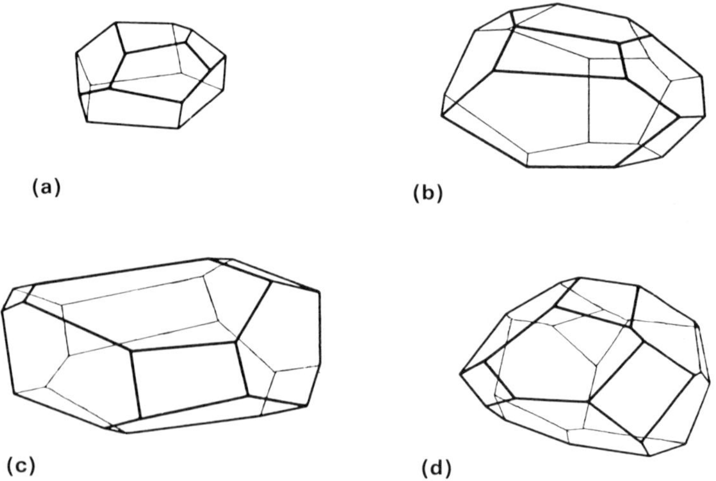

Fig. 5.1 Cells of the pith of *Ailanthus*. (**a**) Small cell with 10 faces. (**b**) and (**c**) Cells with 14 faces. (**d**) Cell with 17 faces. ×200. (From Hulbary,[333] Figs. 1, 5, 7 and 10, p. 564.)

On comparison of his model with real cell populations, however, Korn[387] concluded that the two rules on which it was based, namely that the volume of the cells should be halved by a wall perpendicular to the long axis, and that a wall must not intersect an existing partition between adjacent cells, were sufficient to account for cell shape. The growth pattern of a cell is of fundamental importance. Parenchyma cells may be considerably elongated in one plane, as in the cells of the palisade tissue in the leaf, or they may have several 'arms' or branches, as in the 'stellate' parenchyma cells of the mesophyll of leaves of *Canna* or the pith of *Juncus* (Fig. 5.2). The effect of mechanical stretching during growth on the intercellular spaces between these cells apparently leads to the first stage in the development of the arms of the cells.[264] The arms evidently undergo elongation throughout their entire length, not just in regions close to the spaces.

Intercellular spaces arise either by the splitting apart of the middle lamella region between cells, or, less frequently, by breakdown or lysis of the cells. In certain tissues, notably those of many aquatic plants, intercellular spaces may be exceptionally well developed and form a connected system throughout the entire plant. This tissue is often called aerenchyma, a term reserved by other workers for a tissue in aquatic plants derived from a phellogen. The large vertical air spaces in many aquatic plants (Fig. 5.3) may be occupied by stellate cells and are often transversely intersected at regular intervals by diaphragms, thin plates of cells with intercellular spaces. These diaphragms may be photosynthetic or not; some contain vascular bundles, and there may be a regular sequence of the various types of diaphragm, characteristic of particular species.[376,377,378] In the petioles of water-lily leaves, astrosclereids may branch into the columnar air spaces (Fig. 5.3). These may perhaps be one way of lending strength to the petiole.

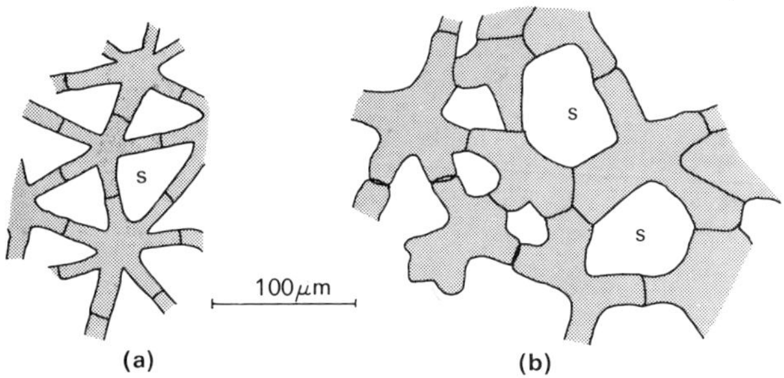

Fig. 5.2 'Armed' or branched parenchyma cells. (**a**) From the pith of *Juncus*. (**b**) From the midrib of the leaf of *Canna*. Large intercellular spaces (s) are present. ×155.

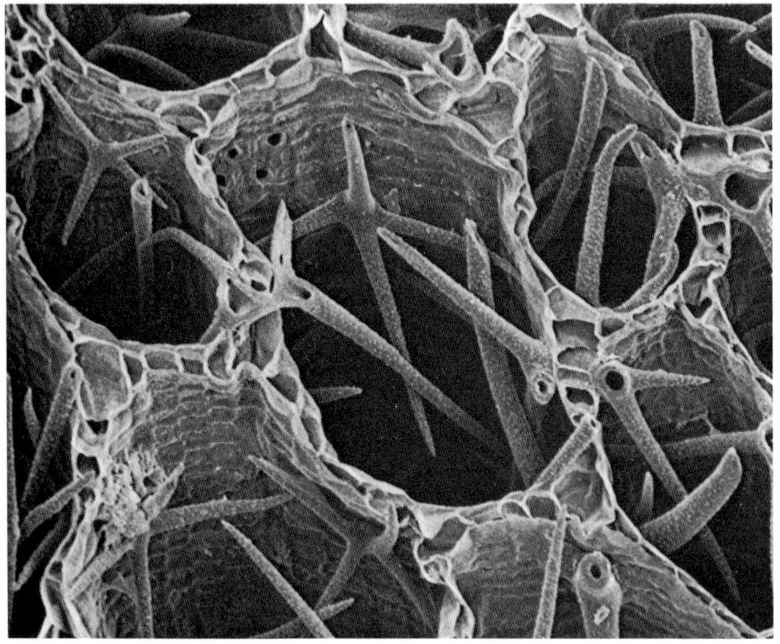

Fig. 5.3 Scanning electron micrograph of part of a section of the petiole of *Nuphar variegatum*, showing numerous astrosclereids branching into the columnar air canals of the aerenchyma. ×125. (By courtesy of Dr. J. N. A. Lott.)

It was formerly believed that this tissue with abundant air spaces functioned only in aeration and in giving buoyancy to aquatic plants; more recently, it has been pointed out that such a system is characterized by exceptional strength for a minimum amount of tissue, and it has been suggested that the honeycomb-like system of intercellular spaces could be an efficient way of withstanding the considerable mechanical stress to which plants in an aquatic environment may well be subjected.[723] However, at least for plants rooted in waterlogged soils, the air space system probably is important in aeration,[116] and this is probably true also for other aquatic and wetland species, especially since it is claimed that loss of oxygen from the roots aids survival by improving poor soil conditions.[17] Despite the constant relationship between the possession of aerenchyma, with its air canals and diaphragms, and an aquatic habitat, nothing is known of the factors influencing its development.

Although relatively unspecialized in the normal development, seeming one of the less interesting tissues, because of their living protoplasts parenchyma cells retain the potentiality for the resumption of meristematic activity, and thus possess striking versatility. Thus, if in horticultural practice an organ is excised from the plant and used as a cutting, it is usually from the parenchyma cells present in that organ that the new root or bud primordia develop. Perhaps the best example of the potentialities of parenchyma cells when removed from their normal environment is the development of whole carrot plants from phloem parenchyma cells of the carrot root, described in Chapter 2. A parenchyma cell, or a small group of such cells, thus possesses the capacity to develop into a whole plant with a complete set of differentiated tissues, but is normally prevented from developing in this way by the restrictions imposed by its position within the plant.

Knowledge of the potentialities of parenchyma cells has recently been enhanced by studies in which viable isolated protoplasts have been obtained from various tissues by removing the cell walls, either by enzymatic or mechanical means. Such naked cells can be grown in culture, and will regenerate new cell walls (see Chapter 4). Under appropriate cultural conditions, which usually include a medium of high osmotic pressure, isolated protoplasts from parenchymatous tissues such as the leaf mesophyll of some species will divide, form callus and ultimately regenerate whole new plants.[112] It is thought that one potentiality of this work is the formation of somatic hybrids by the fusion of isolated protoplasts of different species. By such means various types of genetic incompatibility could perhaps be circumvented. During the process of cell fusion the plasmalemmas of adjacent protoplasts disappear and a new plasmalemma is formed which surrounds the two protoplasts.[524] However, much more work will be required before the main goal is achieved.

COLLENCHYMA

Collenchyma cells have living protoplasts and thickened cellulosic walls. They are extensible cells with a considerable degree of plasticity, and function as supporting tissue in growing organs. They may contain chloroplasts and carry out photosynthesis. They thus differ from parenchymatous cells chiefly in their thick-walled nature, and in being usually somewhat elongated in a plane parallel to the long axis of the organ in which they occur.

Individual collenchyma cells may attain lengths of 2 mm, though rarely. Collenchyma cells differ from sclerenchymatous fibres both in their possession of living contents at maturity and in the cellulosic nature of their walls. Thus collenchyma will not stain red with phloroglucinol and hydrochloric acid, a test for lignin, but will stain blue if treated with a solution of iodine in potassium iodide followed by 66% sulphuric acid, a test for cellulose. In later stages of development, however, collenchyma cells may occasionally become lignified. Collenchyma usually occupies a peripheral position in the organs in which it occurs. In stems it may lie immediately beneath the epidermis, or below a few outer layers of parenchyma. The collenchymatous cells may form a complete cylinder near the periphery of the stem or they may occur in the form of discrete strands, especially in ridged structures such as the petiole of celery, *Apium graveolens*, or many stems (*Calendula*, *Senecio*). Collenchymatous strengthening tissue is commonly found in stems, petioles, peduncles and pedicels; it occurs only rarely in roots, but more frequently in those which have been exposed to the light.[168]

It is often difficult to be certain whether the collenchyma originates from the procambium or from the meristem giving rise to the ground tissue. In the petiole of celery, for example, active periclinal divisions occur on the abaxial side of the procambial strands. These cells become arranged in radial rows and divide, forming a column of elongated, densely staining cells which is separated from the true procambial strand by a secretory duct. This outer column of cells differentiates to form collenchyma.[180] During development, the elongated collenchyma cells may divide transversely, thus having a superficial resemblance to septate fibres.

Usually wall thickening begins in the corners of the cells, but it may spread from there in various ways in different species. In addition to collenchyma cells with more or less uniformly thickened walls, three main types of collenchyma are recognized, according to the disposition of the wall thickening. These are: (i) ***angular*** collenchyma, the commonest type, in which the wall thickening is deposited predominantly at the corners, or angles of the cells, e.g. celery petiole, stems of *Dahlia*, *Datura*; (ii) ***lamellar*** collenchyma, in which the thickening is deposited more heavily on the tangential than on the radial walls of the cells, e.g. stems of *Sambucus*, *Rhamnus*; and (iii) ***lacunar*** collenchyma, in which the thickening is deposited primarily around the intercellular spaces between the cells, e.g. petioles of *Petasites*, aerial roots of *Monstera*. Examples of these various types are illustrated in Fig. 5.4. The mechanism controlling this differential deposition of wall thickening is apparently not understood.

In collenchymatous cell walls there are high amounts of pectin and hemicelluloses. For example, the collenchyma of *Petasites* was found to consist of 45% pectin and 35% hemicellulose, leaving a maximum of 20% cellulose.[550] The micelles of the cell wall are apparently fairly

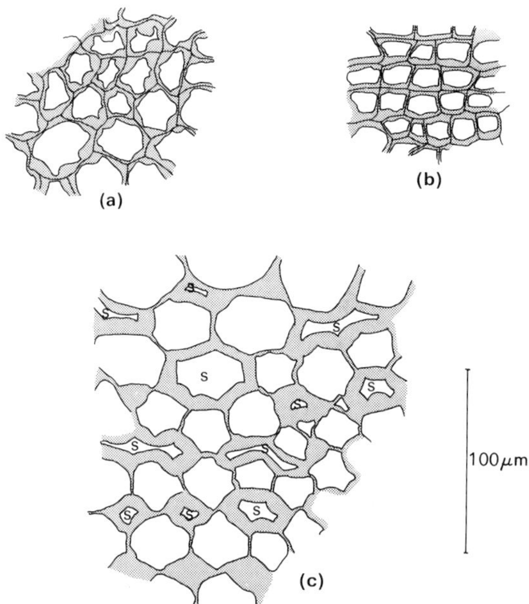

Fig. 5.4 Types of collenchyma. (**a**) Angular collenchyma from the stem of *Cucurbita*. (**b**) Lamellar collenchyma from the stem of *Sambucus*. (**c**) Lacunar collenchyma from the petiole of *Petasites*. s, intercellular space. ×240.

regularly oriented in an axial plane. In celery, studies with the electron microscope show that the wall thickenings are composed of alternate layers of longitudinally oriented cellulose microfibrils and of non-cellulosic material.[48] Visible lamellation of the cell wall has been attributed to alternating cellulose-rich, pectin-poor, and cellulose-poor, pectin-rich layers of material.[526] Such a condition does not seem to occur in all species, however, whereas a study of ten species with the electron microscope has shown a regular alternation of lamellae having transverse orientation of the cellulose microfibrils and lamellae having longitudinal orientation.[96] Further work may be necessary to elucidate this point.

Although there is still abundant scope for a study of the factors controlling the development of collenchyma, a few exploratory experiments have been carried out, and are described below. For example, experiments have shown that mechanical shaking of the plant has a considerable effect on the amount of wall thickening in collenchyma; it does not affect the type of collenchyma formed. In the petioles of celery plants maintained on a mechanical agitator for 9 hours a day over a period of 27 days, 100% more tissue differentiated as collenchyma than in the petioles of control plants. The walls were also 42% thicker.[690] Increases of more than 100% in wall thickness were also obtained in the collenchyma of plants of *Datura stramonium* subjected to comparable treatment for 40 days (Fig. 5.5).[691] It is

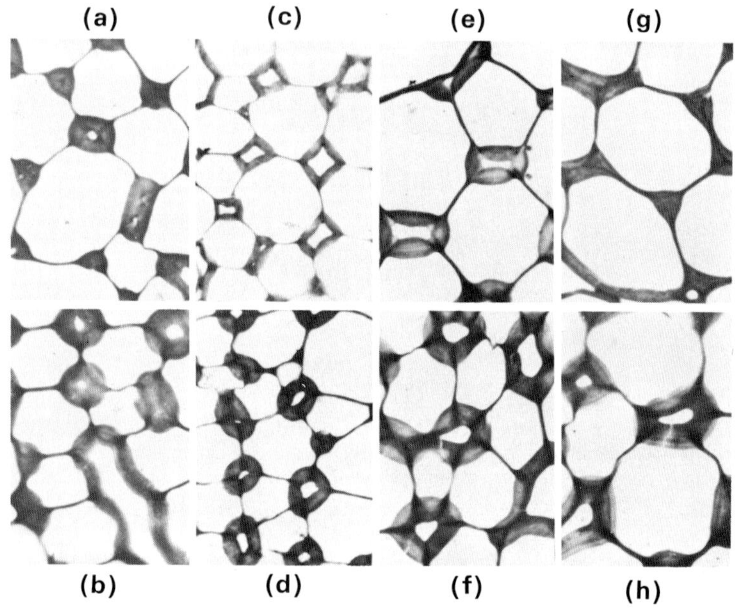

Fig. 5.5 Sections of the stem and petiole of control and agitated plants of *Datura stramonium*. Above, control; below, agitated. (**a**), (**b**) Petiole; (**c**), (**d**) 4th internode; (**e**), (**f**) 3rd internode; (**g**), (**h**) 2nd internode. ×160. (From Walker,[691] Figs. 6–13, p. 720.)

considered that this stimulation of wall thickening is probably accompanied by inhibition of the elongation of collenchyma cells. The development of collenchyma may thus be affected by certain experimental treatments, but the factors controlling its initial differentiation remain to be elucidated. It would be interesting to subject both species that normally do have collenchyma, and some that normally do not, to mechanical agitation during the whole of their ontogeny from the time of germination.

Collenchyma constitutes a living, flexible tissue which possesses considerable tensile stress. Experiments to determine the breaking load of the strands showed that the collenchyma strands in celery were much stronger than the vascular bundles of the same petiole.[180] The plasticity of the walls of fresh collenchyma cells is also higher than that of phloem fibres.[550]

6

Sclerenchyma

This tissue consists of thick-walled elements, which are normally hard and lignified. The cell walls are thickened secondary walls and the cells usually have no living protoplasts when mature. Sclerenchyma is distinguishable from collenchyma both in this lack of living contents and in being lignified; it has a similar function in the plant, however, namely that of support.

Sclerenchyma may be sub-divided into sclereids and fibres; in general, fibres are much more elongated than sclereids, but many intermediate forms occur.

SCLEREIDS

Sclereids, which are sometimes called stone cells because of their hard walls, are usually much more isodiametric in shape than fibres. The gritty texture of the fruit of *Pyrus*, the pear, is attributable to groups of isodiametric sclereids in the flesh. Usually one diameter is not more than three times the size of the other, but some sclereids, called trichosclereids from their superficial resemblance to trichomes, or hairs, are very long (up to ten times as long as broad) and thus do not fit this description. Such sclereids occur in the leaves of *Olea*, the olive. Sclereids are, indeed, extremely variable in shape, and both on this account and because of their distribution within the plant are exceedingly interesting cells. They occur singly or in groups, sometimes associated with the xylem or phloem (e.g. in bark of *Cinnamomum*, cinnamon) but more commonly in parenchymatous tissues, for example the pith and cortex of stems and petioles, e.g. *Hoya*, or of roots, e.g. *Nymphaea* (water lily), the leaf mesophyll, e.g. *Trochodendron*, *Nymphaea*, the flesh of fruits, e.g. *Pyrus* (pear), the seed coat, e.g. *Pisum* (pea), *Phaseolus* (bean). They may occupy a complete layer, as in the seed coat, but more commonly they occur as idioblasts in the tissues mentioned. The question of what causes the differentiation of these often scattered, isolated cells into sclereids is of great interest, and is open to experimental investigation. The distribution of the sclereids may be apparently random, as in the leaves of *Pseudotsuga* and *Trochodendron*[640] and pear fruits,[641] or they may occur in specific positions, for example at the ends of the veinlets, as in leaves of *Mouriria*,[249] *Boronia*[251] and many of the Magnoliaceae.[676] These last are known as terminal sclereids. In the leaves of *Camellia* sclereids occur predominantly near the margins of the leaf.[241] In leaves of *Fagraea* sclereids first differentiate near the tip of the leaf, in the midrib region, and later also in the basal region. Eventually they are formed throughout the leaf.[537]

Types of sclereid

Sclereids may be classified into a number of types, usually based on the extraordinary variation in form found in these cells.

Brachysclereids are shaped like parenchyma cells and are sometimes called stone cells. They occur in the flesh of fruits, e.g. *Pyrus* (pear), *Chaenomeles* (quince), and in parenchymatous tissues or phloem of stems, e.g. *Cinnamomum*, *Hoya*.

Macrosclereids are elongated and columnar in shape, and occur in the seed coat of peas and beans.

Osteosclereids are again columnar but somewhat enlarged at the ends, like a marrow bone, as the name suggests. These again occur in seed coats and in leaves, e.g. *Hakea*.

Astrosclereids are branched and more or less star-shaped. They occur in petioles and leaves, e.g. *Thea* (tea), *Trochodendron*, *Nymphaea*.

Trichosclereids are very much elongated sclereids, somewhat hair-like in form, and sometimes branched. They occur in aerial roots of *Monstera* and in the leaves of *Olea*, olive.

Examples of these types of sclereid are illustrated in Fig. 6.1.

Origin and development

Sclereids which are randomly distributed are usually formed from parenchyma cells which first become distinguishable from adjacent cells by the large size of their nuclei.[24, 261, 640] Subsequently these cells

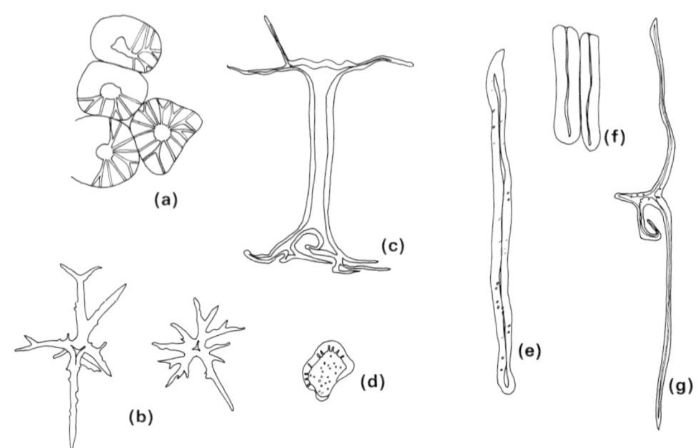

Fig. 6.1 Types of sclereid. (**a**) Brachysclereids from the flesh of pear, *Pyrus*. Note ramiform pits. ×180. (**b**) Astrosclereids from the leaf of *Trochodendron*. ×50. (After Foster, A. S. (1945), *J. Arnold Arbor.*, **26**, 155–162, Pl. III, Figs. 10 and 13.) (**c**) Osteosclereid from the leaf of *Hakea*. ×115. (**d**) Brachysclereid with uneven wall thickening from the cortex of *Cinnamomum* stem. ×115. (**e**) Macrosclereid from the endocarp of apple, *Malus*. ×115. (**f**) Macrosclereids from the testa of pea, *Pisum*. ×180. (**g**) Trichosclereid from the leaf of *Olea*. ×50. (After Arzee,[24] Fig. 8D, p. 685.)

grow very rapidly and may send out branches into neighbouring intercellular spaces. Any part of the wall surface may participate in this growth. In the petiole of *Nymphaea*, the water lily, branches of the sclereid grow in the intercellular spaces (Fig. 5.3), and T-shaped sclereids may develop; if the sclereid initial branches into two intercellular spaces, H-shaped sclereids may result (Fig. 6.2d–g).[261] Astrosclereids in the leaf may bridge the mesophyll from palisade to spongy tissue (Fig. 6.3). In the root of *Nymphaea mexicana*, some cortical cells develop as sclereids. These are cubical cells which develop eight branches or prongs at the corners of the cube (Fig. 6.4). Again, these branches grow out into spaces between the rows of cortical cells. (Being aquatic plants, water lilies have a well developed system of intercellular spaces). The final form of sclereids may thus depend in part on the disposition and ease of penetration of neighbouring tissues. The secondary wall is laid down as the sclereid matures, and may eventually be very thick. Terminal sclereids, associated with the veinlet endings in the leaf mesophyll, are apparently formed from the same cell layer in the meristem that gives rise to the associated procambial strand forming the veinlet.[249] The formation of the sclereid initials thus coincides with the order of differentiation and maturation of the ultimate veinlets.[251]

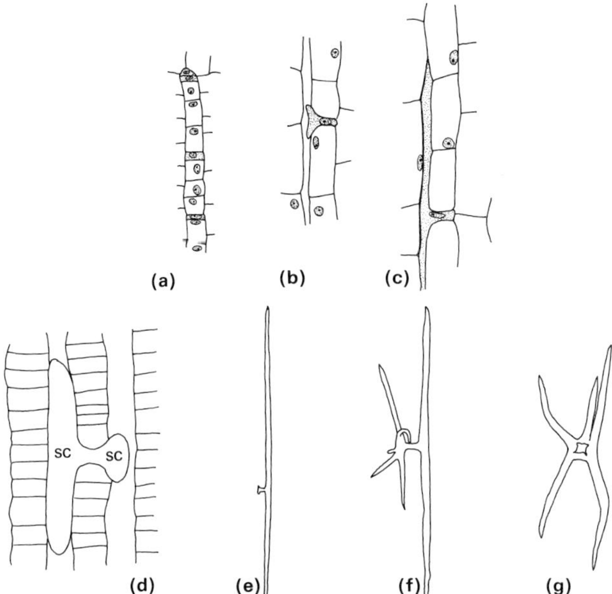

Fig. 6.2 Development of sclereids. (**a**)–(**c**) Longitudinal section of root meristem of *Monstera*. (**a**) Formation of trichosclereids (dotted) at basal ends of cell files. (**b**) 2200 μm from the root apex, showing the outgrowth of processes into an intercellular space. (**c**) Later stage. ×120. (After Bloch,[58] Figs. 5–7, p. 546.) (**d**)–(**g**) *Nymphaea odorata*. (**d**) Longitudinal section of petiole of a young leaf showing elongation of a young sclereid (sc) in the intercellular spaces. ×215. (**e**) Bipolar, (**f**) double bipolar, and (**g**) stellate sclereids from mature petioles. ×67. (Drawn from photographs of Gaudet,[261] Figs 5–7 and 15, p. 527.)

In the aerial roots of *Monstera*, the initials of the trichosclereids are formed by unequal, polarized divisions at the proximal end of files of cortical cells.[58] The small cells thus formed have dense contents and large nuclei and branches from them grow rapidly into the intercellular system (Fig. 6.2a–c). This is another example of the important consequences in differentiation of unequal, polarized cell divisions (see Chapter 2).

Sclereids are thick-walled, but the thickness of the wall may not be uniform. In the brick-shaped stone cells in cinnamon bark, for example, the inner tangential walls are the most strongly thickened (Fig. 6.1d). The walls of sclereids may have numerous pits, which are usually simple; these are often ramiform, i.e. the pit canal is branched. Most sclereids are lignified, and will stain red with phloroglucinol and hydrochloric acid.

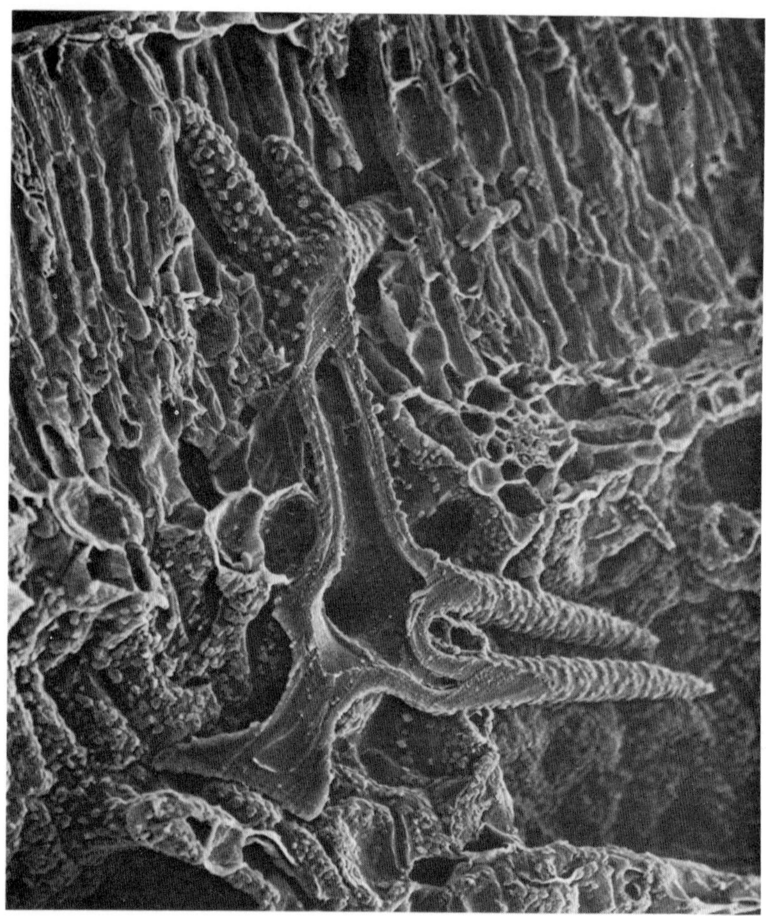

Fig. 6.3 An astrosclereid in the leaf of *Nuphar variegatum*. Part of the wall of the astrosclereid has been cut away, showing the cell lumen. Branches of the sclereid penetrate intercellular spaces in the mesophyll. ×340. (Scanning electron micrograph by courtesy of Dr. J. N. A. Lott.)

Factors controlling differentiation

The facts that sclereids may originate in so many different ways, and may occur in so many different tissues, emphasize the problem of discovering the factors that influence their formation and development. The whole question of what controls the differentiation of

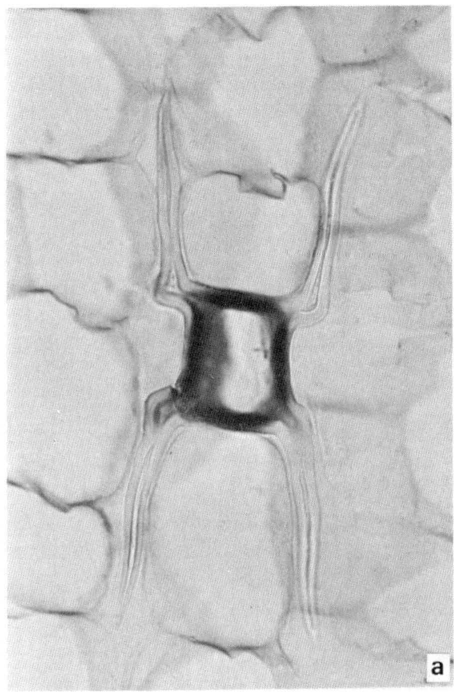

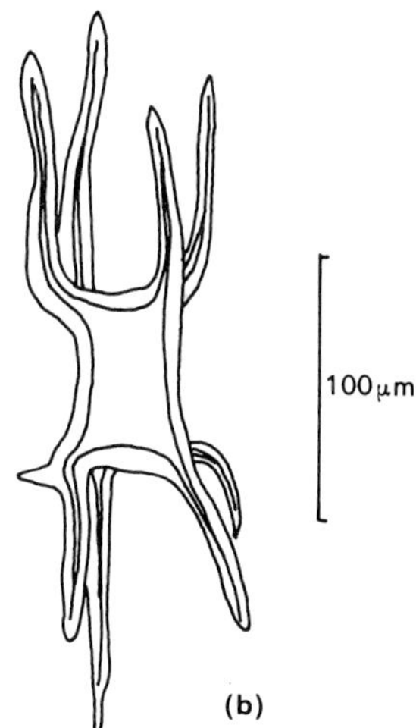

Fig. 6.4 Sclereids from the cortex of the root of *Nymphaea mexicana*. (**a**) Sclereid in longitudinal section of root, showing the position of the outgrowths in the intercellular spaces of the cortical tissue. × 300. (**b**) Drawing of an isolated sclereid, showing the 8 outgrowths from the corners of the cell. × 230.

idioblasts (e.g. crystal-containing cells, tracheioidal cells, sclereids, secretory cells, hairs) is an important one and is the subject of a number of recent investigations.

In some tissues, sclereids apparently differentiate when in close proximity to a surface. This is true of brachysclereids in the aerial roots of *Monstera*,[611] and of astrosclereids in the leaf of *Camellia*, where additional sclereids could be induced to form along new surfaces resulting from experimental incisions in the leaves.[241] Differentiation of sclereids from parenchyma cells was induced by wounding in both leaves and aerial roots of *Monstera*.[57] Sclereids were found to have differentiated from hypodermal cells along the margin of natural wounds in the leaves of two magnoliaceous species.[678]

In leaves of *Fagraea*, however, wounding inhibited sclereid development in a region round the injury, whether this was in apical, lamina or midrib regions of the leaf. This inhibitory influence of wounds was most pronounced during the early stages of leaf development.[535,538] After injury, some of the epidermal cells in *Fagraea* differentiated as brachysclereids, not normally present in the leaf lamina. In excised buds of *Trochodendron* grown in sterile culture, sclereids were found to differentiate first along the margins of developing leaves (Fig. 6.5). The influence of the position of a cell within the plant on its future differentiation, discussed in Chapter 2, is again illustrated by these examples.

Some other observations and experiments seem to implicate other factors. For example, it was found that fewer sclereids developed in detached, cultured leaves of *Camellia* when they were grown in a medium with high concentrations of sucrose, or in medium with the control level of sucrose plus added mannitol,[240] suggesting that osmotic pressure might be a factor affecting sclereid differentiation. In pear fruits, differentiating sclereids appear to stimulate adjacent cells to develop in like manner.[641] The observation that in *Rauwolfia* differentiating sclereid initials at various stages of development showed intensified activity of the enzyme cytochrome oxidase[456] is interesting, but may indicate merely that the cells which later

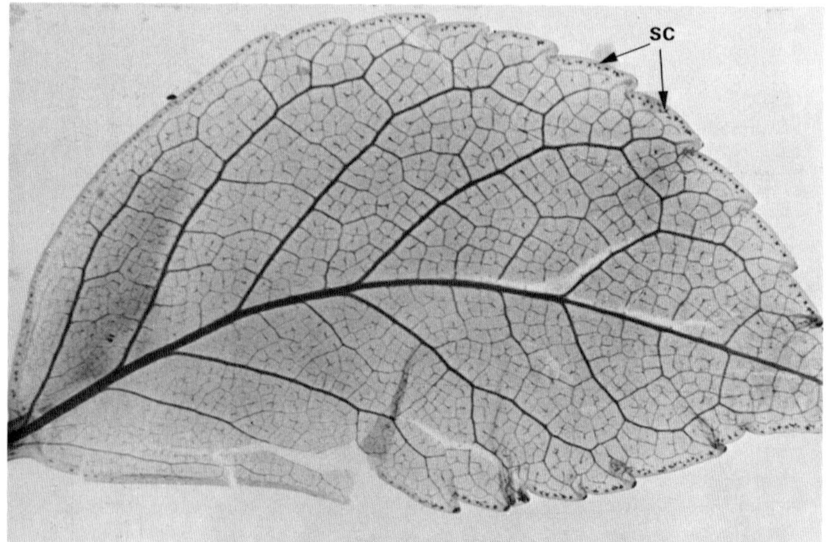

Fig. 6.5 Young leaf of *Trochodendron aralioides*, cleared and stained, from a shoot tip grown in sterile culture. Astrosclereids (sc) are present in the mesophyll at the margin of the leaf only. ×8.

differentiate as sclereids have a higher degree of metabolic activity than their neighbours.

Some important observations and experiments on leaves of Douglas fir, *Pseudotsuga*, suggest that hormonal factors in the developing plant may affect sclereid formation, and play a role in determining when and where sclereid initials will develop.[13] Observation of the leaves along the branches showed that there were more sclereids per leaf in the basal regions of each year's growth than in the terminal regions (Fig. 6.6). This pattern was repeated each year, for example in a 4-year-old branch (Fig. 6.7). These observations suggest that expanding leaves may affect sclereid formation in the younger developing leaves of a branch. It was shown that complete defoliation of a branch, or removal of the leaves from its upper half, caused premature expansion of next year's leaves from the terminal bud. These leaves did not contain sclereids at the time of removal of the outer leaves, but formed them on expansion. They thus provided good experimental material, since the degree of sclereid formation after various treatments could be studied in them. For example, if lanolin paste containing the auxin indoleacetic acid was applied to the defoliated branches, sclereid formation was considerably inhibited in the developing leaves.

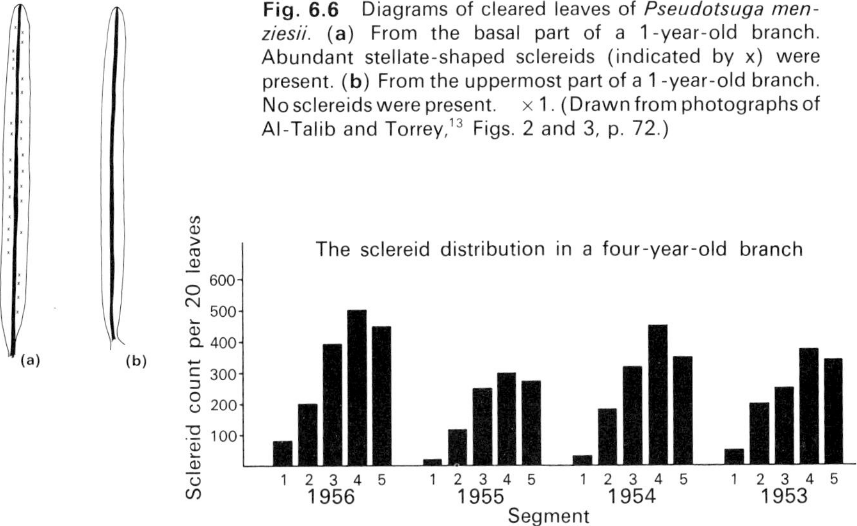

Fig. 6.6 Diagrams of cleared leaves of *Pseudotsuga menziesii*. (**a**) From the basal part of a 1-year-old branch. Abundant stellate-shaped sclereids (indicated by x) were present. (**b**) From the uppermost part of a 1-year-old branch. No sclereids were present. × 1. (Drawn from photographs of Al-Talib and Torrey,[13] Figs. 2 and 3, p. 72.)

Fig. 6.7 Pattern of sclereid distribution along the axis of a 4-year-old branch of *Pseudotsuga menziesii*. Each segment (1–5) contained approximately 20 leaves. (From Al-Talib and Torrey,[13] Fig. 4, p. 74.)

Similar results were obtained in *Fagraea* at certain auxin concentrations. At a lower concentration, sclereids developed but were thin-walled and non-lignified.[536] In other experiments it was shown that in buds of *Pseudotsuga* grown in culture relatively few sclereids developed in the leaves; but in the presence of higher concentrations of various auxins in the culture medium even this number was reduced.[12] It thus appears that auxin may inhibit sclereid formation at certain concentrations, and it seems possible that the auxin produced by expanding leaves may affect differentiation in younger leaves of the bud.

Studies of the order of differentiation of sclereids within a leaf, such as that of *Fagraea*,[537] where the sclereids differentiate first at the leaf tip, may be useful in investigating the factors controlling sclereid formation. For example, it has been pointed out that accumulation of auxin at a leaf tip is a likely consequence of the acropetal movement of auxin in the xylem[593] and perhaps this may be true of other substances also. Such considerations emphasize the possible importance of auxin concentration, and of its interaction with other substances, in the control of sclereid differentiation.

If it is true that hormonal or osmotic factors in the developing leaf play a part in determining how, when and where sclereid initials will develop, then observations on the distribution and development of mature sclereids in the leaves of such plants might tell us something about the auxin relationships and other physiological conditions that prevail during leaf expansion. Interpreted with due caution in this way, plant anatomy could present for us a permanent record of preceding biochemical events—once we have learned adequately to read the code.

FIBRES

Although fibres vary considerably in length, they are typically many times longer than broad. Most fibres are elongated elements with pointed tips, a narrow lumen and thick secondary walls.

Fibres may occur in roots, stems, leaves and fruits, in association with a number of different tissues. They may be present in the xylem or phloem, as a sheath or bundle cap associated with the vascular bundles, especially in leaves, or in the parenchymatous tissues of the pith or cortex. Fibres may occur singly or, more commonly, in bundles. They are sometimes classified[187] in two groups: xylem fibres, and extra-xylary fibres, the latter including all those fibres which occur in tissues other than the xylem, i.e. phloem, cortical and perivascular fibres.

In monocotyledonous leaves fibres may be present not only as a sheath around the vascular bundles but also extending between the bundles and the upper and lower epidermis. It is whole strands of this kind which constitute the 'hard' or leaf fibres used commercially, e.g. sisal (*Agave sisalana*); 'soft' fibres, e.g. flax (*Linum usitatissimum*), are mainly phloem fibres.

Origin and development

Fibres may originate from the procambium or vascular cambium, if they are associated with the primary or secondary xylem or phloem, or from the ground meristem. In the stem of *Linum*, for example, the protophloem comprises a mixture of large and small cells. The large cells are young fibres, the small ones the sieve tubes and companion cells of the phloem.[182] These fibres continue to enlarge and are the source of flax. The fibres of hemp, *Cannabis sativa*,[392] and of ramie, *Boehmeria nivea*,[393] also develop among functional phloem elements from procambial cells.

Primary fibres grow in length with the organ in which they occur.[519] The fibres of *Cannabis* (hemp) and *Corchorus* (jute) extend as the internodes of the stem elongate, but may continue to increase in length after the period of internodal extension.[392] This is true also of the fibres in the fruit of *Luffa*;[608] these networks of fibres are used commercially as sponges. In *Boehmeria*, the fibres elongate faster than the surrounding cells, and at the same time deposition of wall thickening begins at the basal end of the cells.[393] Individual fibres may attain considerable—not to say striking—lengths, e.g. 1–10 cm in hemp, and up to 55 cm in ramie.[7]

The fibres of ramie continue to elongate over a period of months, and may finally attain an increase in length of the order of $2\frac{1}{2}$ million per cent.[250] These cells are thus in a different way no less remarkable than sclereids.

A study[703] of the growth of fibres from the secondary xylem of species with non-storied wood has shown that even fibres derived from the same cambial initial may undergo different amounts of elongation. Such elongating fibres intrude their tips between existing cells; this is called intrusive growth. Intrusive growth occurs only in a region of tissue which is still expanding radially. On the basis of a careful study of developing secondary wood, it is suggested that the fibre tips secrete an enzyme which weakens the middle lamella of adjacent cells. This creates a small space through which the thin-walled tip of the elongating fibre can intrude. On this view the fibre tip would not have to push its way between existing cells.

Deposition of the secondary wall takes place after elongation of the fibre has ceased. In *Boehmeria*, the basal end of a fibre may have a thick secondary wall while the apical end still has living contents and a thin wall.[393] Some fibres are septate, with thin transverse walls, e.g. *Vitis*, *Zingiber* (Fig. 6.8); in such fibres the protoplast may remain living for a long time. These thin transverse septa may be formed after the deposition of the secondary wall material on the longitudinal wall of the fibre. Fibres are usually defined as cells which have no living contents at maturity, but recent evidence[224, 519] indicates that xylem fibres, at least, may in fact retain living contents for several years.

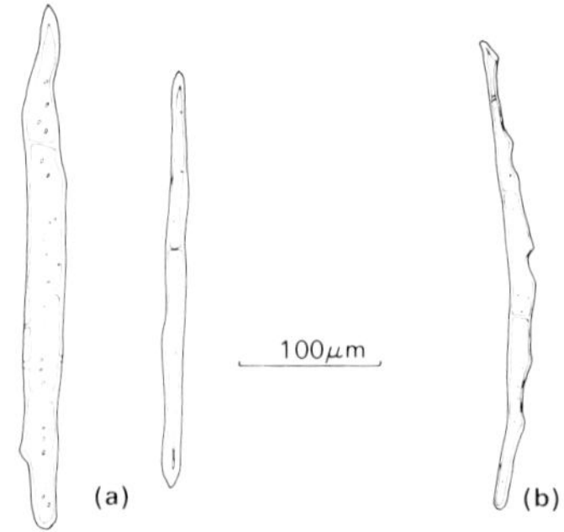

Fig. 6.8 Septate fibres of (**a**) *Vitis* and (**b**) *Zingiber*. ×115.

Fibres may show slight indentations in the wall due to pressure against neighbouring files of parenchyma cells (Fig. 6.8b), or as a result of intrusive growth of the tip.[703] Flax fibres, which may have 90% of the cross-sectional area occupied by the thick wall, are cellulosic, but the walls of many fibres are lignified. Pits are often present in the fibre wall; some of these may be cross pits, i.e. pits with crossed apertures. Formation of the secondary wall is accompanied by vesiculation of the ER and dictyosomes.[520]

Economic uses

Fibres produced by plants have been put to economic use for many centuries. There is evidence that cotton was used between 7200 and 5200 B.C. in the Tehuacán Valley of Mexico;[429] cotton is obtained from hairs on the seed coat, however, and is not a true fibre in the botanical sense (see Chapter 7). About 10 000 years ago the desert peoples of Utah, U.S.A., knew both how to extract plant fibres, perhaps by chewing the plant parts, and how to fashion them into cord; cordage of various kinds has been found in caves at levels dated from

9201 B.C. onwards.[353,614] A complete string bag or net made of the knotted fibres of *Apocynum* has been dated at about 5000 B.C. There is evidence that flax and hemp have been cultivated for fibre for 4000 or 5000 years.[27,250] The fibres of *Salvadora persica*, a prostrate shrub from the Sinai peninsula, are used by the Bedouin as a toothbrush. Pieces of stem with the bark removed at one end are placed on a flat stone and beaten with another stone until the end is frayed into bristle-like projections. The bristles consist of groups of xylem fibres.[211]

At the present time plants from 44 different families are used as sources of fibre. Common commercial fibres may be divided into textile fibres, including flax (*Linum usitatissimum*), jute (*Corchorus* spp.), hemp (*Cannabis sativa*), and ramie (*Boehmeria nivea*), and cordage fibres, including sisal (*Agave sisalana*), bowstring hemp (*Sansevieria* spp.), and New Zealand hemp (*Phormium tenax*). Extraction of most fibres is carried out by a process known as 'retting'. This involves a release of fibres from surrounding tissues by bacterial decomposition of the middle lamellae between the cells. The tissues are left in water for a considerable time while this takes place. Then the retted stems are dried and passed between rollers, which separates the fibres from the other tissues. Finally they are combed, beaten out and placed in bales.[111]

Factors controlling differentiation

Despite the importance of plant fibres, an understanding of what controls or affects their differentiation is largely lacking, though it has long been known that physical stress can stimulate their development. For example, tendrils of *Cyclanthera* which were attached to a support contained more fibres than did tendrils of the same age that were not attached to a support, and the fibres also had thicker walls.[293] In *Cannabis sativa*, the fibres were found to be much stronger in plants from well-watered soil. The physiological factors underlying these observations remain to be investigated. Indeed, the answers to many questions concerning the formation and development of fibres have still to be sought: for example, what factors control their differentiation, not from one, but from several different tissues, what controls the elongation of fibres, and what affects their thickness and strength. Even in the era of man-made fibres, many plant fibres remain economically important and some of these questions would seem to have economic, and perhaps also agronomic, implications. Some studies of the effects of gibberellic acid (GA) on fibre development constitute an interesting attempt to investigate these matters, which may have some economic importance. In jute, treatment with GA increased the amount and percentage of fibre per plant;[612] in jute and hemp, individual fibres of treated plants were considerably longer, wider and more thick-walled.[28, 631] The length of the bundles of fibres which constitute the commercial fibre was increased up to four-fold in GA-treated plants of several species.[630]

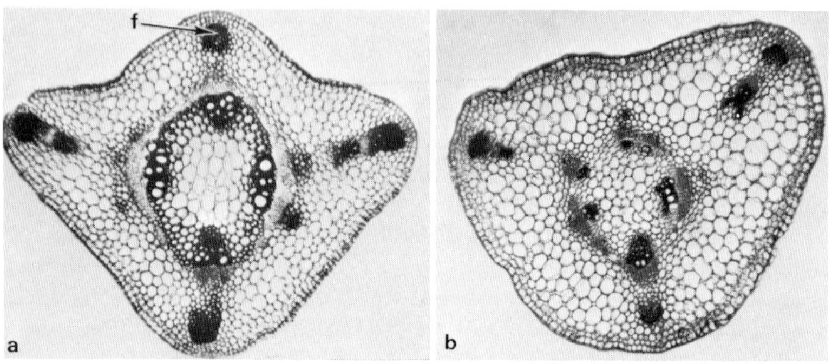

Fig. 6.9 Transverse sections of lateral shoots of pea (*Pisum sativum*), one node below a region which was treated at a primordial stage. (**a**) Control with no treatment. There are two cortical vascular bundles with fibres and two cortical fibre strands. Strand f connected to the third leaf of the shoot, which was the youngest primordium present at the time of treatment. (**b**) A leaf primordium was removed (as well as the apical dome), just after it became visible. The associated fibre strand (corresponding in position to f in Fig. 6.9**a**) almost completely failed to develop. ×70. (**a**, From Sachs,[562] Plate 1A; **b**, by courtesy of Dr. T. Sachs.)

A relationship between substances produced by developing leaves and the differentiation of fibres has recently been demonstrated. In pea stems there is a central cylinder of vascular tissue, but outside this in the cortex there are two vascular bundles including fibres and also two strands consisting only of fibres (Fig. 6.9a). These latter connect to the petioles, which in pea occur in two opposite rows. Sachs[562] showed that if a leaf primordium was removed before it was 25 μm long, its associated fibre strand did not differentiate (Fig. 6.9b). If older leaf primordia were removed, the associated fibre strands did differentiate but were much smaller than normal. Thus it seemed that a stimulus of some kind, formed at an early stage of leaf development, was required for fibre formation. The results of other experiments suggested that this stimulus flowed through certain cells, influencing their differentiation as fibres. This stimulus was apparently not auxin.[562]

Experiments on *Coleus* which combined wounding internodes with the removal of organs distal or proximal to the wound also revealed some interesting results.[9] The shoot was cut off above the fourth internode, and all lateral buds were removed. Some of the treatments are shown in Fig. 6.10; the numbers adjacent to the stem show the number of primary phloem fibres 14 days after the beginning of the experiment, together with the standard error. Fig. 6.10a shows that no fibres were present in the internode below leaf 5 at the beginning of the experiment. If leaves above this internode were removed, no fibres developed in 14 days, whether the internode was wounded or not (Fig. 6.10b,e). When a horizontal wound was made in the fifth internode, and a piece of parafilm inserted to keep the cut tissues apart, fibres developed above the wound on that side of the stem but not below it (Fig. 6.10d). This can be seen also in Fig. 6.11, which shows a section directly above a horizontal wound with well developed primary phloem fibres (Fig. 6.11a), and a section below the wound in which the fibre initials have failed to differentiate into phloem fibres (Fig. 6.11b). These experiments show that mature leaves induce the differentiation of primary phloem fibres in *Coleus*,[9] presumably by the production of some transmissible substance.

Further experiments carried out on beans, *Phaseolus vulgaris*, confirmed that leaves at different stages of development exerted different effects. In this case the fibres of the secondary xylem were studied. Removal of either young leaves or mature leaves showed that both affected cambial activity and the differentiation of its products,

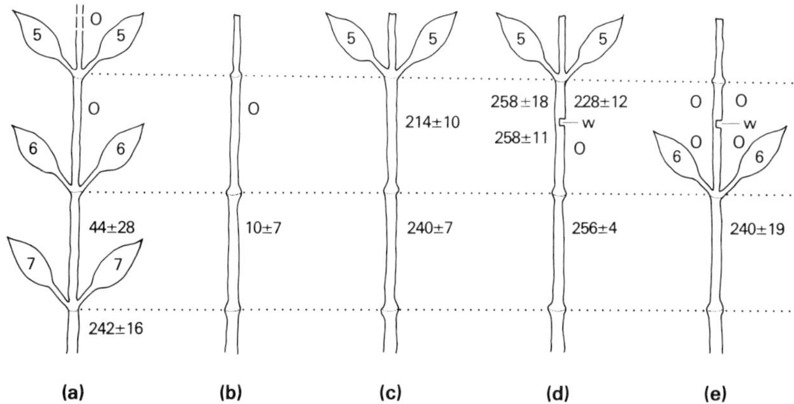

Fig. 6.10 Diagram illustrating the treatments given to internode 5 in stems of *Coleus*. (**a**) At the beginning of the experiment; (**b**) all leaves removed; (**c**) leaves below the fifth internode removed; (**d**) as (**c**) but with a horizontal wound in the fifth internode, the surfaces separated with parafilm; (**e**) as (**d**) but leaf pair 5 removed and leaf pair 6 left on. Numbers beside the stem indicate the average number of primary phloem fibres observed in sections on that side of the internode. (From Aloni,[9] Fig. 18, p. 887.)

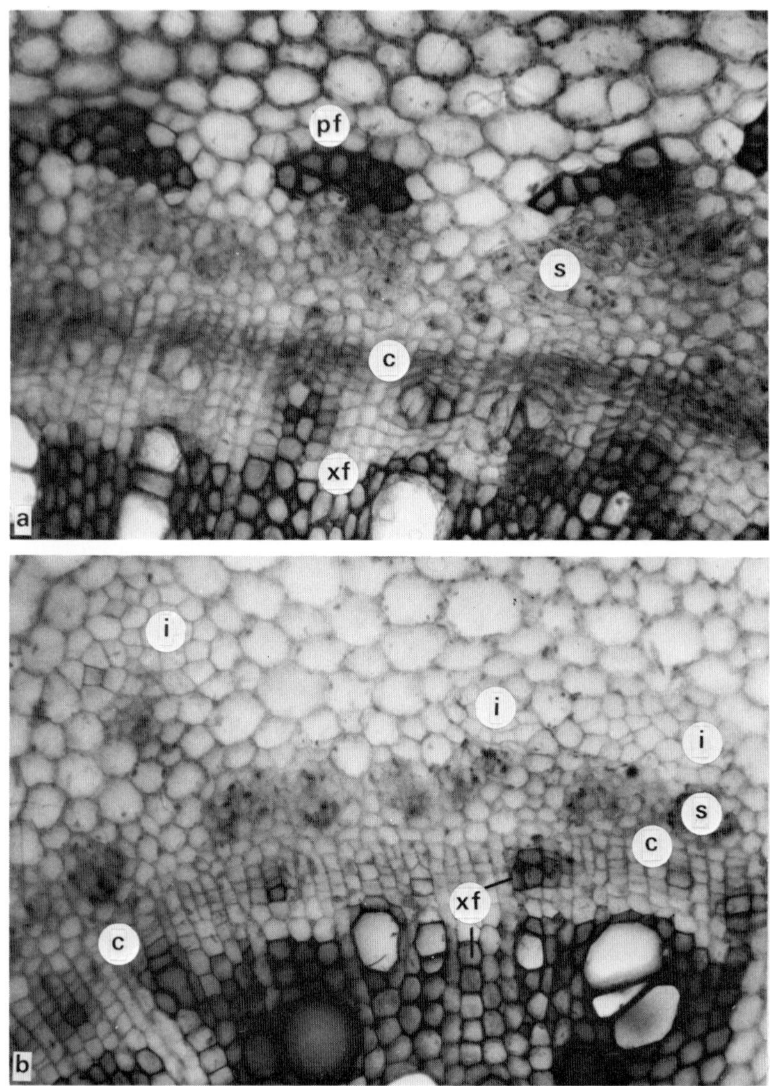

Fig. 6.11 Transverse sections above (**a**) and below (**b**) a horizontal wound in internode 5 of *Coleus*. c, cambium; i, primary phloem fibre initials; pf, primary phloem fibres; s, sieve elements; xf, secondary xylem fibres. ×125. (From Aloni,[9] Figs. 9 and 11, p. 882.)

the proportions of vessels and fibres differing according to the age of the leaf remaining. Although the amount of xylem differentiating below a mature leaf was considerable, it consisted largely of fibres, whereas the xylem developing below young leaves was relatively rich in vessels.[327] In experiments in which excised leaves were replaced by auxin and other hormones, it was found that auxin induced xylem differentiation similar to that associated with very young leaves, whereas gibberellic acid together with auxin led to increased cambial activity and the differentiation of xylem poor in vessels and rich in fibres. Gibberellic acid alone had little effect. It thus appears that bean leaves of different ages produce growth substances in different proportions, which in turn promote the differentiation of different proportions of vessel elements and fibres in the associated xylem. This phenomenon is discussed further in Part 2[136] in relation to seasonal differences in wood structure. Such work certainly makes a stimulating contribution to our understanding of fibre formation and development; interesting as these results are, however, there is still plenty of scope for investigation in this field.

7

Epidermis

The epidermis is the outermost layer or layers of cells on all plant parts during primary growth. It is thus in direct contact with the environment and, as might perhaps be expected, is subject to structural modification by various environmental factors. Both because of its relationships with the environment and because of the conspicuous differentiation that often occurs among the cells of this layer, the epidermis is an interesting tissue and has been much studied. The structure of the epidermis has been fully described by Linsbauer.[417] The development of the scanning electron microscope has allowed direct observation of the epidermal layer in surface view at fairly high magnifications; this instrument also confers a three-dimensional aspect (e.g. Figs. 7.10, 7.11, 7.15).[14,672]

The epidermis of the stem, leaves and floral parts originates from the surface layer of the shoot apical meristem (see Part 2,[136] Chapter 3). That of the root originates from a layer of cells in the root apical meristem that is covered by the root cap; in different species, the epidermis may have a common origin with the cortex or with the root cap (see Part 2, Chapter 2). Usually the epidermis consists of only one layer of cells, but in a few species the cells of this layer may divide periclinally to give rise to a several-layered, or multiple, epidermis. Such a tissue occurs in the aerial roots of some species and is called the velamen. A multiple epidermis also occurs in the leaves of some plants, e.g. species of Moraceae, Piperaceae.

The epidermis of both root and shoot may be differentiated into various kinds of cells. In both instances epidermal cells may elongate at an angle to the surface of the organ to give rise to hairs. In the epidermis of the leaf, and often also of the stem, stomata may be present; in some species cork cells and cells containing silica are also differentiated. Some epidermal cells, or even hairs, may also contain crystals, for example the lithocysts in which cystoliths are formed in the epidermis of *Ficus* (Fig. 3.12). Since these different structures originate from single cells and in a (usually) single-layered tissue which is relatively easy to observe, the epidermis has been used in many studies of cell differentiation and its controlling factors. Recent experiments using tissue culture have revealed additional potentialities of the epidermal cells of some species which are not normally realized when the cells are left *in situ*. This is best seen in the stem epidermis of *Torenia fournieri* (Scrophulariaceae). When segments of internode are excised and cultured *in vitro*, cell division begins in some epidermal cells after 48 hours. At 60 hours, adjacent cells begin to divide also, and gradually numerous 'cell division centres' become recognizable (Fig. 7.1a,b). By the 6th day, about 200 of these are present on any face of the internode.[106] Their distribution is not uniform, but is apparently affected both by endogenous gradients, since the number of cell

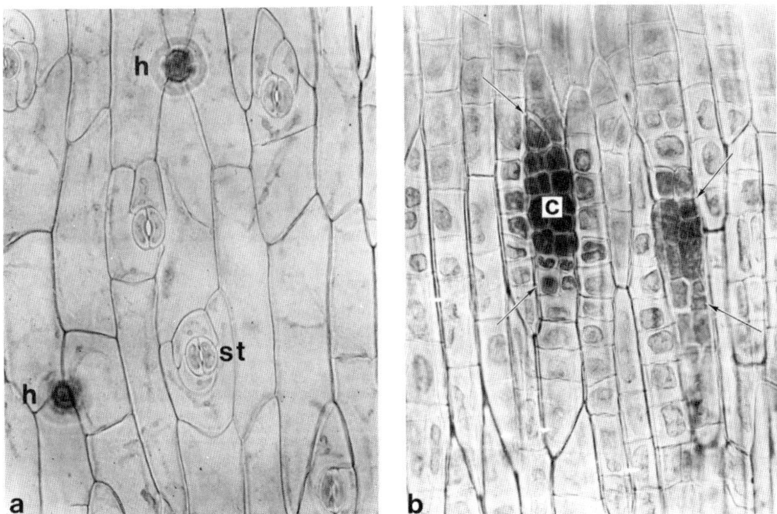

Fig. 7.1 Strips of epidermis from internodes of *Torenia fournieri*. (**a**) Control strip of epidermis, with stomata (st) and basal cells of hairs (h). ×280. (**b**) Part of the epidermis of an internodal segment cultured for several days. Epidermal cells have divided (arrows), c being a central highly divided cell. ×180. (From Chlyah,[106] Figs. 1 and 9; by courtesy of Dr. K. Tran Thanh Van; reproduced by permission of the National Research Council of Canada from the *Canadian Journal of Botany*, **52**, 867–872, 1974.)

division centres increases in the apical to basal direction, and by the underlying tissues, since the epidermis above a median vascular bundle fails to divide and division centres are not formed.[107] Some of the cell division centres subsequently develop as buds. Isolated portions of the lower epidermis of leaves of *Bryophyllum daigremontianum* can be grown in culture; in the presence of a cytokinin and an auxin, the epidermal cells divide to give rise to a flat callus and ultimately to buds.[55] In other species, e.g. tobacco, *Nicotiana tabacum*, small explants of superficial tissue can give rise to floral buds, if excised from certain regions of the plant, or to vegetative buds, roots or callus; and the further development into these various structures can be controlled by the hormones supplied in the medium.[670] In this species, however, these organs originated by division of the sub-epidermal layer.

The epidermis may persist during the life of the plant or, in species which undergo secondary growth, it may be sloughed off along with underlying tissues after the formation of the periderm.

ROOT

In the epidermis of the roots of many flowering plants some of the cells give rise to root hairs. These are merely projections of the epidermal cell, except in isolated instances where multicellular root hairs have been observed.

For a short distance (about 100 μm) behind the tip of the root all the epidermal cells divide and there is little or no evidence of differentiation. Between 100 and 275 μm from the root tip the probability that an epidermal cell will divide becomes progressively less.[178]

Root hairs first grow out at some little distance behind the tip of the root; this distance varies in different species and in the same species under different conditions. In some roots those cells which give rise to root hairs, which are called trichoblasts, are morphologically distinct from those which do not; in other species all epidermal cells are morphologically similar, but only some of them form root hairs. Apparently there is a relationship between length of cell and outgrowth of hair, the shortest cells developing the longest hairs. This suggests that the outgrowth of a root hair indicates, in part, a change in the direction of growth of the cell.[120,121]

Root hairs are generally considered to function by increasing the absorptive surface of the plant. Certainly the area over which absorption could occur is greatly enhanced where root hairs are present. However, it has been shown that the rate of influx of water through the hairless epidermal surface is similar to that through root hairs, and, further, that short hairs are actually more efficient in absorption than long ones.[552] The needs of the plant could be satisfied either by only a few of the root hairs actually present, or by only a proportion of their surface area, and it seems likely that the main biological advantage of root hairs is that by means of their lateral extension they can come into contact with otherwise untapped sources of water.[552] Many aquatic plants possess well developed root hairs, but in others, e.g. *Elodea*, root hairs are formed only, or principally, when the plant grows in soil or mud.[119,724] When grown in water in total darkness *Elodea* produced root hairs, and this was attributed to lack of cuticle, which was present in roots grown in the light and might have acted as a mechanical barrier to outgrowth of hairs. On the other hand, if CO_2 was bubbled through the nutrient medium root hairs were formed on roots grown in light, and a fatty layer did develop on the epidermis. This was shown not to be a true cuticle, however, the CO_2 having prevented the oxidation of the fats to form a cuticle.[142]

A thin cuticle normally covers the cell wall of the root hair, and mucilage may occur superficially.[146] The extreme tip of the root hair is probably the main site of synthesis of cellulose, although work with the isotope ^{14}C indicates that synthesis occurs over a length of about 120 μm behind the tip.[49] If the turgor pressure of the root hair is reduced, by supplying glucose in the external medium, growth in area of the cell wall is inhibited and the polar organization of the cytoplasm is lost.[583] Electron microscopy indicates that the epidermal cell wall consists of two parts, an inner densely staining region and a much wider outer zone. Apparently the root hair represents an outgrowth of the inner region only. Both the root hairs and other epidermal cells contain many vesicles associated with the dictyosomes; these vesicles are much larger than in cells from other parts of the root.[409]

Trichoblasts

In some species, notably many grasses, the cells of the epidermis that will give rise to root hairs (***trichoblasts***) are distinguished in various ways from the other epidermal cells. They are usually smaller and have dense cytoplasm. An interesting example is found in the roots of the aquatic plant *Hydrocharis morsus-ranae*, in which the trichoblasts are easily distinguishable from adjacent epidermal cells by their dense cytoplasm (Fig. 7.2) and other features. For example, in roots grown in the light well developed chloroplasts with grana are present in all epidermal cells except the trichoblasts, in which the chloroplasts progressively revert.[648] In other species, e.g. *Sinapis alba*, white mustard, certain longitudinal files of cells give rise to root hairs. A

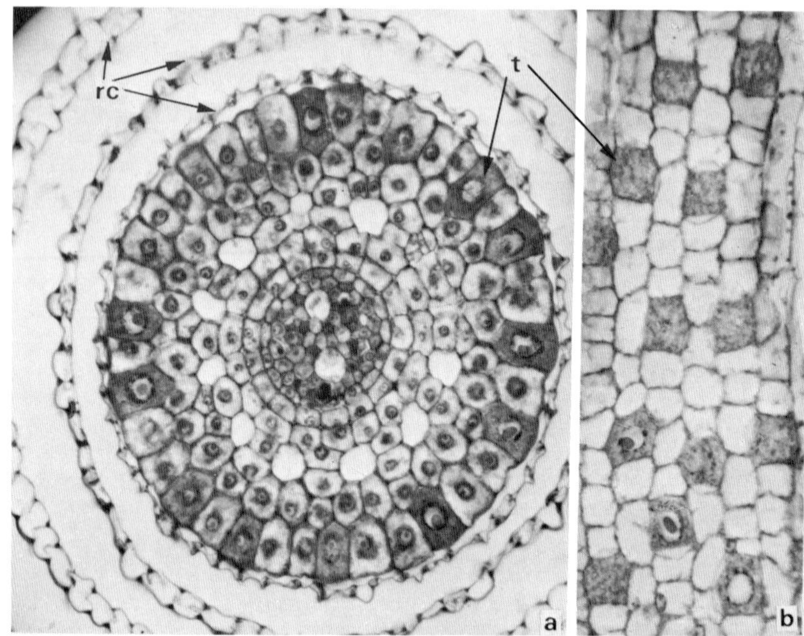

Fig. 7.2 Trichoblasts (t) in the root epidermis of *Hydrocharis morsus-ranae*. (**a**) Transverse section of root, showing the densely cytoplasmic trichoblasts, which also have larger nuclei and nucleoli than adjoining epidermal cells. A three-layered root cap (rc) is present. ×200. (**b**) Tangential longitudinal section through the epidermis, showing the densely cytoplasmic trichoblasts. ×200.

considerable amount of work has been carried out on the differentiation of trichoblasts, including studies of changes in histochemistry and fine structure.

In some plants, including grasses, the trichoblasts are formed by unequal division of an epidermal cell, preceded by an unequal distribution of cytoplasm. In *Phleum* the cytoplasm becomes concentrated at the apical end of the cell (i.e. the end towards the root apex) and later the cell divides to form a small cell in this position and a larger cell proximally. Avers[32] has pointed out that the mitosis itself is asymmetric, and is not merely a symmetrical process occurring in an asymmetrical position at one end of the cell. Thus not merely the distribution of the cytoplasm, but also the mitotic figure itself is asymmetric in these cells. Interesting observations on the roots of species of *Potamogeton* indicate that the trichoblasts may induce the formation of short, densely cytoplasmic cells in the underlying tissues.

Some distance from the root tip, in a region where neighbouring cells no longer undergo much division, an equal division occurs in the cells underlying the trichoblasts; the resulting cells divide again, forming a row of short, densely cytoplasmic cells.[673]

In *Hydrocharis*, it was found that similar unequal divisions took place in the protoderm, but that the trichoblast was the proximal product of the division. Thus a small, densely cytoplasmic cell, the trichoblast, was formed at the proximal end of the original mother cell, and a larger, more vacuolate cell at the distal end (i.e. towards the root apex) (Fig. 7.3a).[137] These observations were confirmed with the electron microscope (Fig. 7.4). Cells in the process of dividing unequally can sometimes be observed; the nucleus is displaced to the proximal end of the cell, and small vacuoles are usually present at the other end. The cell wall between the smaller and larger cells is typically

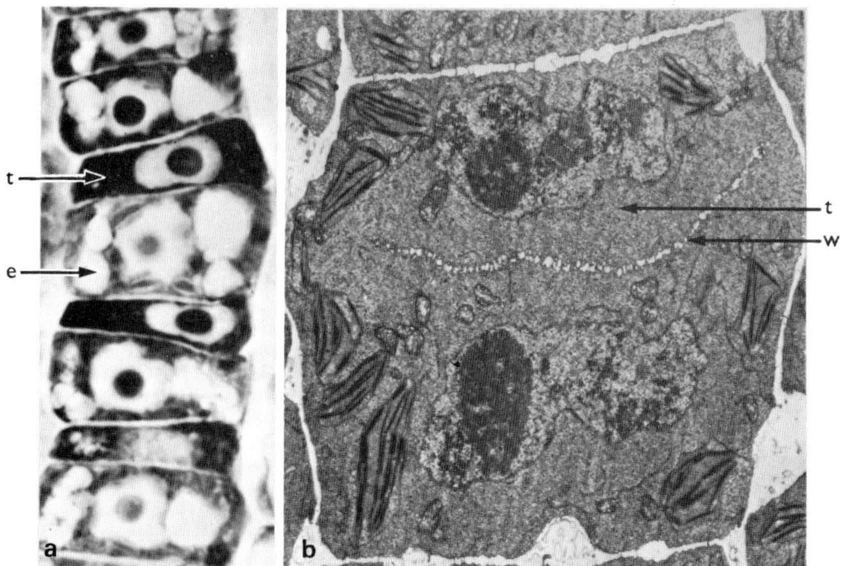

Fig. 7.3 Formation of trichoblasts in the root epidermis of *Hydrocharis morsus-ranae*. (**a**) Light micrograph of part of the protoderm showing the unequal division of the cells. The smaller proximal product of this division, which stains much more densely, is the trichoblast (t); the larger product is an epidermal cell (e). ×1200. (From Cutter and Feldman,[137] Fig. 4, p. 193.) (**b**) Electron micrograph of a protodermal cell in telophase stage of the unequal division giving rise to a trichoblast (t). The wall (w) between the two daughter cells, which is not yet completely formed, is markedly curved, a typical feature of unequal mitoses of this type. ×6000. (From Cutter and Hung,[139] Fig. 6, p. 737.)

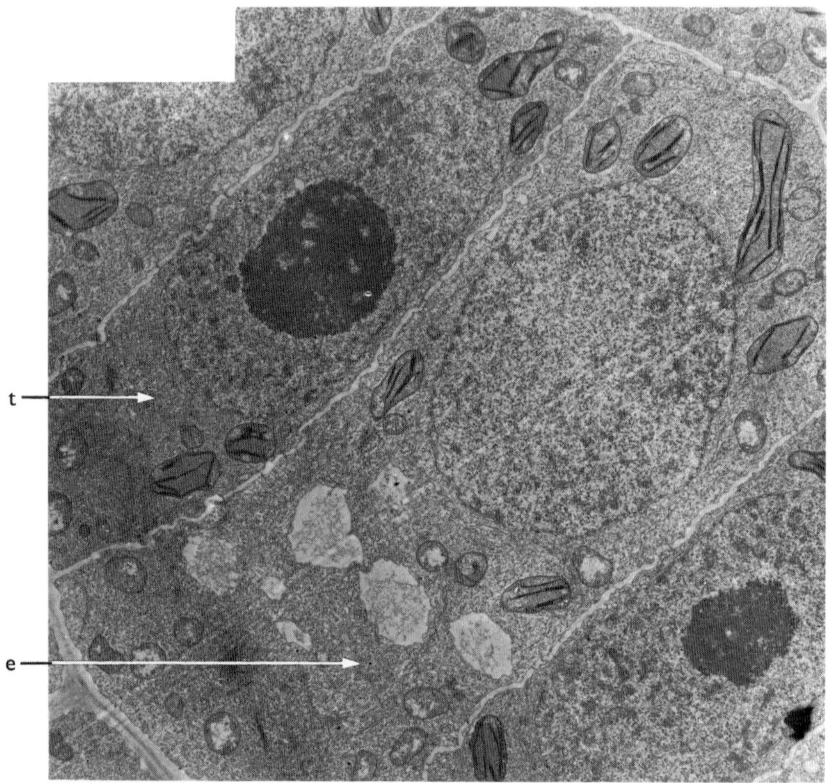

Fig. 7.4 Electron micrograph of a trichoblast (t) and adjacent epidermal cell (e) soon after the unequal division of a protodermal cell in the root of *Hydrocharis morsus-ranae*. The epidermal cell is more vacuolate than the trichoblast. ×6120.

curved (Fig. 7.3b). The unequal division and the appearance of the resulting daughter cells closely resemble those associated with the formation of a guard cell mother cell in leaves and stems of monocotyledons, except that there are less evident cytoplasmic differences between the two daughter cells (compare Figs. 7.3b and 7.4 with Fig. 7.26).[139] Unequal divisions of this kind are important, since they are commonly the prelude to cell differentiation (see Chapter 2).

In *Hydrocharis*, the trichoblasts are very interesting cells. They do not divide, in contrast to the neighbouring larger cell which divides several times, but their nuclei continue to synthesize DNA. This results in their becoming endopolyploid, with much larger nuclei than adjacent cells. Polyploid nuclei have multiples of the normal diploid amount of DNA. At some distance from the root tip the trichoblast nuclei have 8 times as much DNA as adjacent cells. This continued synthesis of DNA was demonstrated both by measuring the light absorption of nuclei stained with the Feulgen reagent with a microspectrophotometer, which enables comparisons to be made between nuclei, and by supplying the roots with ^{3}H-thymidine.[138] In autoradiographs many more silver grains were present over the trichoblast nuclei. It is probable that the polyploid nature of the cell does not directly control its differentiation, but is a factor affecting its considerable growth. Root hairs in this species are especially large cells, about 5 mm long. The polyploid trichoblasts of *Hydrocharis* remain densely cytoplasmic, in contrast to the more vacuolate epidermal cells which intervene between them and are the products of division of the longer daughter cell of the original unequal division (Fig. 7.5).

In species without distinctive trichoblasts the cytoplasm is more or less uniformly distributed and approximately equal division of the cell follows.[607] The proportion of cells that differentiate as hair initials is fairly constant under different conditions, but the number of these cells that actually give rise to root hairs is variable.[724] It seems that the factors that control the outgrowth of root hairs may not be identical with those that lead to their formation. This is, of course, a common phenomenon in morphogenesis. The growth rate of trichoblasts is less than that of other epidermal cells in a region of the root just proximal to the apex[121]; the size relationship between the trichoblasts and other epidermal cells may change during growth.

It is obviously important to establish what distinctive physiological and morphological differences exist between trichoblasts and ordinary epidermal cells. By this means it may be possible to discover the factors that are important in affecting or controlling their ultimate destiny. Early workers suggested that materials that would stimulate root hair formation moved in a polar manner in the root, accumulating at the apical (i.e. distal) end of the cells.[607] Both nuclear and cytoplasmic differences are, indeed, now known to exist between the trichoblasts and other adjacent cells of the epidermis. In the discussion which follows, those cells of the developing epidermis which will not give rise to root hairs will be termed hairless initials, as opposed to the trichoblasts or hair initials. In grasses which possess distinctive trichoblasts nucleolar volume is greater in these than in other epidermal cells (Fig. 7.6).[425, 554] In two species without trichoblasts there was also a difference in nucleolar size—though a less clear-cut one—between those cells which give rise to root hairs and hairless

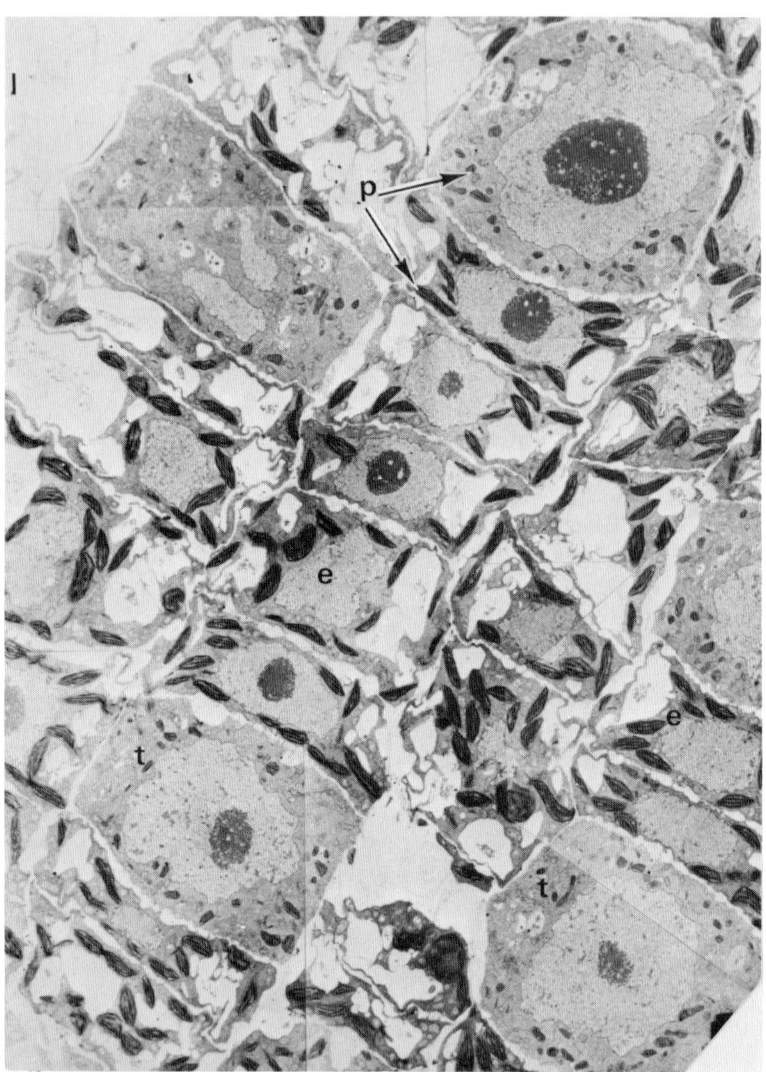

Fig. 7.5 Montage of electron micrographs from a tangential longitudinal section of the root epidermis of *Hydrocharis morsus-ranae*. The trichoblasts (t) are evident as large cells packed with cytoplasm and large nuclei and nucleoli; at this stage in differentiation the intervening epidermal cells have divided to give about 4 quite vacuolate cells (e). Chloroplasts (p) are well developed in the epidermal cells, but small and poorly developed in the trichoblasts. (Cutter and Hung, unpublished.)

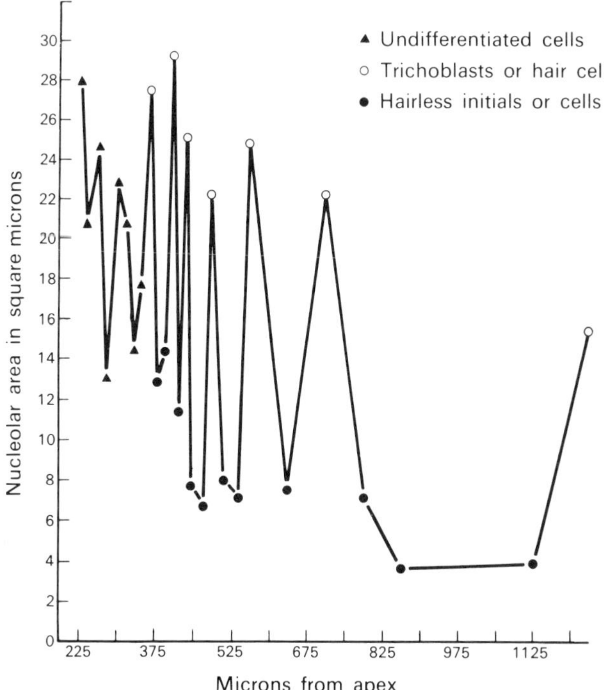

Fig. 7.6 Nucleolar areas of 26 cells in sequence making up a row in the root epidermis of *Festuca arundinacea*. The root was stained with acetocarmine. (From Rothwell,[554] Fig. 8, p. 175.)

cells.[554] Differences in nucleolar size could also be detected in younger epidermal cells which were not yet differentiated. These differences in nucleolar size probably come about at mitosis, since it is known that, following the unequal division of a pollen grain to give the vegetative and generative cells, the nucleolus of the vegetative nucleus has approximately four times the total dry mass and amount of RNA and basic protein of the generative nucleolus; together the total of these components in both nucleoli equals those of the premitotic nucleolus.[442] Thus the nucleolar material seems to be dispersed and reassembled during mitosis. In grasses lacking distinctive trichoblasts hairs are formed from cells which not only have larger nucleoli but also a greater concentration of protein bodies than the hairless cells adjacent to them.[555] Labelling experiments also showed that ^{3}H-thymidine, a precursor of DNA, was incorporated in greater amount in certain of the non-dividing epidermal cells of *Panicum*, hair initials and root hair cells,[2] a result which may be compared with the observations on trichoblasts of *Hydrocharis*. Thus cytological differences may be detectable between cells with different destinies even when morphological differences are not evident.

Intense staining of RNA and ribonucleoprotein was observed in trichoblasts of *Phleum* between 150 and 300 μm from the root apex.[425] Trichoblasts are also the sites of the differential activity of several enzymes. By the use of stains that are specific for certain enzymes it is possible to establish the relative distribution of the enzymes in the cells of a particular tissue, a technique of considerable value in studies of differentiation. Using this technique it was found that both cytochrome oxidase and acid phosphatase occur at a high level in the trichoblasts at an early stage of differentiation (Fig. 7.7).[30, 33] In festucoid grasses, which possess trichoblasts in their roots, intensified activity of the enzyme acid phosphatase in the trichoblasts and loss of activity in the hairless initial cells was evident before full development of these cells. In panicoid grasses, which have no distinctive trichoblasts, no cells inactive in phosphatase were evident during differentiation.[33]

Further work on enzyme activity in epidermal cells of grass roots has revealed an added complexity. It now appears that the activity of

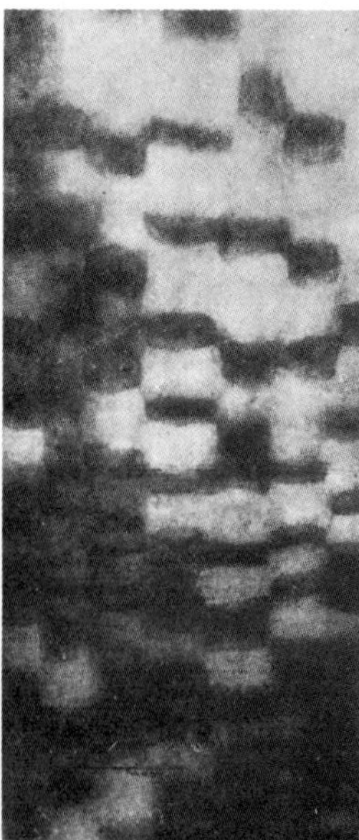

Fig. 7.7 Activity of the enzyme 5-nucleotidase in root epidermal cells of *Phleum*, after histochemical treatment. Active trichoblasts (darkly stained) alternate with inactive hairless initials. ×200. (From Avers,[31] Fig. 8, p. 141.)

enzymes, in this case phosphatases, in the hairless or hair cell initials differs according to the position of these cells along the root. In the cells within 100 μm from the apex of the root little differentiation of the cells was observed; in the region 100–200 μm behind the tip the hairless initials usually showed phosphatase activity while the trichoblasts did not; and from about 200–300 μm behind the root apex the trichoblasts, in general, were active and the hairless initials were not.[141] It thus appears that, superimposed upon differences between future hair-initiating and hairless cells, there are other effects produced by the general gradient of differentiation along the root. It seems possible that this may be related in some way to the progressive differentiation of underlying tissues, which is, of course, proceeding quite rapidly in a basipetal direction. As mentioned in Chapter 2, the nature of the underlying tissues may affect the differentiation of the superficial layer in particular regions.

In interpreting these interesting observations on enzyme activity in the trichoblasts and hairless initials, emphasis is usually placed on the positive, and thus more evident, changes that take place in the trichoblasts. For example, it is considered that high levels of certain enzymes may be necessary for root hair formation. However, an alternative view is that the *loss* of phosphatase activity (or some other change of which this is a symptom) may be important in *restricting* the potentiality for root hair development. Observations on the fine structure of differentiating epidermal cells,[32] which have seemed difficult to interpret, also appear compatible with this viewpoint: namely, that the *restriction* of developmental potentialities may be at least as important in differentiation as the acquisition of positive features such as high enzyme activity. It seems logical to consider the possibility that even in species with trichoblasts all cells are initially capable of giving rise to root hairs; at this early stage of development they are also all rich in enzymes. During differentiation some cells show less enzyme activity, and also lose the capacity to form hairs.

In the roots of some species trichoblasts do not alternate in the longitudinal plane with hairless initials, but form whole rows running longitudinally along the root, other longitudinal rows being composed of hairless initials. The trichoblasts may then have a particular spatial relationship with underlying tissues. For example, in the roots of *Sinapis alba*, the trichoblasts are aligned with the radiating rows of anticlinal walls of cortical cells and the intervening intercellular spaces (Fig. 7.8). Because of this spatial relationship, transport to the trichoblasts of substances from the vascular cylinder and inner layers of the cortex, which are still meristematic and densely protoplasmic at this level, will evidently differ from that to the other epidermal cells.

Some authors[120] argue that substances are more readily transported to the trichoblasts through the intercellular spaces than to other epidermal cells through the cells of the cortex; others maintain that the trichoblasts are, in fact, somewhat isolated physiologically from the central tissues of the root. This latter interpretation is supported by the results of experiments in which the epidermis was separated from the inner tissues by an incision; after this treatment all the isolated

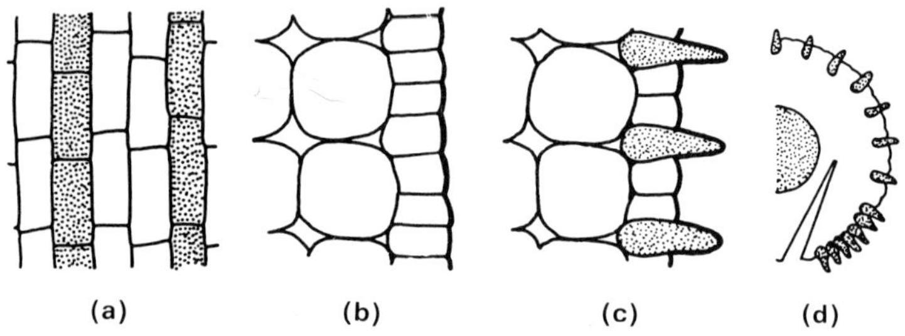

Fig. 7.8 Trichoblasts. (**a**) Distribution of trichoblasts (dotted) and ordinary cells in the epidermis of the roots of some dicotyledons. (**b**) Transverse section showing part of the epidermis and cortex before differentiation. (**c**) Position of emerging root hairs. (**d**) The effect of an incision between the vascular cylinder and the epidermis. All epidermal cells are giving rise to root hairs. (From Bünning,[74] Fig. 4, (IV), p. 115.)

epidermal cells, when adequately nourished, gave rise to root hairs (Fig. 7.8d).[74] Possibly an inhibitor may be radially transported from the central region of the root of some species through the cortical cells to the epidermis, the potential trichoblasts escaping its effects by virtue of their position. Such an interpretation would again place some emphasis on the importance of the restrictive changes taking place in those cells which fail to give rise to root hairs.

AERIAL PARTS

The epidermis of the aerial parts of the plant consists of cells which are more or less tabular in shape, or else with much convoluted anticlinal walls (those at right angles to the surface) (Figs. 7.9, 7.10). In the leaf the cells of the abaxial (lower) epidermis are often more lobed than those of the adaxial (upper) epidermis. Cells lying in costal regions, i.e. over the veins, may be much less sinuous in outline than those in intercostal regions (Fig. 7.9). There are usually few intercellular spaces, or none. In photosynthetic organs such as leaves and young stems the epidermal cells are normally devoid of fully developed chloroplasts, except for the guard cells of the stomata. These are paired cells which surround a pore in the epidermis through which gaseous exchange, including the movement of water vapour, takes place. Epidermal cells are usually thin-walled, but in some species, especially among gymnosperms, they may be thick-walled and even lignified. Pigments, e.g. anthocyanin, may be present in epidermal cells.

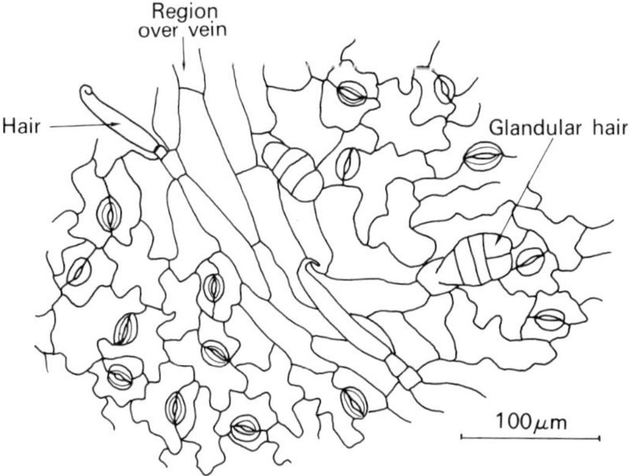

Fig. 7.9 Lower epidermis of the leaf of *Phaseolus vulgaris*, showing two kinds of hair, slightly raised guard cells of stomata, and convolute epidermal cells (except over the veins). ×155.

Cuticle

Very often a layer of fatty material, or cutin, is deposited on the surface of the epidermal cell wall, forming the cuticle. This substance is for the most part impervious to water, and may have a protective function. When the cuticle is of considerable thickness, its chemical nature often varies in different layers, at least proportionally, and may include cutin and wax. The innermost layers may lack cutin, and the outermost may lack cellulose, various proportions of these substances being present in intervening regions.[627] The cuticle is extremely resistant to micro-organisms. Thus in the living plant it affords some protection, perhaps largely mechanical, against infection by pathogens, and in fossilized plant remains it may be extremely well preserved, being resistant to decay. It may persist for millions of years, and can be isolated fairly readily from fossils and studied; indeed, fossil plants can very often be successfully identified in this way.[306] Fossil grass cuticles have recently been used in preference to grass pollen as a means of reconstructing the nature of past vegetation in certain

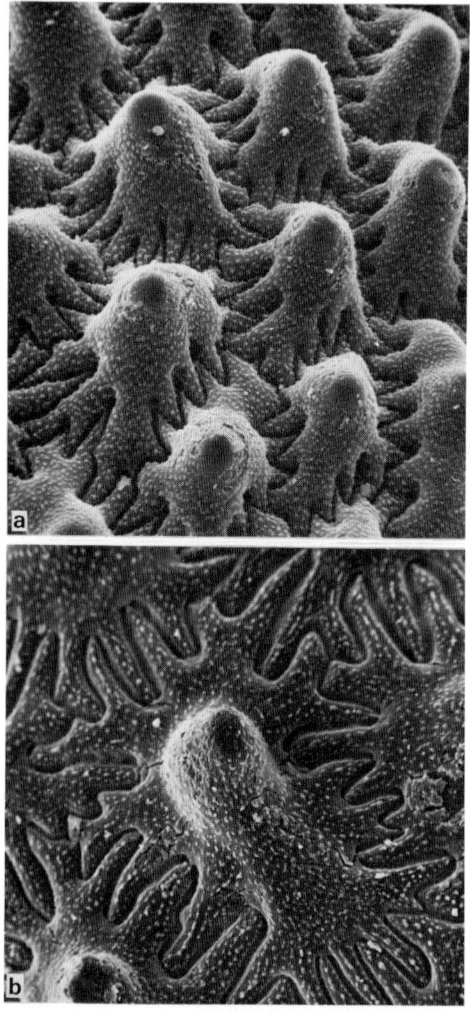

Fig. 7.10 Scanning electron micrographs of the surface of the seed of *Petrorhagia* spp., showing tuberculate epidermal cells with sinuous anticlinal walls. (**a**) *P. nanteuillii.* ×345. (**b**) *P. velutina.* ×365. (Courtesy of Dr. M. Stant. Crown Copyright; reproduced with the permission of the Controller of Her Majesty's Stationery Office and of the Director, Royal Botanic Gardens, Kew.)

regions. Fragments of cuticle at least 28 000 years old deposited in lake sediments in Africa can be identified.[487]

Cuticle deposition can be affected by light intensity, and by the availability of water. In plants under continuous water stress up to twice as much cuticle was laid down as in plants grown under an optimal water régime.[441] Although the cuticle is largely impervious, water can diffuse through it, as occurs in the drying of grapes to form sultanas. The rate of drying of the fruit was found to be inversely proportional to the amount of cuticle present.[443] Recent observations indicate that pores or discontinuities can occur in the cuticle, for example over the secretory head cells of glandular trichomes[325, 447] and on the papillae of the stigma of certain flowers (see Part 2,[136] Chapter 6).[322]

Distinctive cuticular patterns often preserve many of the characteristic features of the underlying epidermis, for example type of stomata and hairs, and their distribution. This may be useful not only in identification of fossils, but in recognizing small fragments of plants, as is necessary, for example, in pharmacognosy, forensic medicine, and studies in animal nutrition. In one investigation 16 species of plants eaten by hill sheep were identified from cuticular fragments in the rumen and faeces, whereas others had been completely digested or were otherwise unrecognizable.[440] Differences in cuticular pattern may be taxonomically useful even at the generic or species level.[3]

In many plants conspicuous deposits of wax are formed on the surface of the cuticle. It is this wax which gives the 'bloom' to some fruits, e.g. grapes, and to glaucous leaves. When viewed under the transmission electron microscope, the wax is seen to form many projections and folds, and by the carbon replica technique complex wax patterns from the surface of the epidermis can be studied. This technique is used mainly for materials which are opaque to electrons, and depends on depositing a replicating material, such as carbon, on the specimen. The replica, which consists of a thin film of electron-transparent material, is then removed and examined under the microscope. Wax and cuticular patterns can now be observed directly on the leaf surface with the scanning electron microscope (Fig. 7.11). Wax may be deposited in large or small flakes or granules (Fig. 7.11a), rods, tubes or sheets, which may then be sculptured into ridges, etc. (Fig. 7.11b). The conditions under which the plants are grown, including altitude, may affect the form of the wax.[173] The wax patterns vary in different species, and to some extent within a single species under different conditions. Wax formation apparently begins at an early stage in leaf development and persists until a late one; the projections seem to reach a fairly uniform density and height and then

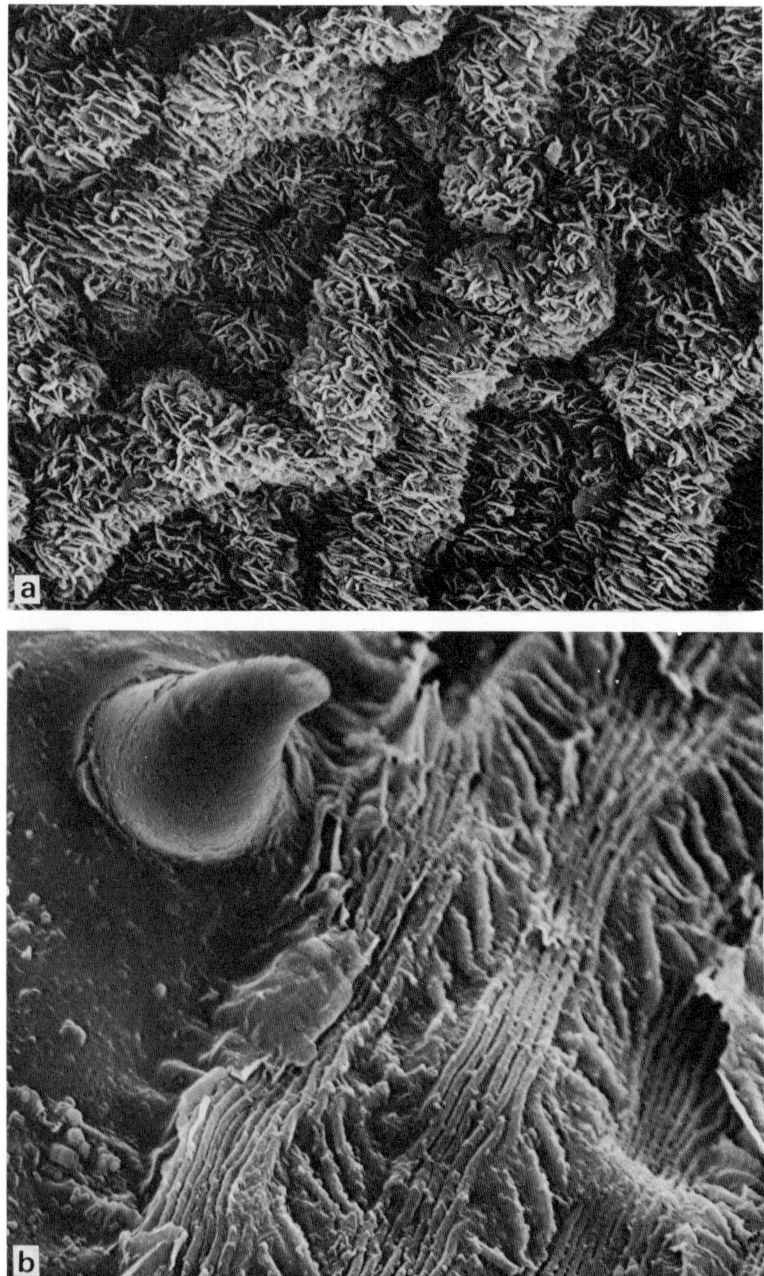

Fig. 7.11 Scanning electron micrographs of leaf surfaces. (**a**) Surface wax on a developing leaf of *Eucalyptus* sp. Stomata can also be discerned. ×1900. (**b**) Adaxial epidermis of the leaf of *Zea mays*, showing ridged wax and cuticle and a unicellular hair. ×1800. (By courtesy of Dr. J. N. A. Lott; (a) from Lott,[423a] *A Scanning Electron Microscope Study of Green Plants*, St. Louis, 1976, The C. V. Mosby Co., Fig. 24, p. 22.)

remain relatively unchanged.[365] If the wax is rubbed off the leaf, it will form again, and may do so within 24 hours. The wax recrystallizes in the original morphological form typical of the species.[296, 297] Some interesting recent work has shown that the surface wax of many species recrystallizes in a form comparable with the original, even when it has been dissolved in various solvents and recrystallized on the surface of porous ceramic dishes of various pore sizes. Thus the wax, in isolation from the plant, can apparently retain properties which determine its morphology, properties which are no doubt associated with its chemical composition. At all events, its fine structure appears to be unaffected by the pores through which it is extruded.[351]

The problem of how the wax reaches the leaf surface, after it is secreted by the cell, is controversial. Some workers maintain that numerous plasmodesmata, called ectodesmata, are present in the outer walls of epidermal cells;[254] others have been quite unable to find pores of any kind through which the wax could have been extruded.[365, 571] On the basis of an investigation using the freeze-etch technique of electron microscopy the existence of microchannels through which the wax is transported is again claimed.[293a] In another such study, no structures resembling ectodesmata could be found in unfixed or glutaraldehyde-fixed material, but in material fixed in sublimate regions extending into the cell wall but not reaching the protoplast were observed. It was suggested that these were sites of specific physico–chemical activity, perhaps regions of wall growth and repair.[427] Ectodesmata-like structures have also been observed in isolated cuticle; the pattern of these was altered by removing the surface wax. It was concluded from this that ectodesmata are not definable cell wall structures.[582] Certainly continuous channels leading from the protoplast through both cell wall and cuticle seem not to have been observed.

These wax patterns on the surface of the epidermis are extremely important in that they affect the degree to which the surface can be wetted. Surface wax resists wetting by sprays, etc., much better than a smooth cuticle.[571] Thus the degree of susceptibility of a plant to herbicides, or the effectiveness of a fungicide, may depend on the extent of development of this surface wax; the selective action of herbicides may be due in part to the amount of surface wax in different species. The development of wax is affected by light, being greater at high light intensities. It has also been noted that a thicker layer of wax is formed in plants which are grown slowly.[365] Thus various environmental factors can affect wax formation, and hence the reaction to various spraying treatments.

Wax usually occurs on surfaces of leaves and fruits, but can occur also on more internal structures such as the juice sacs of citrus fruits, where it is important in maintaining adhesion and thus the integrity of the fruit segment.[227] It is also present on the inside of the modified parts of the leaves which form the pitchers of *Nepenthes*, in the form of a layer of overlapping scales which adhere to insects' feet and make it impossible for them to climb out of the pitcher.[441]

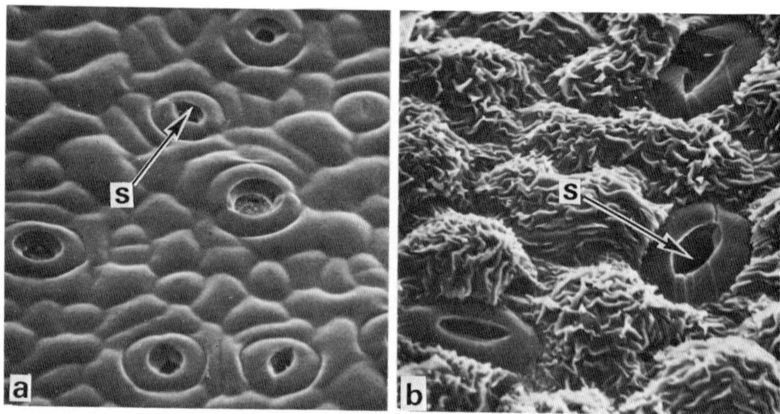

Fig. 7.12 Perspex replicas of surfaces of (**a**) a young leaf of hop, *Humulus lupulus*, and (**b**) an older leaf, both kept in the light. The stomatal pores (s) are wide open. The imprint of surface wax is evident in (**b**). (**a**), ×710; (**b**), ×635. (From Royle and Thomas,[556] Plate 4a and b.)

Wax formed by plants may be commercially useful if it is produced in sufficient quantity. The wax deposited on the leaves of the wax palm, *Copernicia cerifera*, is called carnauba wax and is used in the manufacture of phonograph records and various polishes. Only about 6 ounces (170 g) of wax are obtained from 50 of the large palm leaves.[37]

Various techniques exist for making replicas of a leaf surface, which faithfully reproduce the cellular, cuticular and wax patterns, and can be studied at leisure with the light or the scanning electron microscope (Fig. 7.12). The possibility exists of observing development using a succession of these peels from a single growing surface. One technique requires the making of a negative replica with silicone rubber, followed by a positive replica in nail varnish;[568] others employ Duco cement,[589, 721], Perspex,[556] or a solution of formvar.[255] These replicas can give a good idea also of the shape of stomatal cavities.

Hairs

A hair or trichome is formed by the outgrowth of an epidermal cell. Plant hairs have been described in detail by Uphof.[681] They are formed on all parts of the plant, including stamens, e.g. *Tradescantia*, and seeds, e.g. *Gossypium*, cotton (Fig. 7.13). Cell division may take place, so that the hair becomes multicellular, or it may remain unicellular. Multicellular hairs may consist of a single row of cells, or of many rows. Trichomes are sometimes classified as either glandular hairs, which have a secretory function, or covering hairs, which do not.

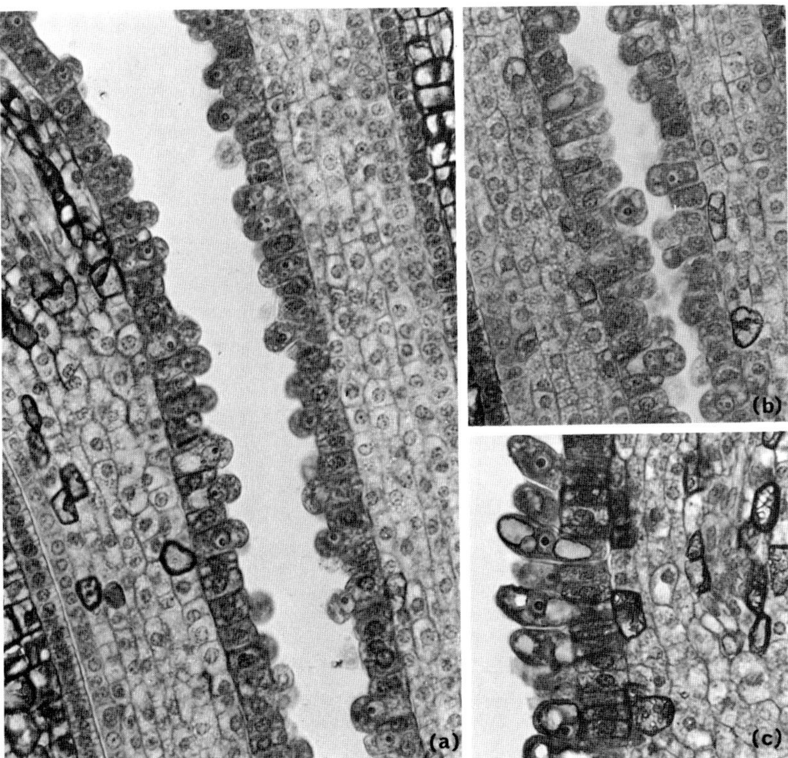

Fig. 7.13 Transverse section of young developing ovules of *Gossypium hirsutum* (cotton) showing early stages in the development of the unicellular hairs (commercial cotton) from the testa. In (**a**) and (**b**) the edges of two adjacent ovules in the ovary are shown. (**a**) Very young stage, with hairs just growing out; (**b**) slightly later stage; (**c**) later stage, hairs becoming vacuolate. ×300.

Scales consist of a discoid plate of cells, usually on a short stalk. Certain leaves, e.g. *Olea*, *Hippophaë*, have a dense covering of scales on the abaxial surface. Scales, like hairs, apparently originate from a single cell. Covering hairs may occur in tufts, as in the leaf of *Hamamelis*, or they may be complex branching structures, as in *Verbascum* (Fig. 7.14). The epidermal cells of many petals have small hair-like projections known as papillae.

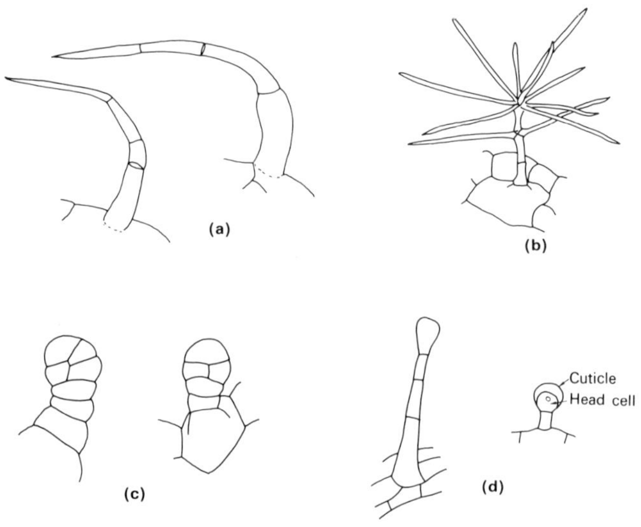

Fig. 7.14 Covering and glandular trichomes. (**a**) Uniseriate covering trichomes from the leaf of *Lycopersicon*. (**b**) Branched covering trichomes from the leaf of *Verbascum*. (**c**) Glandular trichomes with multicellular heads from the leaf of *Cucurbita*. (**d**) Glandular trichomes with unicellular heads from the leaf of *Pelargonium*. (**a**), (**b**) and (**d**) ×115. (**c**) ×180.

Either unicellular or multicellular hairs may be glandular. In such hairs there is a stalk and a head, the head being the secretory region. A cuticle-like structure covers the cell(s) of the head, and the secretion accumulates in the sac formed between the cell or cells and the cuticle (Fig. 7.14d, see also Chapter 11). Oils, resins and camphors may be secreted in this way. Peppermint (*Mentha piperita*) has hairs of this type; they are formed at an early stage in leaf development.[331]

The cell walls of most trichomes are usually thin and cellulosic, but some hairs have lignified cell walls, e.g. those on the seed-coat of *Strychnos nux-vomica*, the source of strychnine. The unicellular hairs on the seed-coat of *Gossypium*, which are cellulosic, are the source of cotton, and are of great commercial importance. In early stages of development the hairs are visible as slight outgrowths from the seed-coat (Figs. 7.13, 7.15), but in some varieties they may be ultimately 1 or 2 in. (2·5 to 5 cm) long. These hairs have been utilized by man for some 7000 or 8000 years, and fabrics woven from cotton are known from 900–200 B.C.[429] The growing of cotton is an important industry in parts of America and Africa, and the economy of the great industrial areas of Lancashire, England, which have a damp climate particularly suitable for cotton spinning, depended for many years on the manufacture of cotton fabrics. The Lancashire cotton industry was built up in the fifteenth and sixteenth centuries and prospered until it encountered serious competition from man-made fibres in more recent times. The industrial and economic importance of this plant structure needs no further emphasis. The hairs on the seed-coat of *Ceiba pentandra*, the source of kapok, are also of commercial importance.

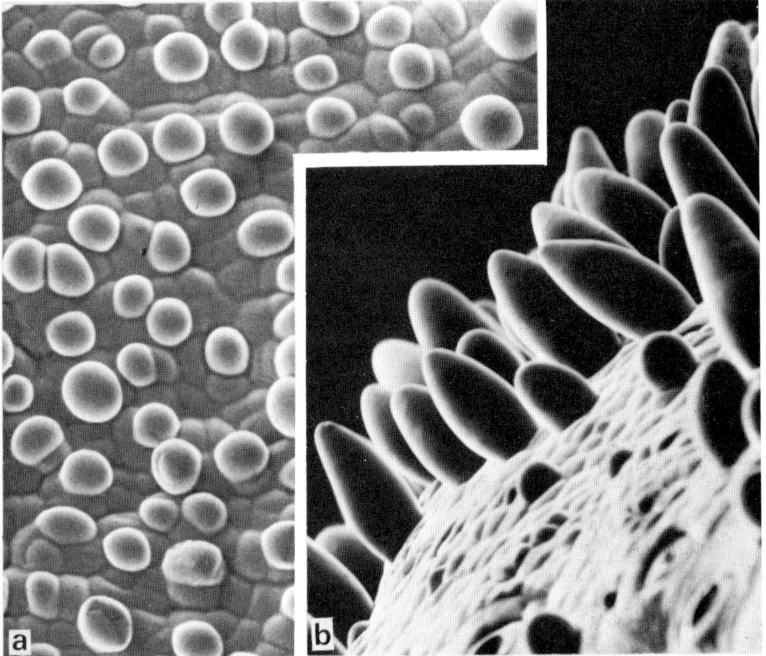

Fig. 7.15 Scanning electron micrographs of developing cotton hairs. (**a**) Emergent hair initials of Pima cotton on the morning of anthesis, at the chalazal end of the ovule. ×740. (**b**) Hairs on an unfertilized ovule, induced to elongate in culture by hormone treatment (see text). Two days after culturing. ×500. (By courtesy of Dr. C. A. Beasley.)

The hairs on cotton ovules begin to elongate on the day of anthesis (i.e., opening of the flower), and continue to elongate for about 20 days (Fig. 7.15). Electron microscopy indicates that one day before anthesis phenolic-type substances are deposited in the vacuoles of epidermal cells. About 16 h prior to anthesis, cells destined to become hairs became distinguishable by the dispersion of phenolic material from the vacuoles, and later by having more ribosomes.[534] The ratio of hair-forming cells to the total number of epidermal cells is about 1 to 3·7; hairs do not occur in a regular pattern.[44,647] A new approach to the production of the commercially important cotton hairs, known industrially as fibres (though botanically this is a misnomer), is to grow isolated ovules in aseptic culture. This affords a means of investigating the factors which promote trichome growth. Since not all epidermal cells that begin to extend produce marketable 'fibres', it seemed possible that growth substances might increase the yield per seed, or alternatively might improve the length or quality of the product.[47] In the event, fertilized ovules removed from ovaries 2 days after anthesis continued to grow on a liquid medium, as did the epidermal hairs (Fig. 7.16a). Addition of GA resulted in a 2- to 3-fold increase in the amount of 'fibre' which developed; IAA also tended to promote fibre production. With appropriate growth regulators the ovules yielded fibres comparable to those on ovules left *in situ*, up to 1 inch (2·5 cm) long and with normal deposits of cellulose in the cell wall.[45] Positive turgor pressure is essential for *in vitro* growth of the hairs,[160] a result not unlike that obtained with root hairs (p. 97). These experiments were later extended to unfertilized ovules. Such ovules may enlarge slightly, but do not form hairs (Fig. 7.16b). In the presence of kinetin, growth of the ovules was promoted but no elongated hairs developed. However, in the presence of IAA or GA, or preferably both together, hairs elongated and fibre production was normal (Fig. 7.16c,d).[43,46]

In other experiments with cotton, isolated protoplasts (see Chapter 5) were obtained from the hairs; about 20% of these survived, formed new walls in culture and divided to form callus. Removal from the cell wall thus led to division rather than expansion of the protoplast.[47]

In some interesting experiments with cultured leaf fragments of *Begonia rex*, formation of unicellular hairs was induced on medium containing the auxin naphthaleneacetic acid.[669] After 6–7 days, 100% of the explants formed hairs, although hairs do not develop in intact plants. Trichomes developed better on liquid medium; apparently the capacity to form hairs was affected by the water balance. Further studies along the lines of the work with *Begonia* and cotton, accompanied by careful anatomical observation, may throw light on the factors controlling hair formation and development.

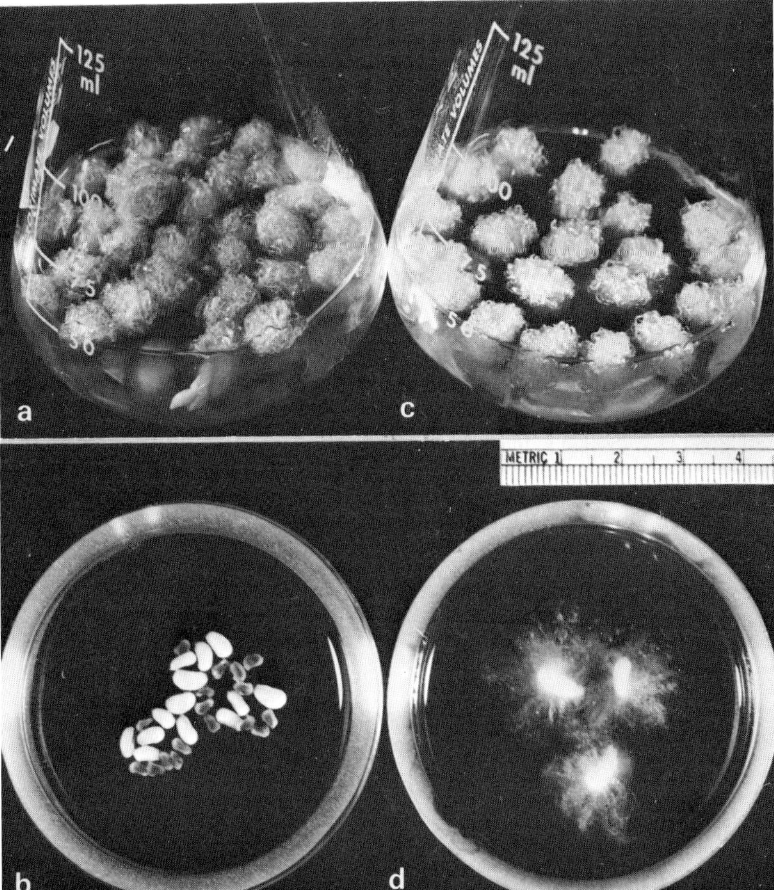

Fig. 7.16 Cultured ovules of cotton, *Gossypium hirsutum*. (**a**) Fertilized ovules in liquid medium. Hairs have continued to elongate, and embryos to develop normally, even to the point of germination. (**b**) Unfertilized ovules from flowers in which fertilization has been prevented. Some ovules have enlarged slightly in culture, but no hairs have developed. (**c**) Unfertilized ovules cultured in medium containing indoleacetic acid and gibberellic acid. With the addition of these hormones unfertilized ovules have enlarged and produced hairs (compare the fertilized ovules in (**a**)). (**d**) Unfertilized ovules from (**c**) treated so as to extend the hairs. (By courtesy of Dr. C. A. Beasley; (**b**) and (**d**) from Beasley and Ting,[46] Fig. 2, p. 190.)

Trichomes are usually of a characteristic form within a species, and may be of taxonomic significance. For example, the numerous species of *Rhododendron*[122] and of the Oleaceae can be identified at least to some extent by their trichomes. Carlquist[87] has emphasized the importance in taxonomic work of considering all the trichomes present; he calls this the 'trichome complement' of the plant. This is borne out by work on *Cannabis sativa* with the scanning electron microscope; the microscopic identification of marijuana depends to a considerable extent on the association of the typical cystolithic hairs with both covering and glandular hairs.[66]

The function of non-glandular trichomes in the plant is obscure, though it is often thought that they are protective and may prevent undue water loss. This view may be attributable to the prevalence of hairs on species from dry habitats. Some culture work is interesting in this connection. If the inhibition of axillary bud growth on isolated segments of pea stems was maintained by applied auxin, and especially if this inhibition was then partially released, there was a conspicuous outgrowth of two kinds of trichomes on the leaves of the buds. No trichomes were present on buds after more complete release of bud inhibition by treatment with kinetin. Also, trichomes were formed abundantly on buds of intact etiolated plants that had been grown for 30 days with a limited water supply. In various other instances it was found that the development of trichomes on pea leaves coincided with some imperfection of the transpiration stream, sometimes caused by incomplete contact between basipetally and acropetally differentiating xylem or by partial destruction of this tissue.[620] These observations are the more interesting in that hairs of this kind are not usually found on pea leaves. Such findings seem to suggest that trichomes may be formed in response to some water deficit (though since they occur in some aquatic plants (Fig. 11.3) this cannot be the complete explanation); whether in fact they protect the plant to an important extent against further water loss remains uncertain. The functions of numerous plant structures, often assigned to them many years ago and cited ever since, are still imperfectly understood and are deserving of further study with modern methods. Sometimes a relationship of a different kind is revealed, just as in this instance there appears to be a causal relationship, rather than a functional one, between limitation of water supply and the presence of trichomes. There is scope for much further observation and experiment in this field.

Another possible function of covering hairs is defence against insects. For example, resistance to leaf-hoppers in both soybean and cotton was found to be related to trichome density. In *Phaseolus* the hooked trichomes often catch the tarsal claw of aphids, and the degree of larval mortality is related to the amount of pubescence.[414] Another example of this is seen in *Passiflora adenopoda*, the leaves of which remain undamaged by the larvae of Heliconiine butterflies although adjacent plants of other species of *Passiflora* suffer extensive injury. The aerial surfaces of *P. adenopoda* were found to be covered with hooked trichomes. Larvae placed on the leaves did not move more than a few millimetres, and in a day were found to be dead and desiccated, owing both to starvation and to loss of haemolymph resulting from puncture wounds from the hooked trichomes.[267] Some glandular hairs also afford protection against insects (see Chapter 11), and plants with stinging hairs[663] are somewhat protected against marauding animals. On the other hand, aphids can transmit viruses through hairs ruptured by their claws.[305]

Silica, cork and crystal cells

In the epidermis of some plants silicon may be deposited in the cell wall or in the lumen. In the internode of oat, *Avena sativa*, some of the smaller daughter cells following unequal divisions of the developing epidermal cells divide by a longitudinal wall and develop as stomata (see p. 131). Others, which lie in intercostal positions, form cork-silica cell pairs by an early, equal division of a short cell by a transverse wall, i.e. one at right angles to the long axis of the stem (Fig. 7.17). Initially the two daughter cells appear the same, but differentiation ensues. The nucleus and cell contents of the silica cell break down, and the lumen becomes filled with fibrillar material, and eventually by amorphous silica bodies. The cork cell becomes highly vacuolate, and its wall becomes suberized.[373, 374] Other short cells in the epidermis do not divide further, but give rise to trichomes, the wall of which also becomes silicified.

In the leaf of *Gibasis* (related to *Tradescantia*), silica cells differ from adjacent epidermal cells in size and shape; they may contain silica bodies. These cells may be solitary and occur scattered randomly over the leaf surface, or may be arranged in longitudinal rows above the veins (Fig. 7.18).[632] In the leaves of barley, also, silica deposits may occur in specialized cells, or trichomes may be silicified.[308]

Development of the technique of electron probe microanalysis has allowed the detection and precise localization of various elements, including silicon. This method involves the detection, analysis and measurement of the X-rays emitted when the beam of electrons hits a circumscribed region of the specimen. The electron beam is focused to a small probe. A multi-channel analyser connected to the electron microscope can plot the distribution of chemical elements present, or

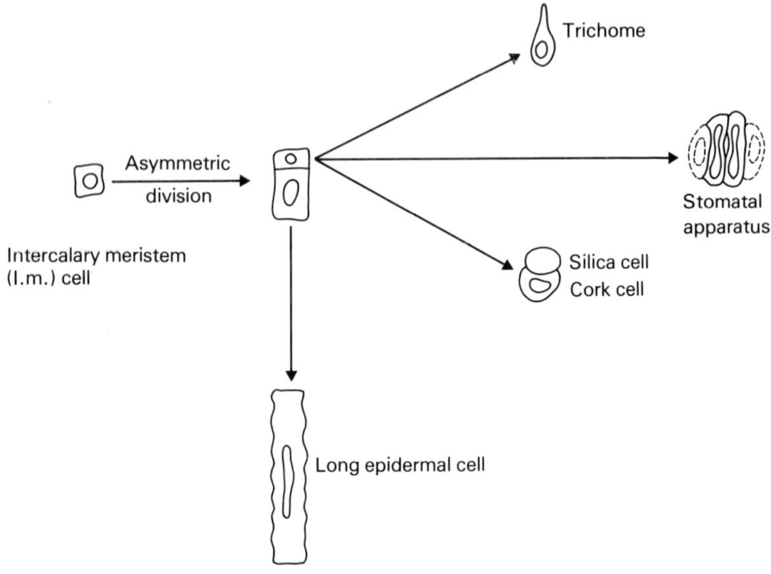

Fig. 7.17 Diagram illustrating the possible pathways of development of the long and short cells resulting from an unequal division of an epidermal cell in the internode of *Avena*. Subsidiary cells (outlined) are derived by division of adjacent epidermal cells. (After Kaufman, Petering, Yocum and Baic,[375] Fig. 1, p. 34.)

X-rays of a particular wavelength can be selected.[98] By use of the scanning electron microscope with this kind of system, distribution of silicon has been studied in the leaf and internode of *Avena*, *Cannabis*, *Cyperus* and *Equisetum*, and in the inflorescence bracts of rice, *Oryza*.[149, 371, 617, 618, 619] Silicon is deposited in marginal trichomes and sclerenchyma cells of oat, where it is thought to enhance the mechanical strength of the leaf sheath and lamina.[617] In the bracts of rice the heaviest deposit is in a layer external to the abaxial epidermis; epidermal papillae would be part of this layer.[619] It was noted in cereals that silica deposits prevented the accumulation of other elements, particularly potassium and calcium.[617] A developmental study of the epidermis of *Cyperus alternifolius* confirmed this decrease in the accumulation of potassium, and also of phosphorus, perhaps as a consequence of the disorganization of cytoplasm and nucleus known to occur in silica cells.[618]

Crystals of calcium carbonate known as cystoliths occur in specialized epidermal cells or lithocysts (Figs. 3.11c, 3.12). These may be

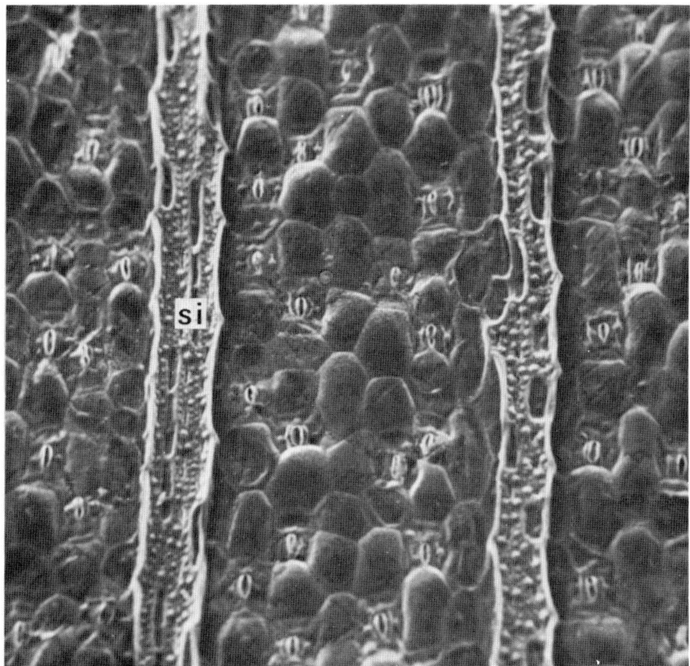

Fig. 7.18 Scanning electron micrograph of the adaxial epidermis of the leaf of *Gibasis karwinskyana* with costal bands of silica cells (si) and intercostal stomata. ×120. (By courtesy of Dr. M. Stant. Crown Copyright; reproduced with the permission of the Controller of Her Majesty's Stationery Office and of the Director, Royal Botanic Gardens, Kew.)

larger than adjacent ordinary epidermal cells. In the leaf of *Canavalia ensiformis* paired crystal-containing cells occur, mainly over the vascular strands.[252] Thus silica, cork and crystal cells all provide further examples of some apparent influence of the underlying cells on differentiation in the epidermis.

When the technique of X-ray microanalysis was applied to parts of the shoot of *Cannabis*, the source of marijuana, silicon was found to be deposited uniformly over covering hairs from all parts of the plant; calcium carbonate occurred at the base of the cystoliths. It is remarkable that both substances could still be detected by this technique in the remains of marijuana ash. The hairs on hop, *Humulus lupulus*, which are rather similar to those on the related genus *Cannabis*, can be distinguished as comparatively poorly silicified.[149]

This technique thus offers one possible means of identifying even small amounts of marijuana.

Stomata

Stomata (singular, stoma) occur on most of the aerial parts of plants, though predominantly on leaves and young stems. A stoma consists of a pore surrounded by two ***guard cells***. The epidermal cells adjoining the guard cells often differ in size or arrangement from the rest of the epidermal cells; such cells are called ***subsidiary cells***. The stoma, together with the subsidiary cells, is sometimes termed the stomatal complex. No plasmodesmata or pits were observed between the guard cells and subsidiary cells of grasses,[72] but they have been detected in leaves of tobacco and *Vicia faba*, largely in the end walls of guard cells.[486] The guard cells usually contain chloroplasts, though electron microscopy indicates that these contain fewer and less well organized lamellae than the chloroplasts of mesophyll cells, resembling early stages in development of normal chloroplasts.[72] However, chlorophyll is present and active assimilation apparently takes place in the guard cells;[734] starch grains are often present in the plastids. Mitochondria, dictyosomes, and ribosomes are numerous, and there is abundant rough endoplasmic reticulum.[606, 713] In some species the guard cells

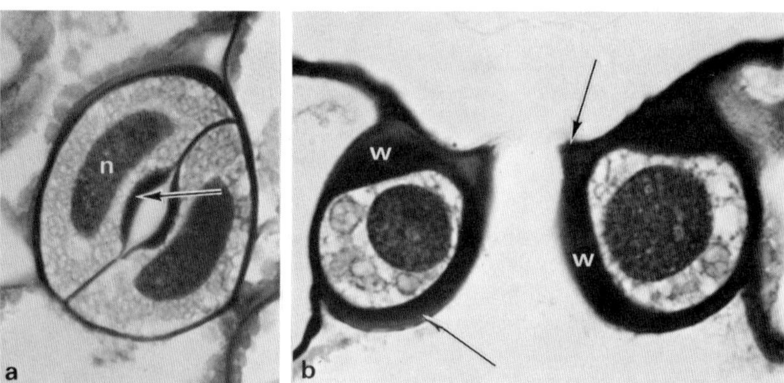

Fig. 7.19 Stomata of *Ophioglossum petiolatum*. (**a**) Paradermal section of material embedded in glycol methacrylate, showing the guard cells. A thickened wall (arrowed) is present adjoining the pore. n, nucleus. ×540. (**b**) Guard cells from T.S. leaf. The cell wall (w) is differentially thickened, and a cuticle (arrowed) surrounds the wall abutting on the pore, ending in a cuticular beak. ×910. (From Peterson, Firminger and Dobrindt,[509] Figs. 17 and 19, p. 1707; reproduced by permission of the National Research Council of Canada from the *Canadian Journal of Botany*, **53**, 1698–1711, 1975.)

may also differ from the other cells of the epidermis in lacking pigments, or crystals or protein bodies.[750]

In most dicotyledons the guard cells are somewhat kidney-shaped, having localized ledges or projections of thick wall or cuticle. In the guard cells of *Allium*, microtubules radiate out in a fan-shaped array from the future pore site; the microfibrils of the thickened region of the wall closely parallel these microtubules, and by regulating the mechanical properties of the wall determine cell shape.[484] The differential thickening may be significant in the opening and closing of the pore.[606,625] In the fern *Ophioglossum* the thickened walls (Fig. 7.19a,b) showed a positive reaction when stained for cellulose; there is a layer of cutin adjacent to the pore (Fig. 7.19b). When the epidermis was stained with aniline blue and viewed with a fluorescence microscope, the thickened walls adjacent to the pore, and also the walls between the two guard cells, fluoresced brightly (Fig. 7.20). Epidermal strips incubated in β-1,3-glucanase enzymes did not fluoresce after

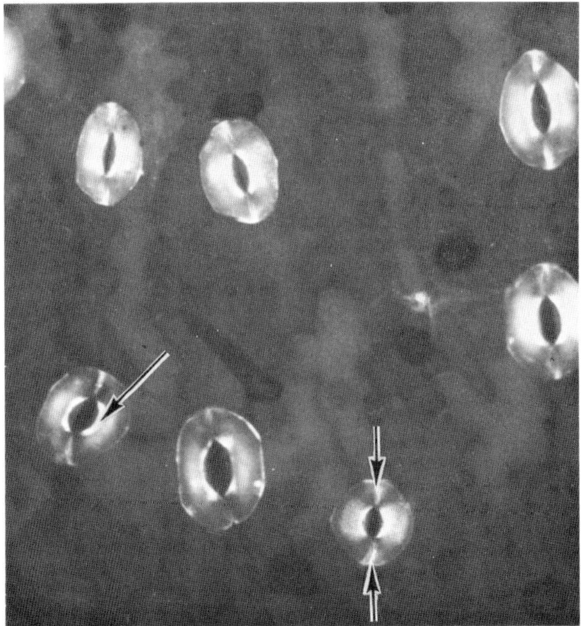

Fig. 7.20 Epidermal strip from a leaf of *Ophioglossum vulgatum* stained with aniline blue in phosphate buffer, pH 8·0, viewed with a fluorescence microscope. Fluorescent regions of the guard cell wall (arrowed) probably contain callose. ×150. (By courtesy of Dr. R. L. Peterson.)

subsequent staining in aniline blue, suggesting that a β-1,3-glucan, probably callose (see also Chapter 9), is a component of the guard cell wall.[509] In many monocotyledons the guard cells are narrow and dumb-bell shaped (Fig. 7.21a), having a narrow central zone with a thick wall and narrow lumen (Fig. 7.21b). This thick-walled region is thought to be important in the mechanism for opening and closing the pore, which is of course of great functional importance, and guard cells of this kind have been compared to a balloon with masking tape along one side.[750] On inflation, such a structure would swell at the two ends, which are thin-walled (or devoid of tape); this swelling would result in the concave bending of the thick-walled (taped) central region. When two adjacent cells act in this way, the pore between the central thick-walled regions would be caused to open. However, observations of the guard cells of *Zea mays* with the electron microscope indicate that the bulbous ends have rather thick walls,[625] which renders the above interpretation less likely.

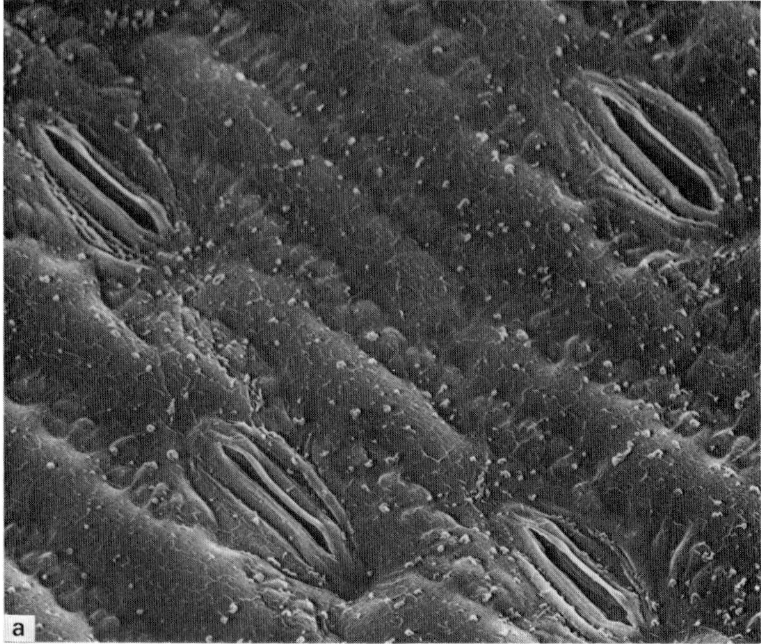

Fig. 7.21a Scanning electron micrograph of the adaxial surface of the leaf of *Zea mays*, showing several stomatal complexes. ×900. (By courtesy of Dr. J. N. A. Lott.)

It is thought that all stomatal movements can be attributed to changes in the turgor difference between the guard cells and the adjoining epidermal cells,[310] although the precise mechanism remains controversial. Active processes may be involved in both opening and closing of stomata. Observations on up to 50 species indicate that guard cells, and even guard cell mother cells, contain a greater concentration of potassium than the surrounding epidermal cells; chloride is sometimes an accompanying anion.[148, 484, 507] The potassium may act osmotically to produce stomatal opening;[484] loss of potassium in the dark apparently brings about closing of the pore.[237, 334]

In leaves stomata may occur on both surfaces, or on only one. In most mesophytic plants—those in a temperate climate with an adequate water supply—stomata are more frequent on the abaxial surface. A mesophytic leaf might have of the order of 180 stomata per mm^2.[310] In aquatic plants stomata may be absent or, in floating leaves, restricted to the upper or adaxial surface. In many xerophytes, i.e. plants which usually have a restricted water supply, the stomata may be deeply sunken below the level of the other epidermal cells, or restricted to grooves or cavities in the leaf surface. In horsetail, *Equisetum*, two subsidiary cells completely overarch the guard cells,

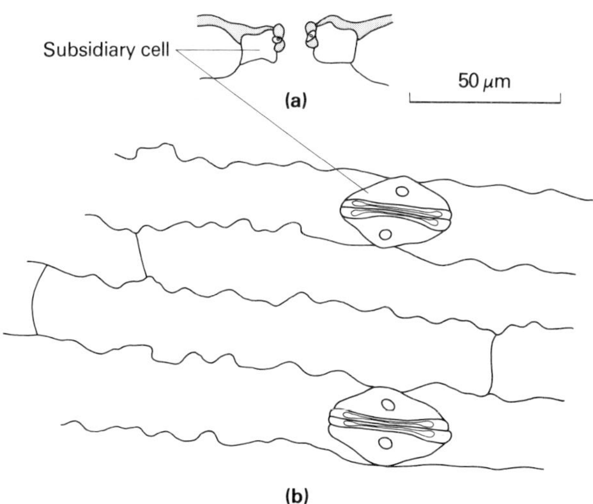

Fig. 7.21b Stomata from *Zea mays*. (**a**) Transverse section of leaf, showing thickened guard cells and subsidiary cells. (**b**) Paradermal section of the leaf, showing guard cells, subsidiary cells and other epidermal cells in surface view. Nuclei of guard cells are much elongated through the narrow part of the cell. Thick walled regions are stippled. ×345.

and control the opening of the pore.[147] In most parallel-veined leaves the stomata occur in rows parallel to the long axis of the leaf, and are formed from apex to base; in most net-veined leaves they are scattered over the leaf surface, and new stomata may be formed between existing ones. A value known as the stomatal index can be readily calculated.

$$\text{Stomatal index} = \frac{\text{No. of stomata}}{\text{No. of stomata} + \text{No. of epidermal cells}} \times 100$$

where the numbers of stomata and of ordinary epidermal cells are measured within a unit area. This value is found to be reasonably constant for any particular species, being affected to any great extent only by humidity. Both the size and frequency of stomata are affected also by the degree of ploidy of the plant.[570] In polyploid plants the stomata are larger and less frequent per unit area of leaf. In a polluted environment, stomatal frequency was reduced on both upper and lower surfaces of the leaves of *Trifolium*, but stomatal size was not affected.[589] A decrease in light intensity also results in lower stomatal frequency.[505] Tomato plants grown under controlled conditions were hypostomatous (i.e. stomata restricted to the abaxial surface); plants grown at high light intensity had amphistomatous leaves (stomata on both surfaces), and had a slightly higher density on the abaxial surface than the plants from low light intensity. Plants which were transferred from high to low light intensity showed a progressive decrease in formation of stomata on the abaxial epidermis; it was concluded that initiation of stomata is affected by the environmental conditions prevailing at the time, rather than those experienced previously.[262]

It is evident from the foregoing that the guard cells are idioblasts, differing in various ways from the surrounding epidermal cells. The distinctive nature of the guard cell mother cell, or initial cell, is evident not only in developing leaf blades but also in stem and petioles.[372,452] For example, in the developing petiole of *Populus* the guard cell mother cell shows an ability to synthesize starch, and other distinctive features, at an early stage of development.[452] The recent isolation of guard cell protoplasts[743] offers a new means of studying their specific physiological properties.

Stomata are clearly of great importance to the physiology of the plant, being sites of gaseous exchange during transpiration, photosynthesis and respiration. They are also the mode of ingress of various pathogens. Observations of the hop leaf with the scanning electron microscope showed that if a suspension of zoospores of *Pseudoperonospora humuli*, the organism responsible for hop downy mildew, was placed on the surface in the light the zoospores settled singly on the

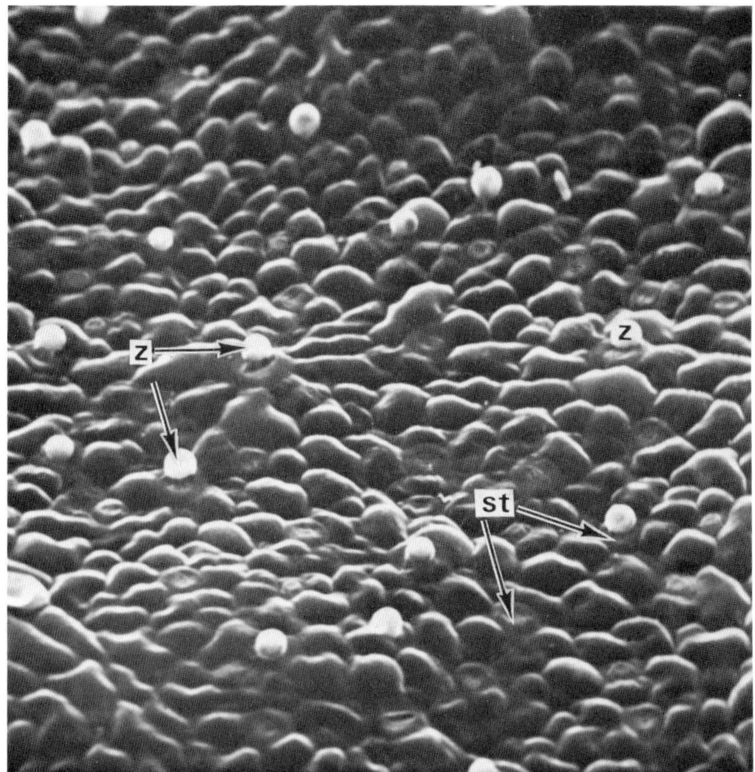

Fig. 7.22 Scanning electron micrograph of the leaf surface of hop, *Humulus lupulus*, in the light, 64 minutes after inoculation with a suspension of zoospores of hop downy mildew (*Pseudoperonospora humuli*). Zoospores (z) have settled on many of the stomata (st). ×378. (By courtesy of Dr. D. J. Royle.)

stomata (Fig. 7.22). A significant proportion had settled after four minutes. Those of *Plasmopara viticola*, the cause of grapevine downy mildew, settled in groups on the stomata of grape leaves; in the light, small groups settled on many stomata (Fig. 7.23a), whereas in the dark, larger groups settled on a few stomata (Fig. 7.23b). If zoospores of *Pseudoperonospora* were placed on Perspex replicas made from a hop leaf in the light, i.e. with open stomata, a large proportion settled on the stomata (Fig. 7.23c,d), whereas on replicas of a leaf from the dark they were usually randomly distributed. It thus appears that both a chemical stimulus emanating from the process of photosynthesis and also a physical stimulus from open stomata (shown by the effect with replicas) are necessary for the complete response of the zoospores to stomata, but each of these stimuli can act independently.[556] These observations also demonstrate the value of replicas of plant surfaces.

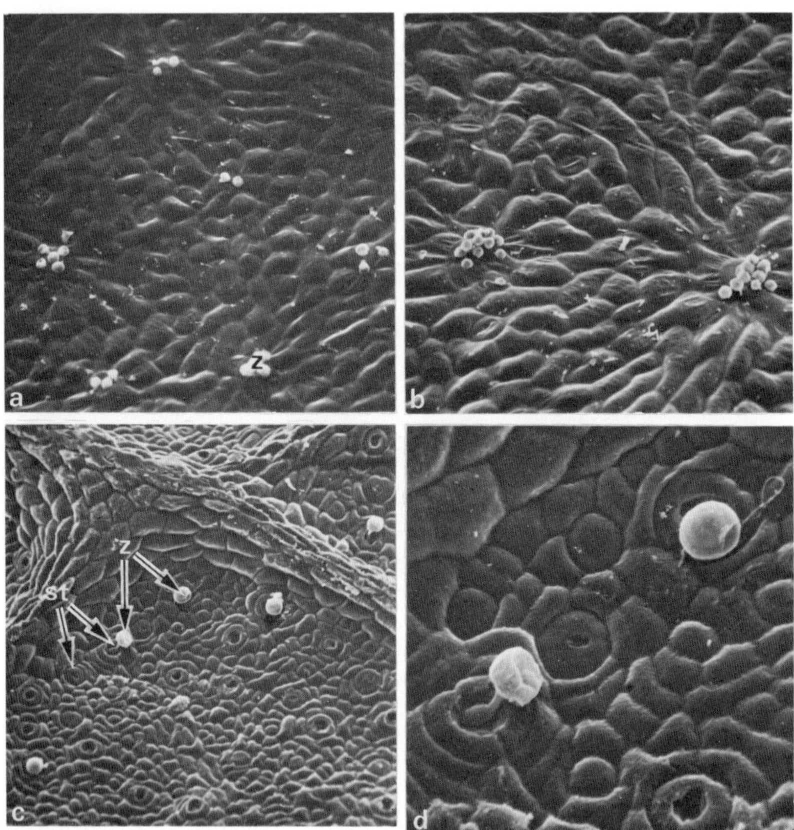

Fig. 7.23 (**a**) and (**b**) Surfaces of young grapevine leaves (*Vitis vinifera*) with zoospores of grapevine downy mildew, *Plasmopara viticola*, 64 minutes after inoculation. In the light (**a**), small groups of zoospores (z) have settled on many of the stomata; in the dark (**b**), larger groups have settled on a few of the stomata. (**a**), ×205; (**b**), ×245. (**c**) and (**d**) Perspex replicas of a young hop leaf (*Humulus lupulus*) equilibrated in the light, 6 hours after inoculation with zoospores of hop downy mildew, *Pseudoperonospora humuli*. In (**c**) all 5 zoospores (z) have settled on stomata (st). ×255. (**d**), enlargement of two zoospores in (**c**). ×810. (From Royle and Thomas,[556] Plate 3a, b, and Plate 4d and e.)

Formation of stomata

Stomata in dicotyledons may originate by a division resulting in an oblique wall in an epidermal cell.[491] The smaller cell resulting from this division functions as the guard cell mother cell. In many monocotyledons the guard cell mother cell is formed by an asymmetric division of an epidermal cell; the process is similar to the formation of some trichoblasts in the root epidermis. A differential distribution of cytoplasm first takes place, the cytoplasm accumulating at the apical, or distal, end of the cells. Subsequently the cell divides asymmetrically, and the distal small, densely cytoplasmic cell functions as the guard cell mother cell.[75,77] In the leaves[634] and stems[372] of grasses, as already mentioned, the small idioblast thus formed may give rise to a hair, a pair of guard cells, or a silica cell plus a suberous (cork) cell; its final development depends in part, at least, on its position in the epidermis (Fig. 7.17). Because of the divergent fates of their products, these small cells are of great interest in studies of differentiation.

In grasses and other monocotyledons the guard cell mother cell (gcmc) is the smaller, more densely cytoplasmic product of the unequal division of a protodermal cell (Fig. 7.24a). In some monocotyledons the guard cell mother cell induces divisions in the adjacent epidermal cells, probably in part by setting up an osmotic gradient. The smaller cells produced from these induced asymmetric divisions in the adjoining epidermal cells constitute the subsidiary cells (Fig. 7.24b).[636,638] After the formation of the subsidiary cells the guard cell mother cell divides by a wall parallel to the long axis of the leaf to give rise to the two guard cells (Fig. 7.24c) Temperature changes have differential effects on the division of the guard cell mother cells and on the preceding formation of the subsidiary cells by induced mitoses, suggesting that these are quite distinct processes.[16] During the formation of the gcmc in *Allium*, the metaphase plate usually forms at an oblique angle to the longitudinal axis of the cell. In late anaphase–telophase the spindle apparatus and both daughter nuclei rotate, until the cell plate is brought into a longitudinal position, a process which usually takes 15–20 minutes.[482] Microtubules are apparently essential for spindle re-orientation, since antimicrotubule agents such as colchicine prevent it.[483]

Studies of the epidermis in developing leaves of wheat with the electron microscope have shown that before the unequal division the nucleus becomes displaced towards one end of the cell, and vacuoles to the other end, other organelles being uniformly distributed. Before prophase, a band of microtubules was evident round the nucleus.[518] After formation of the guard cell mother cell, the nuclei of adjacent epidermal cells move towards the guard cell mother cell and divide to

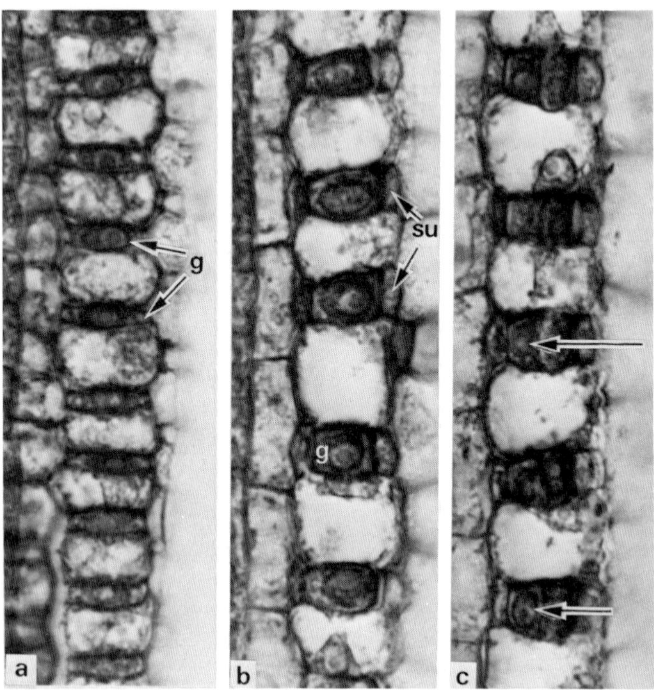

Fig. 7.24 Stages in development of the stomatal complex as seen in paradermal sections of the young leaf of *Cyperus alternifolius*. (**a**) Guard cell mother cells (g, arrowed) have been formed by unequal divisions of the protodermal cells. (**b**) Subsidiary cells (su) have been formed lateral to the guard cell mother cells (g), formed by unequal division of the adjoining epidermal cells. (**c**) Guard cell mother cells have divided equally by a longitudinal wall (arrowed) to form two guard cells. All ×1000. (Slide by courtesy of Dr. J. B. Fisher.)

form the subsidiary cells. A hemispherical cell plate forms, apparently aligned by microtubules. These pre-prophase microtubules indicate the location where the new cell plate will join the mother cell wall. Later work on wheat and *Commelina* indicated that this pre-prophase band of microtubules was not, in fact, responsible for positioning the nucleus prior to mitosis.[515,516] It is believed that the gcmc in some way induces the asymmetric division of adjacent epidermal cells to form the subsidiary cells; this view is supported by observations on barley leaves, where epidermal cells that were adjacent to two gcmc's divided asymmetrically twice to form the subsidiary cells of both stomatal complexes.[742] Subsequently the guard cell mother cell divides symmetrically to give rise to the guard cells, the intervening wall being incomplete with one or more holes at either end.[518] Similar events occur during the formation of stomata in the internodal epidermis of oat, and again there are well defined openings between the two guard cells.[375] Stages in the formation of guard cells in the stem epidermis of oat are illustrated in Figs. 7.25–7.29. The movement of the nucleus to one end of the cell prior to the unequal division leading to formation of the guard cell mother cell is evident (Fig. 7.25), as is the concentration of cytoplasm at that end of the cell and in the smaller product of the unequal division (Fig. 7.26). The larger product of the division, which develops as an ordinary cell of the epidermis, is much more highly vacuolate. These illustrations afford a striking example of polarized cellular differentiation in a single layer of tissue. Figs. 7.25 and 7.26 may be compared with Figs. 7.3 and 7.4, illustrating the unequal division of cells in the root epidermis in the formation of trichoblasts. The contrast between the larger and smaller product of the unequal division, both in size and density of cell contents, is greater in the guard cell mother cells illustrated than in the example of trichoblast formation.

After the formation of the two guard cells in oat, the nuclei become much elongated and plastids with a few grana become evident. A pore develops between the two guard cells (Fig. 7.29). As already mentioned, microtubules become apparent along the common wall between the guard cells, and the wall forms a thick pad in this region.

In the leaves of *Galtonia candicans* (Liliaceae), after the unequal division the gcmc had a smaller nucleolus than the larger product, about $\frac{3}{5}$ of the volume of the latter. It was also noted that, whereas the gcmc's merely doubled their DNA content in preparation for mitosis, the epidermal cells showed an increase from the 2C to the 20C level of DNA, accompanied by a 10-fold increase in length.[588] These interesting observations may be compared with those on the trichoblasts of *Hydrocharis* (p. 100), in which it was the smaller product of the unequal division that showed a larger nucleolus, endopolyploidy and extensive elongation. In four members of the Liliaceae, the leaf epidermal cells reached tetraploid levels of DNA before the formation of the guard cells, and it is suggested[588] that the induction of mitosis in adjacent epidermal cells by the gcmc, to form the subsidiary cells, may fail because polyploid nuclei divide less readily than diploid ones. These species have no subsidiary cells. These relationships between nucleolar size, DNA content, cell elongation and cell division merit further investigation.

In dicotyledons such as *Dianthus*, the guard cell mother cells also develop from the smaller product of an unequal division,[490] but it is more difficult to follow their development in detail. In grass leaves the

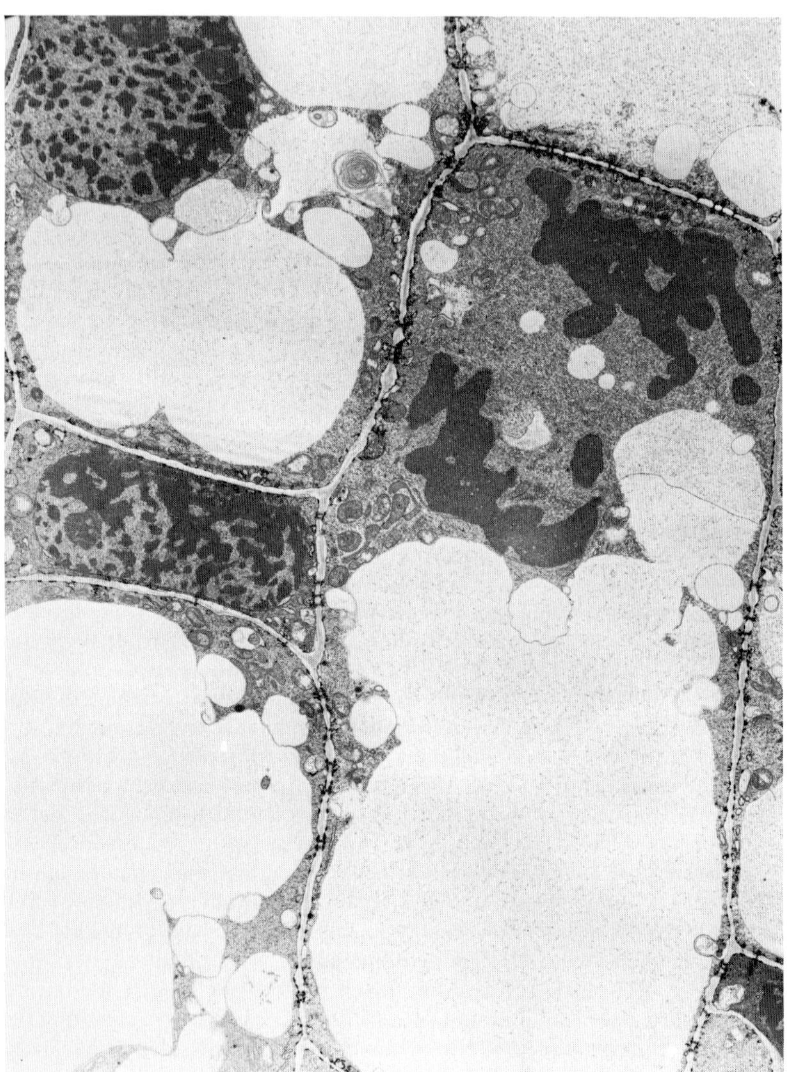

Fig. 7.25 Unequal division of an epidermal cell in the developing internode of oat, *Avena sativa*, in the intercalary meristem region. Prior to mitosis the nucleus and much of the cytoplasm has moved to the end of the cell (on the right). Mitosis is in progress. On the left a small densely cytoplasmic cell and a large, highly vacuolate cell, the products of such an unequal division, are seen. The small cell will probably develop as a guard cell mother cell (in this region, however, some small cells develop in other ways, e.g. as trichomes). $\times 1880$. (From Kaufman *et al.*,[374] Fig. 3, p. 284.)

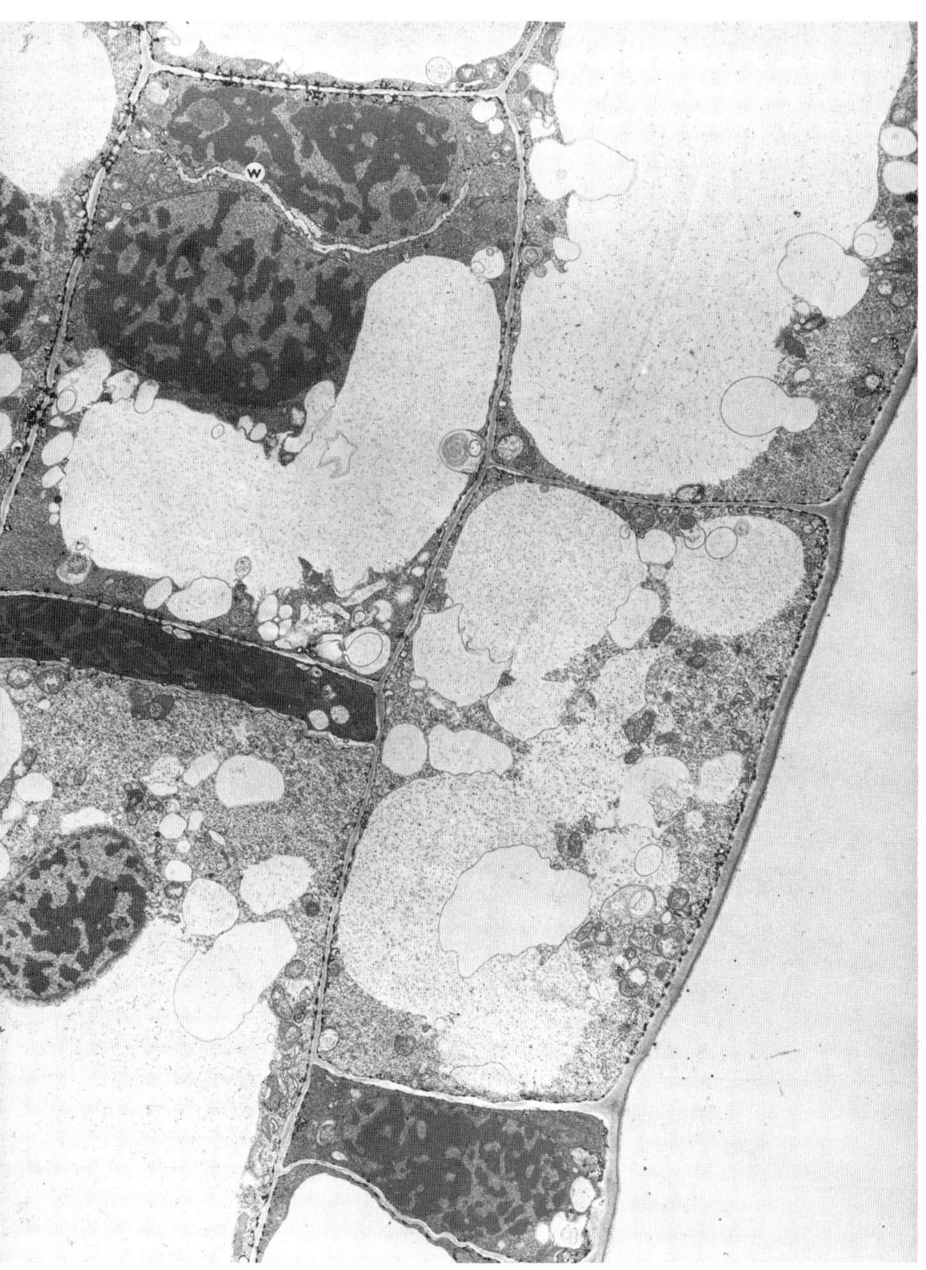

Fig. 7.26 Short and long cells in the meristematic region of the internodal epidermis of oat. Note the dense cytoplasm of the short cells, presumptive guard cell mother cells, and the much greater degree of vacuolation of the long cells. The wall (w) between the guard cell mother cell and its sister cell is markedly curved. × c. 3800. (From Kaufman *et al.*,[373] Fig. 11, p. 177. © 1970 by The University of Chicago. All rights reserved.)

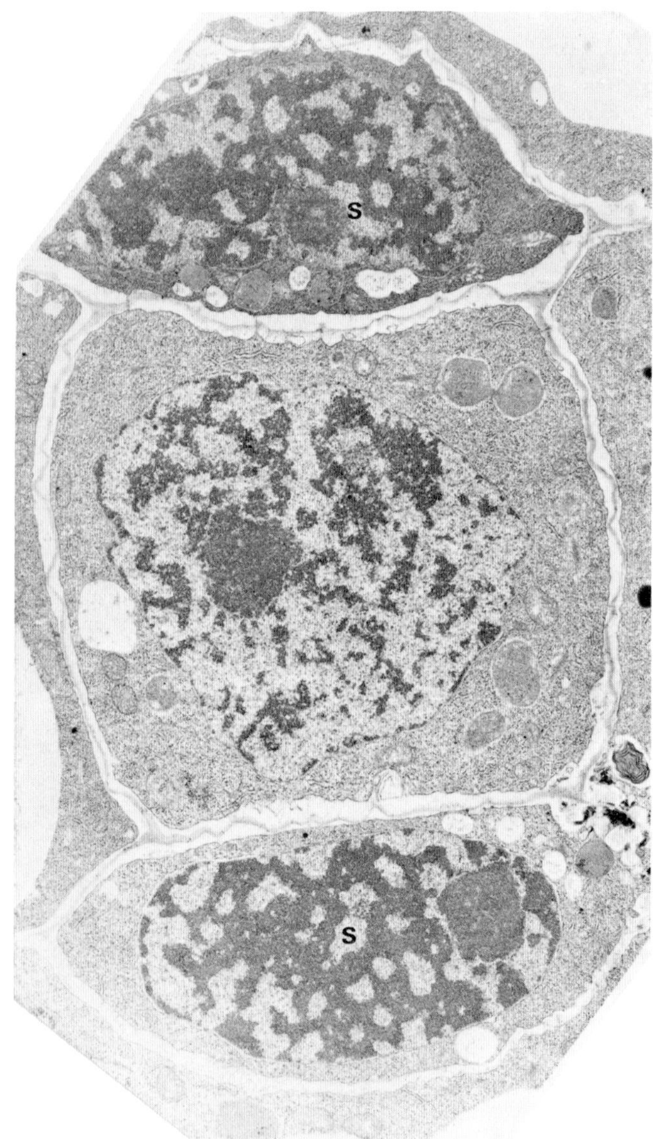

Fig. 7.27 Guard cell mother cell with two subsidiary cells (s) from internodal epidermis of oat. The orientation of this figure and of Fig. 7.28 is at right angles to the orientation of Figs. 7.24–7.26. ×c. 2700. (From Kaufman *et al.*,[374] Fig. 7, p. 288.)

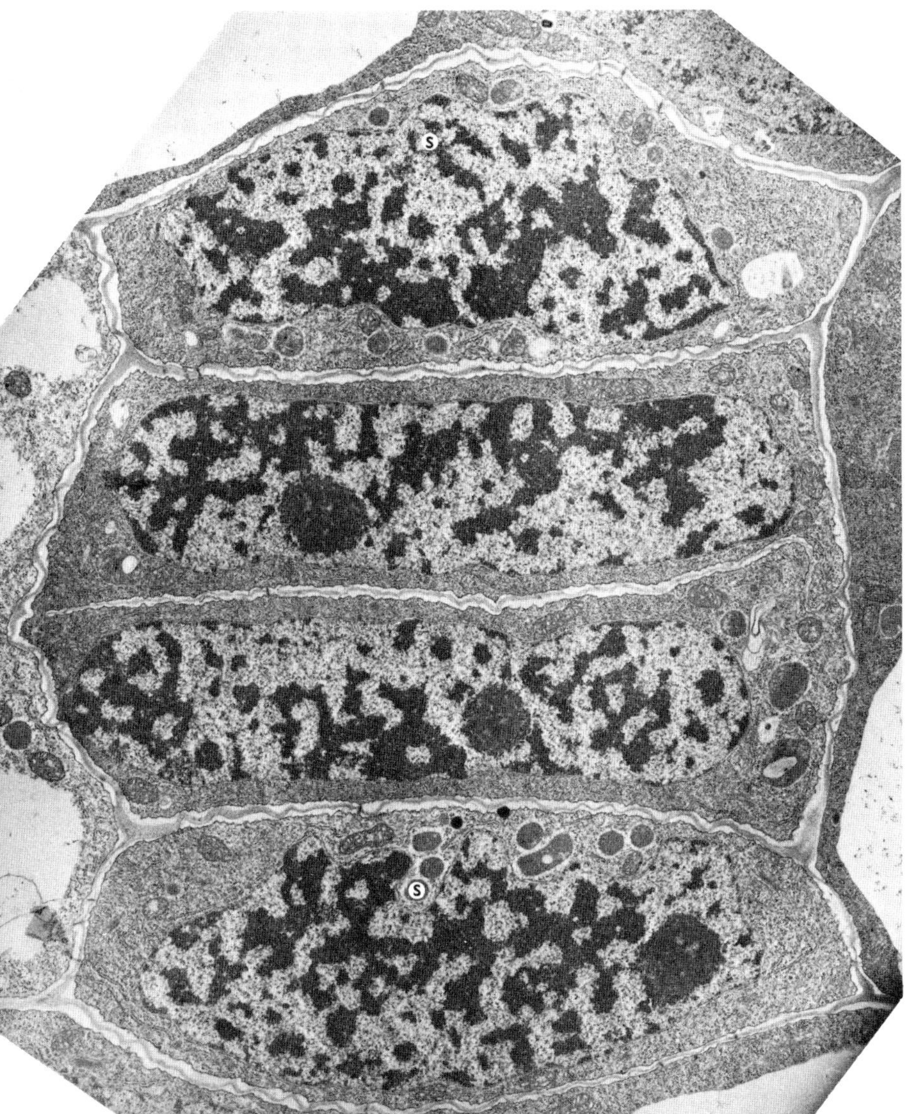

Fig. 7.28 Early stage in development of stomata in the internodal epidermis of oat. The guard cell mother cell has divided by a wall parallel to the long axis of the stem to give the guard cells. s, subsidiary cell. × 7500. (From Kaufman *et al.*,[375] Fig. 16, p. 39.)

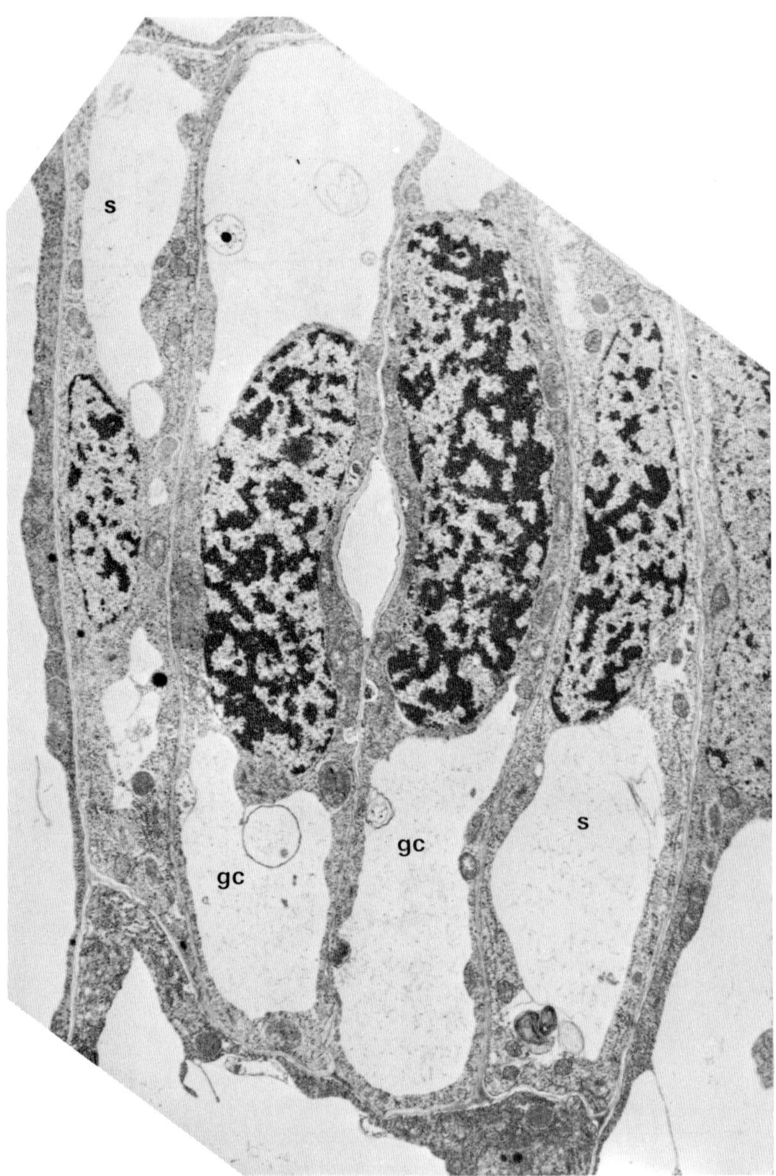

Fig. 7.29 Stomatal complex in the internodal epidermis of oat. The guard cells (gc) and subsidiary cells (s) are now somewhat vacuolate. A pore has formed between the guard cells. ×5120. (From Kaufman *et al.*,[375] Fig. 23, p. 44.)

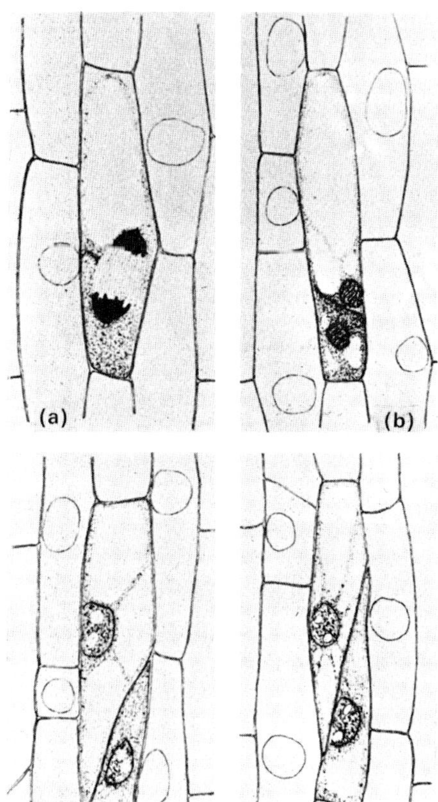

Fig. 7.30 The effect of centrifugation on stomatal development in *Allium cepa*. A stomatal mother cell is apparently formed at the basal cell pole. (**a**) Displacement of the nuclear spindle after centrifugation. (**b**) Cell wall formation in the basal part of the cell after centrifuging for 3 hours. (**c**) Outgrowth of the cell formed towards the base, 12 hours after the end of centrifugation. (**d**) The same, 24 hours after the end of centrifugation. (From Bünning,[75] Fig. 23, p. 23, after Bünning and Biegert.)

stomata develop sequentially in longitudinal rows, because of the activity of the basal intercalary meristem, but stomata of dicotyledonous plants are more scattered. However, studies have been carried out on the ontogeny of stomata in a number of dicotyledonous groups;[335–338] ontogeny is often somewhat variable even in a single species.

Experiments on stomata

A number of experiments have been carried out in an endeavour to discover the factors controlling stomatal formation and distribution. Bünning and Biegert[77] showed that after centrifugation the normal accumulation of cytoplasm at and the movement of the nucleus towards the apical pole of the epidermal cells of onion leaves (*Allium cepa*) was altered, and an initial-like cell was formed at the basal (proximal) pole instead (Fig. 7.30). This did not, however, give rise to guard cells. Later observations on centrifuged wheat seedlings showed that pre-prophase nuclei were rather resistant to sedimentation, perhaps in part because of the incorporation of many pre-prophase microtubules into the developing spindle.[515] Nuclei of epidermal cells dividing asymmetrically to form subsidiary cells also resisted sedimentation.

These experiments interfered with the distribution of the cytoplasm in differentiating cells. Treatments affecting the nucleus also gave some interesting results. For example, in barley seedlings treated with 2-mercaptoethanol, a substance that interferes with spindle formation, a large number of stomata had one subsidiary cell missing.[638] This substance blocks mitosis temporarily, and after growth is resumed some guard cell mother cells divide by a wall perpendicular to the long axis of the leaf, i.e. at right angles to the normal. Similar results were obtained by simply removing the sheaths of culm leaves from the plant, without further treatment. In such leaves mitoses occurred again in guard cell mother cells after a period of quiescence consequent upon removal of the leaf; about 88% of such mitoses produced guard cells abnormally oriented at right angles to the long axis of the leaf.[634] The effects that these treatments have in common are a temporary delay in the onset of mitosis and the frequent absence of subsidiary cells as well as the reorientation of mitoses in the guard cell mother cells. Before mitosis the latter synthesize both DNA and RNA very actively.[639] Somewhat similar results were obtained by removing the calyptra from the developing capsules of two species of mosses. The long axis of the stomata is normally parallel to that of the capsule, but if the calyptra was removed prior to the division of the gcmc an essentially random orientation of stomata resulted.[255]

Working with the leaves of various dicotyledons, Bünning and Sagromsky[78] showed that treatment of developing leaves with alkali stimulated the formation of stomata, whereas treatment with acid was inhibitory. One half of a young leaf was treated, while the other half was left untreated and served as the control. Treatment with the hormone indoleacetic acid stimulated cell division but inhibited formation of stomata, suggesting that the production of such a substance by stomata and other meristemoids might be, or might resemble, the normal mechanism by which the formation of other similar structures in their immediate vicinity is prevented (see Chapter 2 and below). Wounding the leaf also inhibited stomatal development in proximity to the wound. Although guard cells rarely divide, and, for example, remain undivided during the formation of cell division centres in the epidermis of *Torenia* (p. 95),[106] divisions were found in guard cells of 26 Magnoliaceous taxa in 7 genera after natural or experimental wounding.[677] For example, in wounded leaves of *Magnolia omeiensis* each guard cell seemed to have enlarged and then divided once or several times. The influence of the wound was very localized.

Treatment of leaves with extracts of growing tissues was also tried. For example, when a paste consisting of the tissues of young pods which were forming stomata was smeared on the leaves of *Theobroma cacao*, stomatal development in the leaves was completely prevented.[78] When seedlings of *Vicia faba* were grown in the presence of apples, and hence in the presence of ethylene gas, the number of stomata per mm^2 was greatly enhanced.[391] With 6 apples, leaf expansion was greatly inhibited, rendering the results more difficult to interpret, but with 3 apples the stomatal index was increased, i.e. the number of epidermal cells per stoma was decreased. Several stomata often occurred adjacent to one another,[391] suggesting that any mutually inhibitory influence was diminished or nullified by the ethylene. Experiments of this kind seem worth pursuing with modern more precise techniques.

Distribution of stomata

In most parallel-veined monocotyledons stomata occur in rows parallel to the long axis of the leaf. In horsetail, *Equisetum*, two rows of stomata are present in each scale leaf.[147] In dicotyledons, however, stomata are usually scattered over the leaf surface, and, as already mentioned in Chapter 2, observations suggest that they may be mutually inhibitory. Thus it is possible to derive mathematically a curve relating stomatal number to leaf area in tomato that is the same as one plotted from actual data if two assumptions are made, one of which is that new stomata are inhibited by existing ones.[262] Computer models of epidermal development in *Pelargonium zonale* and *Sedum stahlii*, based on observed quantitative features and the assumption that stomatal initials inhibit adjacent cells from developing also as stomatal initials, produce a probability distribution frequency for inter-stomata distances similar to that for real cells.[386] In the leaf of *Ilex crenata* var. *convexa*, large stomata arise early in development and differ from the more typical form. Computer models of several possible modes of origin of these stomata were tested for their similarity to actual observations of their distribution. The only model which proved satisfactory was one in which each newly formed large stoma produced a zone of inhibition that prevented further similar structures from forming.[388] These recent quantitative studies thus bear out the hypothesis that existing stomata inhibit the formation of new ones. Work with the more ordered epidermis of a monocotyledon, *Crinum*, however, demonstrates a method for measuring the relation of stomatal frequency to the distance from a central stoma which may be applicable to other species. Here again a region was found in the

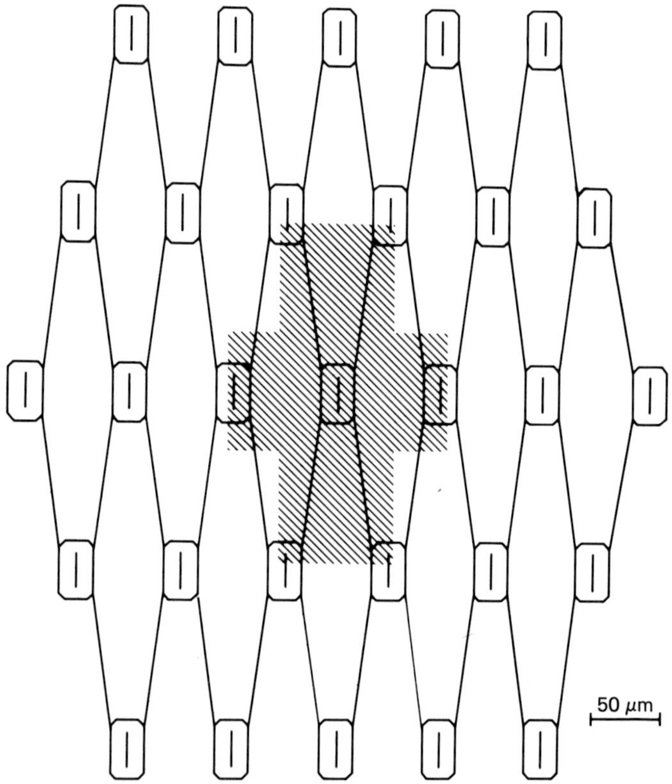

Fig. 7.31 Diagram showing stomata (oval with a central line) in every possible location in the epidermis of *Crinum*. An orderly arrangement results. The cross-hatching represents the region for which neighbouring stomata frequency was found to be below that of a random distribution (thus indicating the area of inhibition round the central stoma). (After Sachs,[564] Fig. 5, p. 317. © 1974 by The University of Chicago. All rights reserved.)

vicinity of each stoma in which other stomata did not occur. This region seemed to extend through one epidermal cell in each direction, and thus, like the epidermal cells themselves, had a longer dimension parallel to the long axis of the leaf. Stomata had no effect on other stomata over greater distances. The pattern of stomata in *Crinum* thus is not the result of inhibition between the stomata, but simply of the alternating arrangement of the epidermal cells and the formation of a gcmc distally by the unequal division of each epidermal cell, except those in costal regions.[564] In Fig. 7.31 the distribution of stomata in every possible location is shown diagrammatically, together with the region in which the frequency of stomata in the vicinity of the central stoma was found to be below that of a random distribution (cross-hatched area). This corresponds well with the actual arrangement. Preliminary work with dicotyledons suggests that the phenomenon may be complicated by additional factors.[564]

Types of Stomata

In most species the number and arrangement of the subsidiary cells around the stomata are relatively constant. Various classifications of stomata according to these arrangements have been drawn up, and are sometimes useful to taxonomists. Not all the stomata present on a leaf are good examples of a single type, and fairly large numbers of stomata should be examined to determine the most prevalent type. The early classifications were based on mature structure only, but it is now believed that ontogenetic studies of stomatal development are important.

Solereder[616] described four types of stomatal complex that occurred especially in certain families of dicotyledons, and to which he gave family names. Metcalfe and Chalk[451] gave rather more descriptive names to these types, as follows:

Anomocytic or irregular-celled (Ranunculaceous): the surrounding cells are indefinite in number and do not differ from the other epidermal cells (Fig. 7.32a);

Anisocytic or unequal-celled (Cruciferous): usually three subsidiary cells surround the stoma, one cell being considerably smaller or larger than the other two (Fig. 7.32b);

Diacytic or cross-celled (Caryophyllaceous): two subsidiary cells surround the stoma with their common wall at right angles to the guard cells (Fig. 7.32c);

Paracytic or parallel-celled (Rubiaceous): one or more (often two) subsidiary cells are present, with their long axes parallel to the guard cells (Fig. 7.32d).

In addition to these types, three others are sometimes recognized: ***actinocytic***, with four or more subsidiary cells elongated radially to the stoma; ***cyclocytic***, with four or more subsidiary cells arranged in a narrow ring round the stoma;[627] and ***tetracytic***, with four subsidiary cells, two polar and two lateral, as in many monocotyledons.[448]

This classification takes no account of the ontogeny of the stomata, and it is known that the various types can be arrived at by different

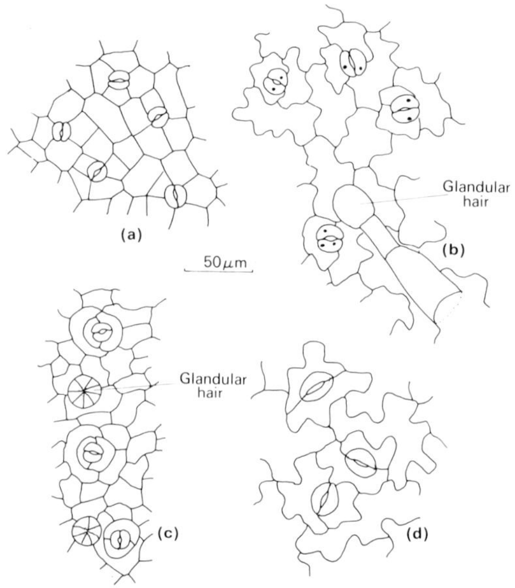

Fig. 7.32 Types of stomata of dicotyledonous plants. (**a**) Anomocytic, having no special subsidiary cells (*Cucurbita*); (**b**) anisocytic, having 3 subsidiary cells, one larger or smaller than the other two (*Petunia*); (**c**) diacytic, having 2 subsidiary cells with their common wall at right angles to the guard cells (*Hygrophila*); (**d**) paracytic, having 2 subsidiary cells with their long axes parallel to the guard cells (*Phaseolus*). ×180.

developmental paths. Relatively minor anomalies during ontogeny, such as fewer mitotic divisions or a slight difference in the position of the cell wall, can result in formation on the same leaf of stomata of different types.[489] Accordingly, a new type of classification based on ontogeny has been proposed.[488] Stomata are divided into the following types:

Mesogenous: the subsidiary cells have a common origin with the guard cells, developing from the same meristemoid as the guard cells;

Perigenous: the subsidiary cells do not have a common origin with the guard cells, but are formed by cells lying around the meristemoid that divides to form the guard cells;

Mesoperigenous: at least one of the subsidiary cells has a common origin with the guard cells, but the others do not.

These terms are not easily applicable to monocotyledons,[485] in which four categories of stomatal complex have been recognized.[636,637] More recently, many additional types of stomata have been described, including some known only in ferns.[257,682] One suggestion is to assign descriptive names in which a prefix indicates the nature of the adult stoma, and a suffix indicates the mode of origin, e.g. dia-mesoperigenous, dia-mesogenous, for diacytic stomata that originate in a mesoperigenous or mesogenous way.[257] The naming of stomatal types, though systematically useful, seems in danger of becoming over-complicated.

8

Xylem

The vascular system of the plant is made up of xylem, the water-conducting tissue, and phloem, the food-conducting tissue. Since an adequate supply of water and food is obviously essential for growth, it is at once evident that the vascular system is functionally very important within the plant. The possession of vascular tissues separates the higher plants from some of the more primitive groups of plants which are devoid of comparable conducting tissues. Many of the components of the xylem are hard and thick-walled, and are thus usually better preserved in fossil materials than the soft-walled phloem. Consequently, the phylogeny of xylem is better understood than that of the phloem. For other reasons also, the xylem is easier to study than the phloem, and will be considered first.

Hardwood and softwood timbers, many of which are of great commercial value, are made up of the secondary xylem of dicotyledons and gymnosperms respectively (see Part 2,[136] Chapter 4). Timber accounts for approximately 20–25% of all photosynthetic matter produced on earth.[623] Wood is not only used for various constructional purposes, but is also the raw material of paper. In the U.S.A. alone, 25 million tons of paper and paperboard are made each year.[37] Consequently xylem is of considerable economic importance, as well as being of great functional importance to the plant.

Origin

During the primary growth of the plant, xylem differentiates from the procambium, situated below the growing root or shoot apex or associated with a leaf primordium. The procambium consists of meristematic, densely cytoplasmic cells elongated in the longitudinal plane of the organ in which it occurs. The differentiation of primary xylem and phloem from this tissue is described in Part 2, Chapters 2 and 3. The first elements of the primary xylem to differentiate and become mature are known as the ***protoxylem***; those which mature later are called ***metaxylem***. In plants which undergo secondary growth, i.e. most gymnosperms and dicotyledons, the vascular cambium gives rise to secondary xylem. The patterns in which the elements of the xylem and phloem differentiate differ in the various organs of the plant. Differentiation of vascular tissues in these organs, and the controlling factors involved, are considered in Part 2.[136]

Differentiating metaxylem elements are known to become polyploid,[422] and recent work in which the DNA of immature metaxylem vessel elements was hybridized with labelled ribosomal RNA indicates that these elements contain substantially more rDNA than neighbouring cells, i.e. that ribosomal cistrons become considerably amplified during differentiation of the metaxylem.[29] Work with cultured cortical explants of pea roots indicated, however, that since both tracheary elements and cortical parenchyma showed polyploid levels of DNA, differentiation of xylem elements could not be wholly a consequence of the level of DNA.[514]

Elements of the xylem

Xylem is a complex tissue, composed of the conducting or tracheary elements, fibres and parenchyma (Fig. 8.1). Xylary fibres are elongated elements with pointed ends and are thought to have evolved from tracheids. They usually have thicker walls and the pits may have smaller borders than the tracheids of the same species.[187] Some evidence[221,224] suggests that wood fibres may retain living protoplasts for as long as twenty years, although they are usually defined as nonliving elements. Further studies of xylary fibres from a considerable number of species seem advisable. There may be transitional types between fibres and tracheids. Living parenchyma cells are present in both primary and secondary xylem. They may contain starch or crystals, and fulfil a storage function; eventually they too may become lignified. The proportion of fibres and axial parenchyma in the wood can be modified, for example by inhibitors of auxin transport.[462]

Tracheary elements are of two kinds, tracheids and vessel members. Both are elongated cells, thick-walled and usually devoid of living contents at maturity. The cell contents may be discernible up to the time of lignification of the wall. The secondary wall thickening is laid

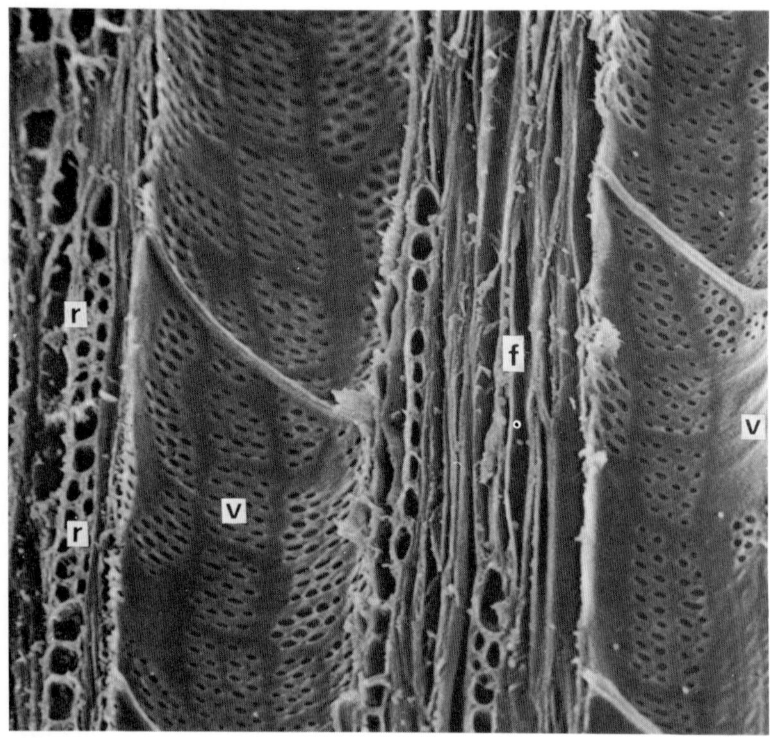

Fig. 8.1 SEM view of a tangential longitudinal section of the secondary xylem of *Juglans*, walnut. Broad vessel members (v) with groups of pits, fibres (f) and ray parenchyma (r) are evident. ×400. (By courtesy of Dr. J. N. A. Lott.)

down in various patterns and usually becomes lignified. ***Tracheids*** originate from single cells, are normally elongated and pointed at both ends, and are imperforate, i.e. the primary wall is always continuous, pit-closing membranes being present in the regions of pits. Tracheids are present in all groups of vascular plants, and the tracheary elements of most pteridophytes and gymnosperms consist exclusively of tracheids (Fig. 8.2).

By contrast, ***vessel members*** are perforate elements aggregated longitudinally into files of cells connected through the pores or perforations. These chains of cells are ***vessels***, and may vary from two cells to an indefinite but considerable length, perhaps 1 m or more. More information on vessel length is required. Vessels of the primary and secondary xylem are thus formed from a longitudinal file of procambial or cambial cells respectively. Vessels are present in the wood of nearly all angiosperms; the exceptions are certain members of the Ranales, an order of plants usually considered to be relatively primitive, which have very uniform wood composed almost entirely of tracheids. Vessels are absent from the wood of most gymnosperms and pteridophytes.

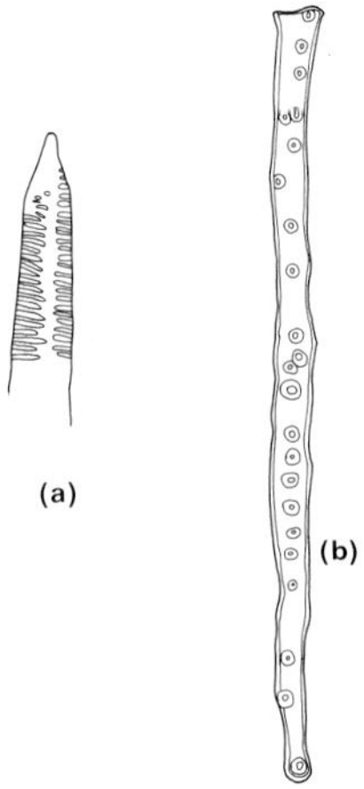

Fig. 8.2 Tracheids. (**a**) Part of a tracheid of *Polypodium*, showing scalariform pitting. ×180. (**b**) Tracheid of *Pinus*, showing circular bordered pits. ×115.

Xylem elements which differentiate during early phases of growth, i.e. protoxylem elements, usually have a thin primary wall with rings or helices of secondary wall thickening deposited upon it. These annular or helical (spiral) xylem elements (Fig. 8.3a, b) are extensible and often become much stretched during the elongation of the organ in which they occur. During this process the original rings or helices of secondary wall material become more widely separated. The later-formed elements of the metaxylem or of the secondary xylem have much more extensive regions of secondary wall; these are reticulate or pitted elements (Figs. 8.1, 8.3c, d). A comparative study of the primary xylem from some 1,350 species of angiosperms indicated that these later-maturing elements are in fact ontogenetically derived from the helical elements by the deposition of additional secondary wall material between the gyres of the existing helix. Bierhorst and Zamora[54] interpret the secondary wall system of tracheary elements as consisting of a first-order framework (the helical system) and a second-order framework (additional secondary wall deposited between the gyres of the helix). The fact that elements in intermediate stages of development are observed only relatively infrequently is attributed by these authors to the occurrence of such elements in only some vascular

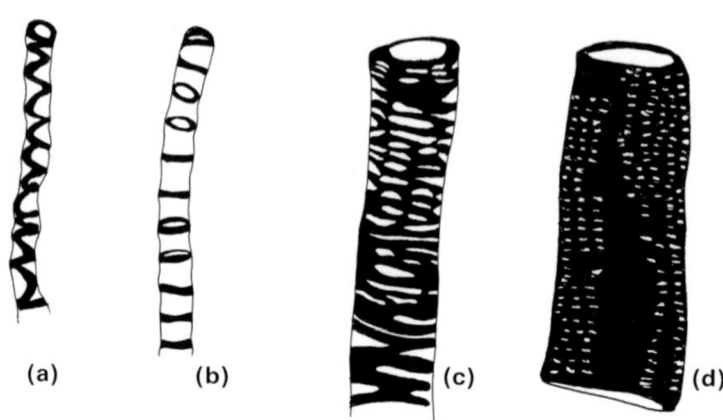

Fig. 8.3 Vessel members from the primary xylem of *Phaseolus* showing different types of secondary wall thickening. (**a**) Spiral or helical; (**b**) annular; (**c**) partially pitted and partially reticulate; (**d**) pitted. Secondary wall shown black. Only parts of a vessel member are shown in (**a**)–(**c**). (**a**)–(**d**) ×240.

bundles of certain internodes of the stem. Occasionally, however, all stages of development of the secondary wall may be observed in a single rapidly-elongating element. In terms of this view of the development of tracheary elements, annular and helical components could be considered to be juvenile forms showing a type of arrested development. It is interesting to consider why the second-order framework should be deposited only on the later-maturing elements. It is known that exposure to light accelerates secondary wall deposition,[270, 628] but the mechanism involved is not fully understood.

In a number of genera of the family Commelinaceae the development of the second-order framework can be seen particularly clearly.[54] The metaxylem elements may be helical or annular, and these thickenings of the primary framework become interconnected by a very uniform system of vertical strands, the developing second-order framework (Fig. 8.4). These vertical strands are themselves occasionally interconnected by transverse strands or sheets of second-order framework (Fig. 8.4d). It is not known how the cell determines which pattern of wall thickening to deposit.[477]

The development of the scanning electron microscope has allowed useful three-dimensional views of the various components of xylem (Fig. 8.1); see also Meylan and Butterfield.[453]

Perforation plates

The region of the wall of a vessel member in which a pore or perforation occurs is known as the perforation plate. These plates are usually terminal in position, but may be sub-terminal or lateral; the

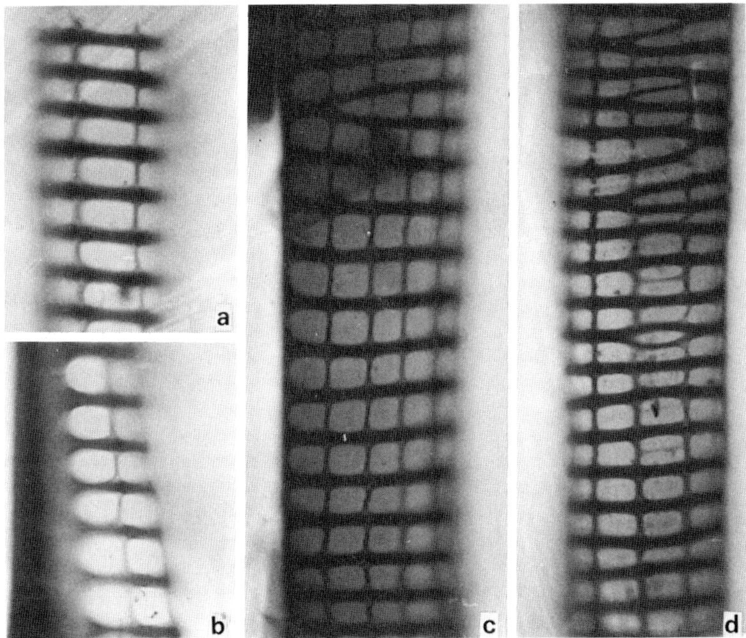

Fig. 8.4 Portions of the metaxylem of *Aneilema vitiense* (Commelinaceae), showing presence of a primary framework of wall thickenings and progressive formation of the secondary framework. ×650. (From Bierhorst and Zamora,[54] Figs. 128–131, p. 669.)

perforation may be ***simple***, with one pore (Fig. 8.5c) or multiple, with more than one. In multiple perforation plates, the pores may be arranged in various ways. When the pores occur in a ladder-like arrangement, with intervening bars of thickening, the perforation plate is ***scalariform***, e.g. *Liriodendron*, the tulip tree (Figs. 8.5a, 8.6). When the perforations are more or less circular and are grouped together, the perforation plate is ***foraminate***, e.g. *Ephedra*, one of the few gymnosperms that possesses vessels (Fig. 8.5b). If numerous small pores are separated by a network of secondary wall thickening, these form a ***reticulate*** perforation plate. Simple perforation plates are thought to have been phylogenetically derived from the multiple type, by loss of the bars of thickening. Occasionally, vessel members may have different types of perforation plates at each end, e.g. simple and scalariform. Adjoining vessel elements may also have different types of plate.[455] Such observations serve to emphasize how variable

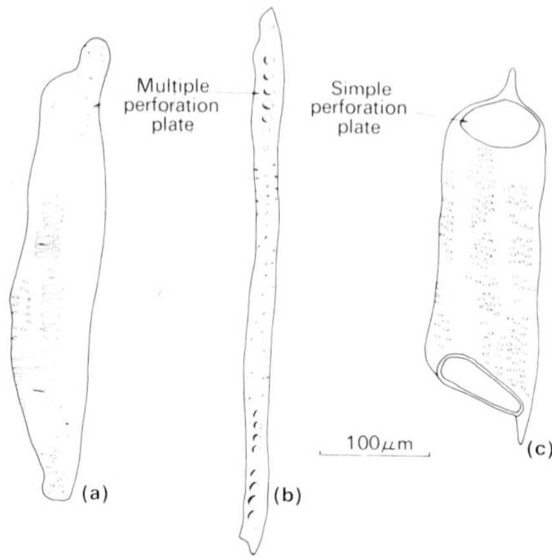

Fig. 8.5 Perforation plates in vessel members. (**a**) Scalariform (*Liriodendron*); (**b**) foraminate (*Ephedra*); (**c**) simple (*Quercus*) perforation plates. Pits are also shown. ×115.

the formation of perforation plates is and how little we know about the factors controlling their development, or the patterns of wall material. During differentiation, secondary wall material is laid down over the whole of the primary wall, except for regions that will become pits or perforation apertures. At a late stage of development these regions of primary wall are broken down by enzyme action.[80, 454] At least in later-formed xylem elements, a slight thickening may be present on the region of the primary wall that later becomes the perforation; subsequently it disappears gradually.[54, 181]

Fine structure of xylem elements

In an electron microscopic study of parenchyma cells of *Coleus* which had been induced to differentiate into thickened xylem cells as a result of wounding (see below), Hepler and Newcomb[314] reported a concentration of organelles and vesicles along the cytoplasmic bands occupying the sites of subsequent wall thickenings. Dense strands of cytoplasm which foreshadowed the positions of the secondary wall had been reported much earlier from studies with the light microscope.[134, 610] In studies of the differentiation of the normal xylem of *Acer*, *Beta*, *Cucurbita*, and of the *Avena* coleoptile, however, no indication of any regular distribution of cytoplasm and organelles in relation to wall thickenings was observed with the electron microscope.[132, 202, 729] However, microtubules have been observed in close

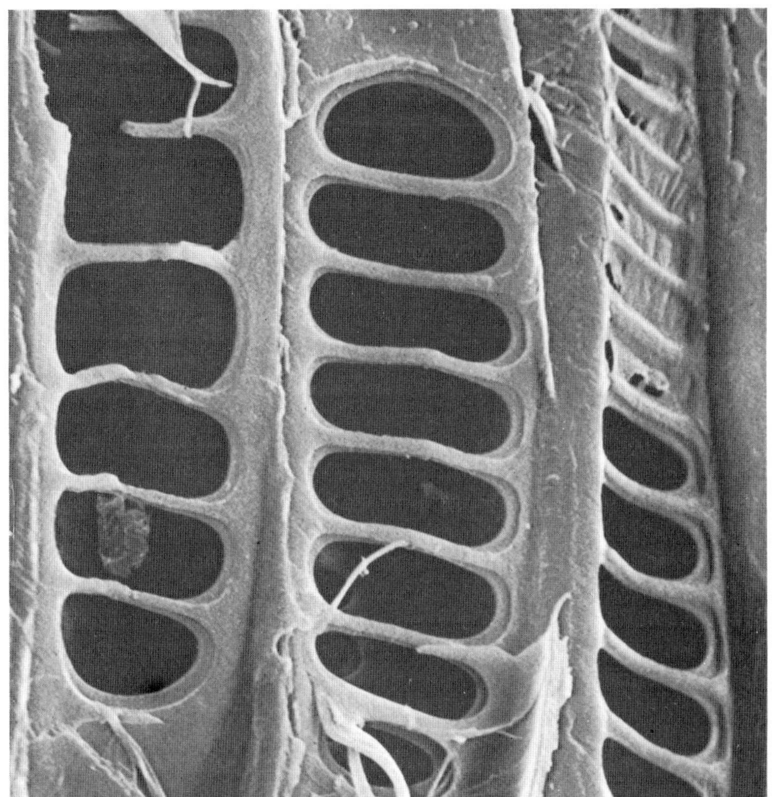

Fig. 8.6 SEM view of scalariform perforation plates in vessel members of the secondary xylem of the tulip tree, *Liriodendron tulipifera*. See also Fig. 8.5a. ×2000. (By courtesy of Dr. J. N. A. Lott.)

association with the developing secondary wall thickenings, both in normally differentiating tracheary elements of *Acer*,[729] *Avena*[132] and *Cucurbita*[203] and in xylem cells induced to regenerate from parenchyma in *Coleus*[315] and *Nicotiana*.[129] Microtubules are thought to play a major role in defining the pattern of secondary wall deposition, and treatment with colchicine, which is known to interfere with microtubule assembly, results in deformed secondary walls, which are more or less smeared over the primary wall.[313,668] A histochemical study of tracheary elements formed after wounding showed that the enzyme peroxidase and lignin occurred together in the secondary wall thickenings, suggesting that peroxidase is involved in the process of lignification.[317] It has been suggested that the hormonal control of xylem differentiation may act through the effects of auxin on peroxidase.[542]

In young tracheary elements, dense cytoplasm is present, and dictyosomes and rough endoplasmic reticulum are especially prominent. In the later stages of differentiation, after the deposition of wall thickenings, both the endoplasmic reticulum and various organelles break down.[205,729] These degenerative changes are comparable with those that occur in differentiating sieve elements (see Chapter 9). Xylem parenchyma cells are also rich in cytoplasm, with well developed mitochondria and abundant rough endoplasmic reticulum. Numerous plasmodesmata in the walls suggest a possible route for symplastic transport.[404]

Phylogeny

A considerable amount of work has been done on the phylogenetic trends shown by the xylem elements, since these elements are relatively conservative and also often remain well preserved in fossil materials.

Vessel elements are considered to have evolved from tracheids, and to have developed independently and in a parallel fashion in the various groups in which they occur.[35,100] In the dicotyledons vessels appeared first in the secondary xylem, and only subsequently in the primary xylem; in monocotyledons, which usually have no secondary xylem, vessels occurred first in the oldest parts of the primary xylem. It is believed that they occurred first in the roots, and only later in the stems and aerial parts of the plant.[101,103] In this connection, the report[716] that in ferns the most highly evolved tracheids (and, in species in which they occur, vessels) mainly occur in the roots rather than the aerial parts is of considerable interest.

In general, the specialization of vessel elements has gone from long narrow elements with tapering ends, to short, wide, squat elements with approximately transverse end walls. Scalariform perforation plates are considered to be the most primitive type and simple perforations the most highly evolved.[100] A statistical correlation has been shown between long vessel elements, very oblique end walls, and perforation plates with many bars.[87] All of these are considered to be primitive or unspecialized characters. The arrangement of pits is also considered to have progressed from scalariform through opposite to alternate, but Bierhorst and Zamora[54] have recently pointed out that only slight modifications during ontogeny are required to transform opposite into alternate pits. From their observations on the xylem of a large number of living species, these authors also drew the conclusion that there is a general trend of specialization in the xylem which expresses itself in the ever earlier appearance during ontogeny of advanced features and the elimination of primitive ones.

CONTROL OF DIFFERENTIATION OF XYLEM ELEMENTS

One of the most interesting problems of differentiation is the development of procambial or cambial cells into both xylem and phloem. These are tissues with distinctive physiological and anatomical characteristics and very different functions in the plant; yet they differentiate from the same precursors and in close spatial relationship. A considerable body of interesting experimental work is accumulating which throws some light on the factors controlling xylem differentiation, but much still remains to be done. Most of this experimental work involves the induction of tracheary elements from parenchyma, either following wounding or in cultured parenchyma tissue or callus. The problem of xylem differentiation and its control has been recently reviewed.[546,547]

Genetic aspects

Differentiation of xylem, as of other tissues, is clearly under overall genetic control, and some interesting mutants exist which have been investigated from an anatomical standpoint. For example, a recessive mutant of maize, designated *wilted*, shows severe signs of wilting during most of the growing period, even when the soil is kept well moistened. Anatomical studies showed that the wilting was attributable to the greatly retarded differentiation of the two large metaxylem elements in the vascular bundles of the stem (compare Fig. 8.7b and c with 8.7a).[523] During later stages of development most of the vessel elements did differentiate, and the wilting symptoms were alleviated. In another single-gene mutant, *wilty-dwarf* tomato, the water deficit leading to wilting symptoms was again attributable to anomalous vessel development. In this instance, secondary wall material was deposited across the primary end wall prior to its breakdown to form the perforation plate. This was considered to be due to the failure of a control mechanism during development.[8]

These examples illustrate the effects of single gene changes on the differentiation of a xylem element once this has been initiated. What controls the initial differentiation of a cell as a tracheary element? The

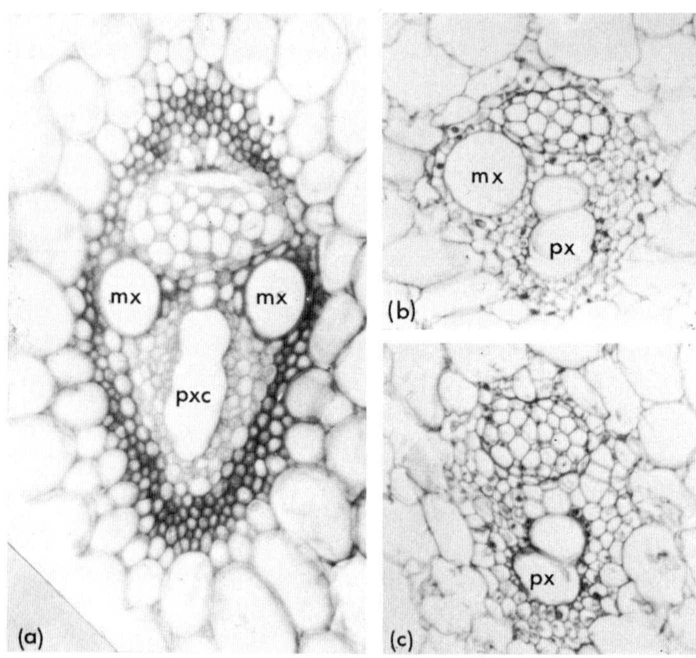

Fig. 8.7 Vascular bundles from transverse sections of stems of *Zea mays*. (**a**) From a normal plant. (**b**) and (**c**) From *wilted*, a recessive mutant. In (**b**) one large metaxylem element (mx) is present, but only slightly differentiated. In (**c**) no large metaxylem elements have differentiated. px, protoxylem; pxc, protoxylem canal. ×200. (Material of *wilted* by courtesy of Dr. S. N. Postlethwait.)

results of many experiments indicate that auxin is closely implicated, and that other growth substances may also be involved.

Induction of wound vessel elements

Some sixty years ago, Simon[605] found that if incisions were made in the vascular strands of a stem, so that these were interrupted, new xylem cells differentiated from the parenchymatous cells of the pith, and these formed between two vascular bundles. This work was later extended by Sinnott and Bloch,[609,610] who used plants of *Coleus*, which has pairs of opposite leaves. They showed that vascular strands did not always regenerate between the apical and basal ends of a single vascular bundle, but sometimes did so also between this bundle and another across the stem from it. In the pith cells in the path of this future vascular strand divisions took place and oblique walls were formed, apparently along a new axis of polarity (Fig. 8.8a). Later, bands of lignin were laid down on the walls of these parenchymatous cells. Pores were present in some of the cells (Fig. 8.8b). Sinnott and Bloch[610] noted that the first sign of the lignified thickenings was a reorientation of the cytoplasm into granular bands around the cell, occupying the position of the future bands of lignin. Twenty years later Hepler and Newcomb,[314] working with the electron microscope, reported the aggregation of cell organelles in these positions in

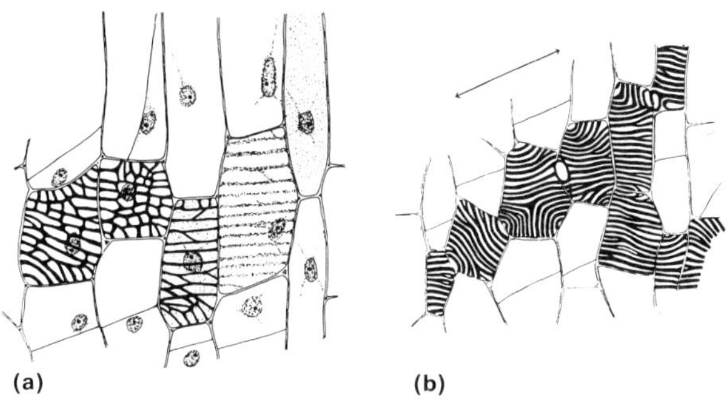

Fig. 8.8 Parts of regenerating strands of xylem elements in the pith of wounded stems of *Coleus*. (**a**) Recent cell divisions in the pith cells are evident, and stages in the differentiation and lignification of the ringed and reticulate patterns in the cell walls. On the right, bands of cytoplasm indicate the places where lignified thickenings will later differentiate. ×212. (**b**) Pattern of lignified bands in the cell walls and the position of pores, 8 days after wounding. The new cell walls and the line of pores are parallel to the course of the new strand (arrow). ×150. (From Sinnott and Bloch,[610] Figs. 7 and 9, pp. 152 and 154.)

regenerating xylem cells; however, as already mentioned, such aggregations have not been observed in normal xylem.[132,729] Sinnott and Bloch[609] also pointed out that the pattern of wall thickening was often continuous from one cell to another, and drew the conclusion that the cytoplasmic changes occurring in a cell during its differentiation do not take place in isolation but are an integral part of the changes taking place in a group of related cells.

More recent work with stems of both *Cucumis sativus* and *Coleus* has shown that, in fact, continuity between the damaged vessels above and below the wound is not restored by the regenerating strands, but that these connect only with newly formed elements lying parallel and contiguous with the original vascular strand.[52]

It seems likely that some sort of gradient is established along the new axis of polarity of the regenerating strand. After considering all the information available at the time, Jacobs[340–342] designed a series of experiments to investigate the view that an auxin gradient might be involved and, further, that auxin was a limiting factor in xylem regeneration (see Appendix). Jacobs wounded internodes of *Coleus*, and observed the regeneration of xylem under various conditions. In some experiments leaves and buds distal or proximal to the wound were removed. He established that neither auxin transport nor xylem regeneration in the stem was wholly basipetal (i.e. from apex to base), as had been believed, but that there was a quantitative relationship

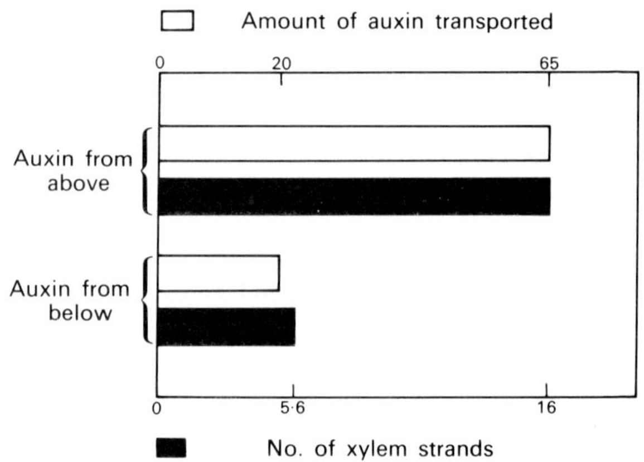

Fig. 8.9 Histogram showing the relationship between the amount of auxin transported in isolated sections of the stem of *Coleus* and the number of strands of xylem regenerated when the leaves above and below are excised. (From Jacobs,[342] Fig. 2, p. 165. By permission of the University of Chicago Press.)

between the amount of acropetal or basipetal auxin transport and the amount of acropetal or basipetal xylem regeneration (Fig. 8.9). Jacobs[340] also showed that a young leaf distal to the wound had a stimulatory effect on xylem regeneration, and that this effect could be simulated by applying indoleacetic acid (IAA) to the cut stump of the petiole of an excised leaf or to a decapitated internode (Fig. 8.10). In pea, also, a mature leaf above the wound was essential for xylem differentiation.[52] These findings indicated that auxin was a limiting factor in the regeneration of xylem under these experimental conditions. The capacity of the internode for transporting auxin is also implicated. This view is supported by experiments in which the

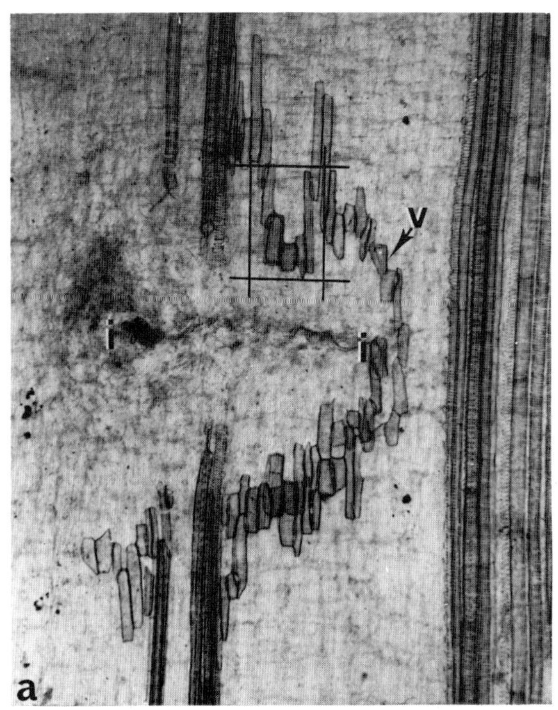

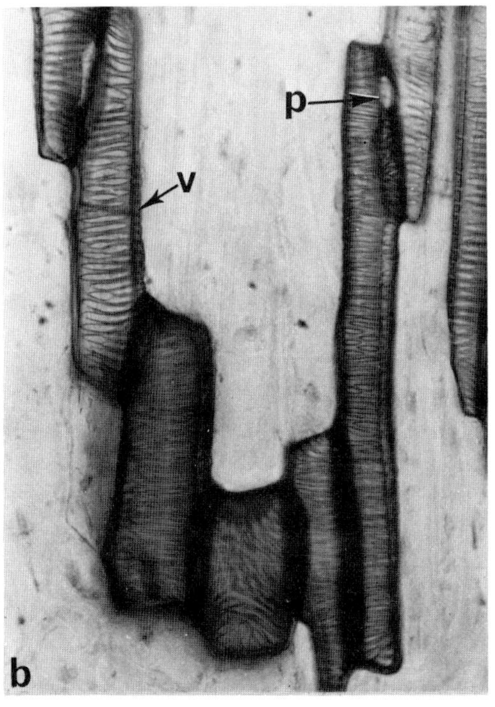

Fig. 8.10 Regenerated xylem. (**a**) Part of a cleared and stained fifth internode of *Coleus* in which the small vascular bundle has been incised (i–i, incision). The plant was otherwise intact. Wound vessel members (v) have regenerated around the wound, from parenchymatous cells. One week after treatment. *See* Appendix. ×45. (**b**) Part of **a** enclosed in rectangle, enlarged. p, pore. ×270.

amount of regeneration was reduced when tri-iodobenzoic acid (a substance which blocks auxin transport) was applied between the wound and the applied IAA.[658] By a careful study of normal differentiation in the shoot of *Coleus* it was later shown[344] that auxin was a limiting factor in the differentiation of normal xylem also. There is a precise quantitative relationship between the rate of production of diffusible auxin and of xylem cells in a particular internode in the normal development (Fig. 8.11). Excision of a developing leaf resulted in cessation of primary xylem differentiation in the internode below. Applied auxin compensated for the absence of the leaf as regards the number of vessels, but not in all respects.[692] The amount of auxin required to induce the differentiation of a parenchyma cell as a regenerated tracheary cell is approximately 14 times that required for normal differentiation from a procambial cell.[344]

In support of his thesis that auxin is produced as a consequence of cell death (and thus by xylem elements undergoing autolysis), Sheldrake[592] has argued that the correlation between auxin production and xylem differentiation in *Coleus* (Fig. 8.11) could indicate that auxin is being produced by the differentiating xylem itself.

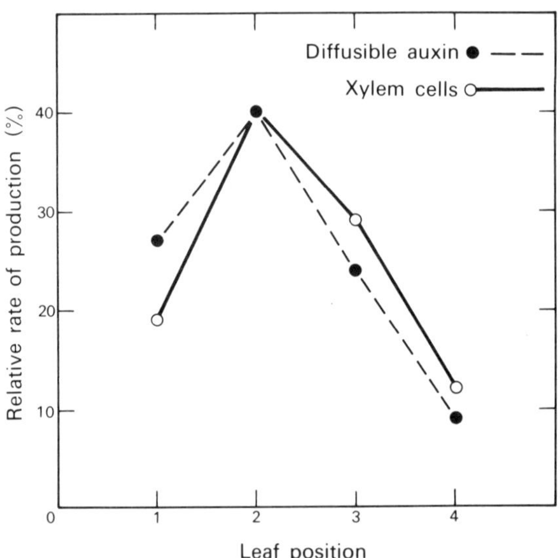

Fig. 8.11 Relative rates of production of diffusible auxin and xylem elements in normally differentiating unwounded plants of *Coleus* in relation to leaf position on the plant (the production of leaves 1–4 inclusive is taken as 100%). (From Jacobs and Morrow,[344] Fig. 16, p. 838.)

However, this view fails to account satisfactorily for the effects of applied auxin.

In further work on xylem regeneration in *Coleus*, it has been shown[548] that severance of the primary xylem only does not result in the regeneration of xylem elements from pith cells, but it is not certain whether the factors involved resulted from the severance of the primary phloem only or from that of both the phloem and xylem.

Using grafts between the root and shoot of pea seedlings, Sachs[558] showed that xylem regenerated across the graft union, but that if the tissues were kept slightly separated, xylem differentiated only on the shoot side of the cut. The effect of the shoot could be replaced by IAA in lanolin. From these and other grafting experiments Sachs concluded that the shoot produces a xylem-inducing factor which moves towards the root through the vascular system. After an incision severing the latter, the stimulus moves instead through neighbouring parenchyma cells, causing them to differentiate as tracheary elements.

An interesting effect of auxin concentration was demonstrated in experiments in which 2 mm stem segments of *Coleus* were grown in aseptic culture.[247] Xylem elements differentiated in the slices even on the control medium; addition of indoleacetic acid (IAA), 2,4-dichlorophenoxyacetic acid (2,4-D), tri-iodobenzoic acid or kinetin inhibited differentiation of these elements, but low concentrations of IAA or 2,4-D led to over 100% increase in the number formed. It is possible that high concentrations of these substances may inhibit the polar transport of auxin, or may simply be supra-optimal for differentiation.

Cultured stem segments of *Coleus* have been used in a number of other experiments. Fosket[246] found that wound vessel members were not discernible until the third day of culture, and that the greatest number differentiated between the third and fourth days (Fig. 8.12). In an experiment in which the incorporation of a labelled precursor of DNA, ^{3}H-thymidine, was measured, it was found that the greatest amount of DNA synthesis occurred during the second day of culture and subsequently declined (Fig. 8.12). Addition of 5-fluorodeoxyuridine, an inhibitor of DNA synthesis, to the medium resulted in approximately 80% reduction in the number of wound vessel members; this inhibition could be overcome by supplying thymidine.[246] These results thus indicate that DNA synthesis is necessary for wound vessel member differentiation. Experiments on cortical cells of pea roots also showed that either prevention of DNA synthesis or the synthesis of an abnormal DNA prevented the induction of tracheary elements by cytokinin.[597] In further experiments on cultured stem segments of *Coleus*, differentiation of xylem elements was inhibited by colchicine, which interferes with spindle formation, suggesting that

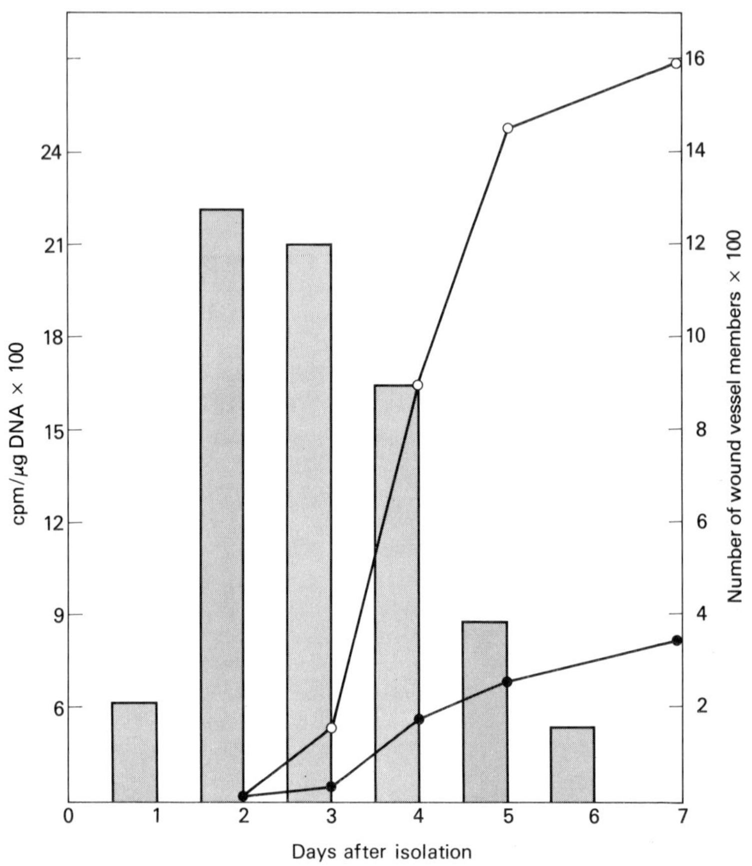

Fig. 8.12 Histograms showing the amount of ^{3}H-thymidine incorporated by cultured stem segments of *Coleus* up to seven days. The curves show the mean number of wound vessel members that differentiated on control sucrose-agar medium (black circles) and on the same medium supplemented with indoleacetic acid (open circles). The major amount of DNA synthesis (as shown by incorporation of ^{3}H-thymidine) precedes the differentiation of any substantial number of wound vessel members. (From Fosket,[246] Fig. 1, p. 65.)

completion of mitosis is necessary in order for differentiation to occur.[313]

In these cultured stem segments, transference to a medium containing IAA led to a four-fold increase in the number of wound vessel members present after seven days of culture. It was noted that supplying auxin after three days' culture on medium without auxin resulted in the formation of a new population of vessel elements,[246] leading to the conclusion that auxin does not regulate xylem differentiation by its effect on DNA synthesis.[246,547] However, some change in the cells may have occurred during the three-day period. In wounded tobacco stems cell division always occurred before differentiation of parenchyma into tracheary elements, and it was concluded that auxin may limit differentiation of xylem (and phloem) by limiting the number of cells that have divided and are thus available for differentiation.[651] In some culture experiments with isolated cells, however, differentiation into tracheary elements took place without prior division,[384,667] though polyploidy may have occurred.

Experiments in which IAA labelled with tritium was supplied to the apical end of wounded internodes of *Coleus* showed that the label, as revealed by histoautoradiography, occurred in the secondary walls of differentiating tracheary elements, suggesting the incorporation of IAA derivatives into cell wall precursors.[557] In other experiments carbon-labelled IAA was loaded into anion exchange resin beads only 430 μm in diameter. One of these beads was pushed into each wounded *Coleus* stem above the wound, giving a more localized source of labelled auxin. Prospective wound xylem elements did not show accumulation of radioactivity, compared with adjacent parenchyma cells, up to three days, when early stages of lignification were discernible. Thus it seems unlikely that accumulation of IAA in the region of differentiation is the stimulus, though there may be a greater flux of IAA through the cells.[263] Sachs[563] also concluded that polar transport of auxin is likely to be much more significant than any local gradients, though it seems likely that these must occur.

Using pea seedlings, Sachs[559,560] showed that if IAA in lanolin was applied to the side of a decapitated plant a strand of tracheary elements differentiated from the cortical cells and joined up with the central vascular cylinder. If the plant remained intact, however, or auxin was applied also to the existing severed vascular strand, little xylem differentiated and it failed to join the existing vascular tissue (Fig. 8.13). Thus a strand of xylem appears to differentiate towards existing vascular tissue if this is devoid of a source of auxin, but is inhibited from forming in the vicinity of a strand already well supplied with auxin. Sachs[559,561] has applied these interesting observations to an interpretation of the linkages between the vascular strands of developing leaf primordia and of axillary buds with existing vascular strands in the shoot (see also Part 2,[136] Chapter 3). However, as O'Brien[477] has pointed out, the fusion of mature vascular strands is preceded by the fusion of differentiating strands of procambium, and little is known

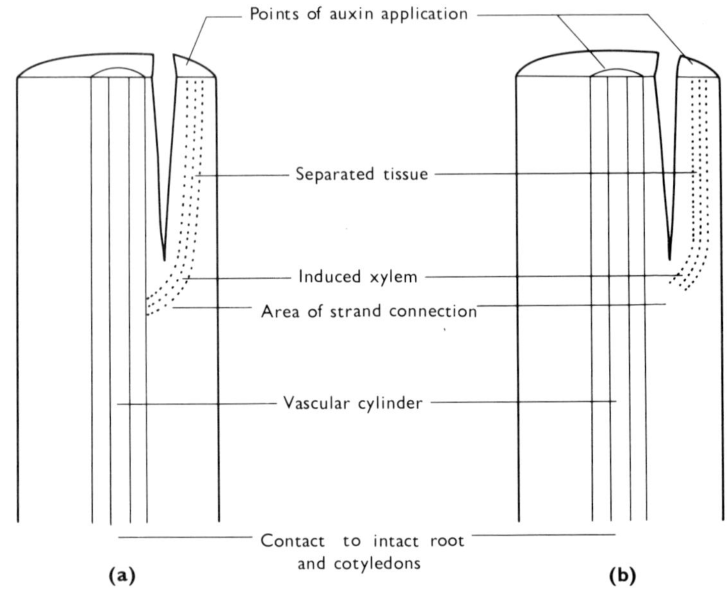

Fig. 8.13 Diagram of an experiment designed to test the influence of the concentration and direction of transport of auxin on the differentiation of xylem, using the pea epicotyl. (**a**) Auxin was applied only to the partially isolated flap of tissue. (**b**) Auxin was applied both to the flap of tissue and to the vascular cylinder. The induced strand of xylem is shown dotted. (From Sachs,[560] Fig. 2, p. 269.)

about the factors controlling either the direction of differentiation of procambial strands or their fusion. It is not yet clear to what extent it is reasonable to apply the results of observations on xylem differentiation from parenchyma cells to the differentiation of procambial strands and its control. From other experiments, Sachs[565] has concluded that in the early stages of vascular differentiation, the axis, but not the direction, of the movement of stimuli that induce differentiation is determined.

Explants of organized parenchyma (pith or cortex)

The importance of auxin in controlling the differentiation of tracheary elements is further indicated by experiments in which pith excised from herbaceous stems was grown in aseptic culture, and sterile micro-pipettes containing aqueous solutions of IAA, kinetin or both these substances were inserted into the pith. In tobacco pith small areas of dividing cells developed just below the tip of the pipette, and xylem tracheids occurred among them in a haphazard manner. No cell divisions were associated with empty pipettes or those containing only sterile water.[109] In cabbage pith various results were obtained, including the differentiation of xylem elements, depending on the ratio of kinetin to IAA.[110] Further work with excised tobacco pith confirmed these results, and showed that around pipettes with 2 mg/l IAA cell divisions occurred on the third day, tracheary elements beginning to differentiate on the sixth day. The size of the region of dividing cells, and the number of lignified cells formed, both increased with increasing concentrations of IAA.[651]

When segments of pith of *Coleus* stems were cultured, addition of IAA to the medium resulted in little growth but induced the differentiation of large numbers of tracheary elements. Omission of sucrose from the medium prevented this. Xylem elements were single, or in clusters or strands. Addition of kinetin had no effect on xylem differentiation or was inhibitory.[171] On the other hand, in cultured cortical explants of pea roots, i.e. segments of pea root from which the central cylinder had been removed, leaving only parenchyma, cell division did not occur on a medium containing auxin as the only hormone but took place from the third day when kinetin was also present. Tracheary elements began to differentiate on the seventh day, and continued for as long as 3 weeks.[513]

In cultured pith discs of Romaine lettuce (*Lactuca sativa* var. *Romana*), also, both an auxin and a cytokinin were required for the differentiation of tracheary elements, which took place within 4 days, preceded by cell division. A complete ring of strands differentiated at the periphery of the explant, at right angles to the medium, and tracheids also occurred scattered or in small clumps. Slightly different patterns of differentiation were observed according to whether the cytokinin accompanying the IAA was zeatin or kinetin.[144] With lettuce pith, the concentration of cytokinin supplied was a limiting factor for xylem differentiation in the presence of an optimal amount of auxin, but cytokinins alone could not induce xylem differentiation. With the optimal combination of IAA and zeatin, 6·9% of the cells differentiated as tracheids.[143]

In proliferating cultures of tubers of Jerusalem artichoke (*Helianthus tuberosus*), however, up to 30–35% of the cells differentiated as tracheids, in the presence of appropriate concentrations of naphthaleneacetic acid (NAA, an auxin), benzyladenine (a cytokinin) and soluble starch. Histochemical studies of this tissue showed that several enzymes often considered to be associated with xylem differentiation did not occur in proportion to the number of differentiating cells. It was also observed that deposition of lignin was not specifically associated with the secondary walls of tracheids, but occurred on all kinds of cells.[457] In suspension cultures of sycamore (*Acer pseudoplatanus*) cells, also, the lignin content per cell could be affected by the concentration of sucrose or 2,4-D, but electron microscopy confirmed the absence of secondary walls, such as occur in tracheary elements.[86] It has been suggested that lignification and the differentiation of xylem elements may be less closely associated in tissue culture than they are in the intact tissues of growing organs.[457]

Experiments with callus

In addition to pith, undifferentiated callus tissue has been used for studies of vascular differentiation. Callus is unorganized tissue in which parenchymatous cells are randomly arranged and usually no other cell types are present. When a growing shoot apex of lilac (*Syringa vulgaris*) was grafted into callus of the same species and

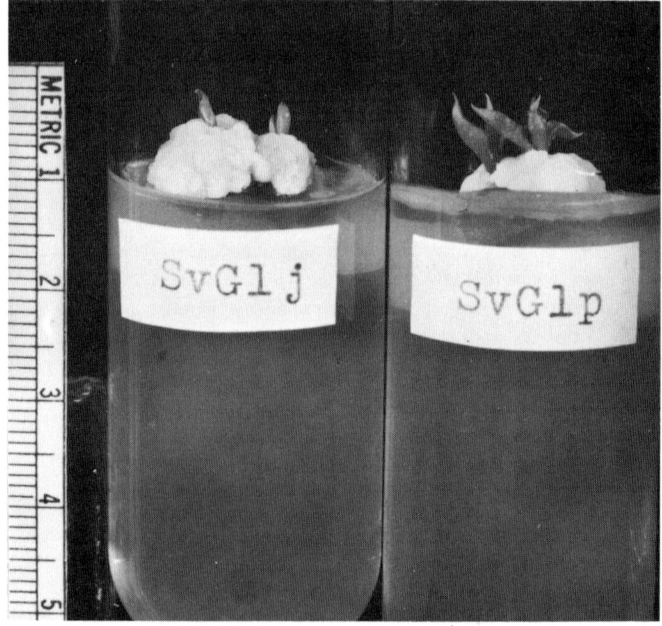

Fig. 8.14 Shoot apices of *Syringa vulgaris* grafted into callus of the same species, after 7 weeks in culture. (From Wetmore and Sorokin,[708] Pl. 1, Fig. 1.)

maintained in aseptic culture on a defined medium, vascular tissue differentiated in the callus (Figs. 8.14, 8.15).[83, 708] Substitution of the scion by a block of agar containing physiological concentrations of auxin gave similar results; nodules of vascular tissue were formed in the callus at some distance from the graft, the distance depending on the concentration of auxin. This work was subsequently extended to five additional species; sucrose and/or IAA were supplied in the agar placed in the cut.[707] When sugar and auxin were placed in the cut on

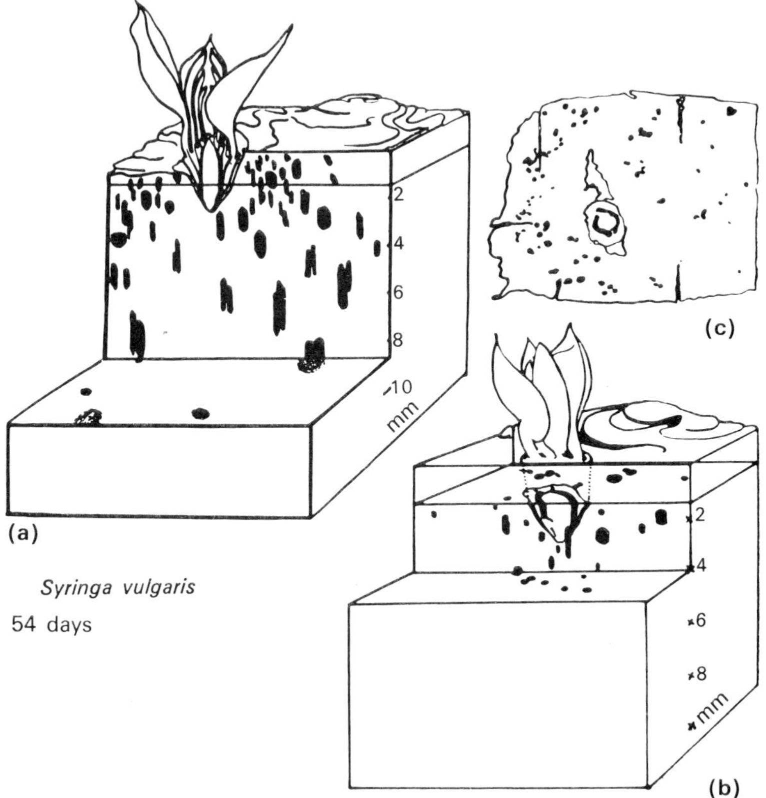

Fig. 8.15 (**a**) and (**b**) Three-dimensional diagrams of a piece of lilac callus (*Syringa vulgaris*) into which an apex of the same species with two or three pairs of leaf primordia was grafted. 1% agar containing 0·05 mg/l. naphthaleneacetic acid was placed in the incision. This is a composite drawing constructed from serial sections taken 54 days after the beginning of the experiment. (**c**) Transverse section at about the 2 mm mark. Vascular tissue black. (From Wetmore and Sorokin,[708] Figs. 1 and 2, p. 309.)

top of the callus and neither was present in the medium, nodules of vascular tissue were formed in a circle below the site of application. When auxin and sugar were present in the culture medium as well, another set of nodules was present towards the base of the callus, below the level of the medium (Fig. 8.16). Both xylem and phloem were present in the nodules. Low concentrations (1·5–2·5%) of sugar were found to favour xylem formation, high concentrations (3–4%) phloem; and at intermediate concentrations (2·5–3·5%) both xylem and phloem were formed, usually with cambium in between. Individual nodules showed the normal orientation of vascular tissues found in a stem, i.e. phloem towards the periphery and xylem towards the centre of the callus. Wetmore and Rier[707] further showed that when sucrose and auxin were supplied through a micro-pipette in the callus, as in the experiments with pith already described, a complete

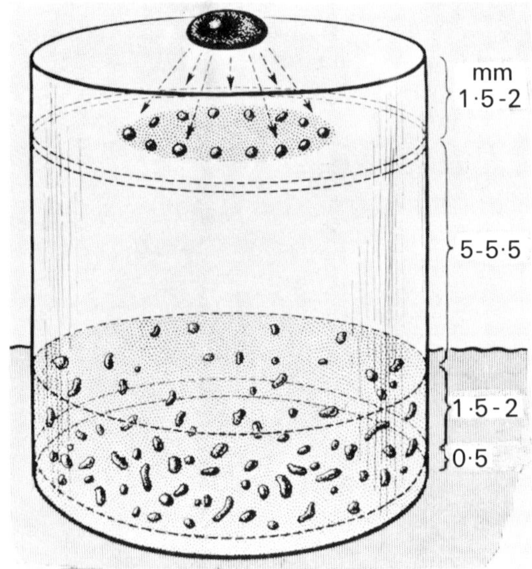

Fig. 8.16 Three-dimensional diagram of a piece of callus of lilac (*Syringa vulgaris*). 1% agar containing 0·1 mg/l. naphthaleneacetic acid and 3% sucrose (or different concentrations of these substances) was applied to the top of the callus, which was grown on a culture medium which also contained these substances. A circle of nodules of vascular tissue (tracheids surrounded by sieve elements) differentiated below the agar block. The diameter of the circle depends on the concentration of auxin. Additional nodules (with phloem towards and xylem away from the medium) were formed below the medium, usually in greatest abundance 1–2 mm from the callus surface. (From Wetmore and Rier,[707] Fig. 12, p. 426.)

ring of xylem, or of xylem with phloem, differentiated around the pipette. In some experiments cambium was present between the xylem and the phloem.

In comparable experiments with callus of bean, *Phaseolus vulgaris*, nodules of xylem surrounded by a cambium-like layer and peripheral phloem were again formed. Experiments with various sugars showed that sucrose and certain other disaccharides, including maltose and trehalose, had a specific action in inducing these vascular nodules.[352] Rier and Beslow,[541] working with callus of *Parthenocissus*, obtained the maximum number of tracheary elements with 8% sucrose. This may have been due, at least in part, to the osmotic properties of such a high concentration of sucrose, since Doley and Leyton[164] found that xylem formation in callus of *Fraxinus* increased with a decrease in water potential. At each water potential used there was an optimal concentration of IAA for xylem differentiation. Thus physical factors, as well as auxin and other hormones, seem to play a role in xylem differentiation.

With cultured soybean callus, neither NAA nor kinetin alone could induce the formation of tracheary elements: both were required. Without cytokinin there was no cell division and no differentiation. Xylem differentiation could be stimulated by 2,4-D, however, in the absence of added kinetin. Nests and strands of xylem were formed.[248,668]

Induction of secondary xylem

Secondary xylem differentiates from the vascular cambium in an organized way, and the factors controlling this are quite fully dealt with in Part 2,[136] Chapters 2 and 4. It is shown there that auxin originating in the growing shoot is again an important factor influencing the differentiation of secondary xylem in both shoot and root. This is supported by girdling experiments on *Robinia*, in which cambial reorientation was found to depend on contact with the shoot; much less new xylem was formed in the absence of contact with the shoot, the effect of which could be partially replaced by a source of auxin.[383] In stems of *Fraxinus* and other woody species, as is already known (see Part 2,[136] Chapter 4), IAA affects the sites and differentiation of vessels in the secondary xylem. Changes in water potential again affected the level of IAA required, a decrease in water potential lessening the effect of IAA and *vice versa*.[163] Water potential may influence xylem differentiation indirectly through its effect on the rate of cell division.[547]

In some interesting experiments on *Phaseolus vulgaris*, Hess and Sachs[327] excised young or mature leaves and observed the differentiation of cambium and secondary xylem in the subjacent stem in the presence or absence of these leaves. They found that below a young leaf the cambium was active and the xylem was rich in vessels. Below a mature leaf, cambial activity was again evident but the secondary xylem was poor in vessels, consisting mainly of fibres. The effect of a young leaf could be simulated by applied auxin, that of a mature leaf by GA + IAA. Indeed, most of the experimental evidence implicating gibberellins in the control of xylem differentiation seems to come from studies of secondary xylem.[546] For the differentiation of normal secondary xylem, a complex combination of auxin, cytokinins, gibberellins, carbohydrates and ethylene may be required.[547]

Some workers consider that auxin is not only involved in the control of cambial activity and vascular differentiation, but is also produced as a consequence of these processes.[591,592] Experiments in which tobacco stem segments cultured in the presence of tri-iodobenzoic acid (TIBA), an inhibitor of auxin transport, showed considerable cambial activity and the production of serried ranks of tracheids, are considered to show a positive feedback mechanism in the vascular cambium, since auxin would normally be removed by polar transport.[592] Some auxin might, however, have been present initially in these segments, from outside sources.

Conclusions

The more experiments that are carried out on the control of xylem differentiation, the more involved the situation seems to become. This is probably in part because of the use of different species and tissues, in which the endogenous hormones are largely unknown but may be expected to differ. Moreover, there may be a sequential requirement for different substances.[366] Thus it may not be surprising that many experiments with *Coleus* have shown that the addition of auxin alone can lead to the differentiation of tracheary elements from parenchyma, whereas in pea cortical explants, soybean callus and lettuce pith a cytokinin is usually also required. It has been suggested that the cytokinin requirement is in fact one for cyclic AMP, but more evidence is required.[547] Sucrose is also necessary, and this may be related to the effects of water potential.

The experiments of Wetmore and Rier,[707] in particular, come close to inducing, in unorganized callus tissue, not only the kinds of tissues, but the arrangements of tissues, normally found in stems (see Part 2,[136] Chapter 3). In the normal development the shoot apex (probably including the young leaf primordia; see Esau[188]) would no doubt be the source of the auxin and sugar required for the differentiation of the vascular tissues. It has also been suggested that auxin may be produced as a consequence of cellular autolysis, and that the differentiating xylem elements themselves may be an important source. Auxin is present in the guttation fluid of *Avena*, lending some support to this view.[594] Clearly, however, already differentiating xylem elements cannot be the primary source of auxin. If the shoot tip is the source of auxin, because of its continuous vertical growth away from proximal regions of the stem, strands of vascular tissue, instead of nodules at a certain fixed distance from a static source, could result. These various experiments have shown, in addition, that not only procambial cells, but any cell supplied with the necessary substances in appropriate concentration, can give rise to xylem or phloem.[707] This contention is supported by the successful induction, with sucrose and an auxin, of tracheids in fern gametophytes which never normally contain such cells.[151] Cytokinins are also effective in inducing xylem differentiation in the prothalli of ferns and lycopods.[546]

To the procambium, however, is attributed the important function of giving pattern to the vascular system; the characteristic pattern is blocked out during the procambial phase of development.[706] This pattern is not only under genetic control, but differs fundamentally in the root and the shoot of the same plant (see Part 2,[136] Chapters 2 and 3, where some of the factors involved are discussed).

Thus in recent years knowledge of the factors controlling the differentiation of xylem has been greatly advanced. Further work on xylem is still required, however, not only in this field but in that of phylogeny, comparative anatomy, and others.

9

Phloem

The function of the phloem as the food-conducting tissue of the plant, and the still controversial nature of the mechanism by which this function is fulfilled, have resulted in intensive studies of the anatomy and physiology of this tissue. In particular, the fine structure of phloem elements has received much attention. Because of the peculiar nature of the phloem elements, they are technically difficult to study; and although much is now known of their structure and ultrastructure many aspects are still incompletely understood, particularly those relating to function. Mechanisms of phloem transport have been recently reviewed.[749]

The material translocated in phloem sieve tubes consists largely of carbohydrates, sugars constituting about 10–25% of the exudate obtained.[189] In addition, amino acids may be present in small quantities,[130] and also hormones, including indoleacetic acid,[295] gibberellin-like substances, abscisic acid and the unidentified floral stimulus.[694] Virus and mycoplasma particles are also carried in sieve tubes.

That the sieve elements of the phloem can transport a large volume of material is evidenced by the often rapid growth of tubers such as the potato, or of fruits, such as those of the sausage tree (*Kigelia pinnata*), each of which can weigh up to 10 kg.[84] The relatively slender stolon attaching the potato to the axis is characterized by abundant phloem from an early stage of development, and the fruit stalk of the sausage tree develops abundant secondary phloem.[84] Little or no attention seems to have been directed towards a study of the factors that promote this development, which might prove a promising field of study.

Origin

As in the case of xylem, the primary phloem differentiates from procambium, and the secondary phloem from the vascular cambium, during primary and secondary growth of the plant respectively. The first elements of the primary phloem to mature are known as protophloem; subsequently differentiating components constitute the metaphloem. In plants with internal as well as external phloem, the internal phloem may differentiate from the innermost elements of the procambium, or by division of adjacent pith parenchyma.[61,258]

In early stages of differentiation, there may be discontinuities between the sieve elements.[346] Later, a continuous system is formed, and there may be many anastomoses between different phloem strands, both between internal and external phloem,[61,258] and between primary vascular strands of stems which lack internal phloem. Such anastomoses were found in 18 out of 26 species examined.[11] These connections, which were shown to be affected by growing conditions, may be important in the physiology of the plant, and in the interpretation of many physiological experiments.

As already indicated, the differentiation of two tissues—phloem and xylem—so dissimilar in structure, physiology and function, from the same precursors, and in close spatial proximity, poses a number of developmental problems which have yet to be solved.

Elements of the phloem

Like the xylem, phloem is a complex tissue, composed of several different types of elements. These comprise sieve elements or sieve cells, companion cells, parenchyma, fibres and sclereids. Secretory cells or tissues of various kinds may also be associated with the phloem, e.g. the laticifers in *Hevea*, from which rubber is derived, and oil cells in the secondary phloem of *Cinnamomum*, the source of cinnamon. It has been aptly remarked that the xylem elements themselves are of economic importance, since they constitute the bulk of the wood, whereas in the main the phloem elements are less important economically than are the other tissues associated with them. However, many phloem fibres (e.g. flax) are of considerable commercial value.

The proportion of the cross-sectional area of the phloem region that is occupied by sieve elements is sometimes important in considering the mechanism of translocation, and is often taken as about $\frac{1}{5}$. Recent observations on the secondary phloem of tropical trees, however, suggest that this figure should be higher, and is closer to the proportion of $\frac{2}{3}$ originally suggested by Münch.[406] Difficulties are experienced in identifying sieve elements with certainty with the light microscope; sections of resin-embedded material stained with aniline blue and viewed with ultraviolet light seem to give the best results.[85]

Fibres may be present in both primary and secondary phloem. They have thick secondary walls with simple or slightly bordered pits; the walls are often lignified. Phloem fibres are elongated elements with overlapping ends. Protophloem elements often differentiate into fibres in later stages of development. Observations[458] of rotation of cytoplasm in phloem fibres suggest that they may conceivably play a role in transport of materials, in addition to providing support. Phloem fibres may be septate in certain plants, e.g. *Vitis* (Fig. 6.8a). Sclereids also often occur in association with the phloem, e.g. in many barks (*Quercus*, oak; *Rhamnus*, cascara; *Cinnamomum*, cinnamon).

Parenchyma cells are present in both primary and secondary phloem. These cells have primary walls with primary pit fields, and living contents; in inactive phloem the cell walls may later become thickened and lignified. Some phloem parenchyma cells may develop wall ingrowths and function as transfer cells (see Chapter 10).

Parenchyma cells may store starch, resins, etc., and crystals are often present; crystal-containing parenchyma cells may form a sheath around fibres or bundles of fibres, especially in the secondary phloem.

Parenchyma cells may be physiologically related in some way to the sieve elements, like the companion cells, but less closely. Sometimes parenchyma cells die at the same time as a neighbouring sieve element.

SIEVE ELEMENTS

Like the tracheary elements of the xylem, sieve elements are of two kinds. ***Sieve cells*** occur in pteridophytes and gymnosperms; they are single, somewhat elongate cells with specialized sieve areas in their lateral and sometimes also in their terminal walls. ***Sieve tubes***, which occur in angiosperms, are longitudinal files of cells, each of which is a sieve tube member. In such cells, one or more of the sieve areas, usually in or near an end wall, is more specialized, and forms a sieve plate. This is a region of the wall comprising a number of pores, through which strands connect one member with another, usually vertically; these are called connecting strands.

Structure

Sieve elements are very remarkable cells, indeed unique, for they are living cells which at maturity usually contain no nucleus. The sieve element has been aptly termed a nonconformist cell.[184] A necrotic nucleus has been observed in the mature sieve cells of *Pinus*,[468] and nuclei have also been observed in mature sieve cells of other gymnosperms, and in sieve elements of woody dicotyledons,[213] and a monocotyledon.[179] In palms examined, nuclei had typically degenerated in mature sieve elements,[496,497] as also in a number of ferns,[214] and indeed it is considered that electron microscopic studies have confirmed, in general, the breakdown of sieve element nuclei at the time of other degenerative changes.[189,498] An ultrastructural study of the differentiating sieve elements of the sensitive plant, *Mimosa pudica*, followed the changes that occurred in the nucleus, revealing that the nuclear envelope first became dilated; this was followed by breakdown of the interior of the nucleus. Portions of the nuclear membrane persist for a while, but even these are not present in fully differentiated sieve elements.[192]

Although the mature sieve element is typically devoid of a nucleus, in most angiosperms sieve elements are closely associated with living, nucleate cells, the ***companion cells***; these originate from the same mother cell as the sieve tube member, and are thought to have a very close physiological and functional relationship with it throughout life. A careful study of secondary phloem in members of the Calycanthaceae revealed that most sieve elements had one or two, and occasionally up to five, companion cells associated with them ontogenetically; nearly 5%, however, had no associated companion cells.[102] Much information has recently stemmed from studies of phloem elements with the electron microscope, but despite this careful analytical work the development and mode of functioning of the sieve element and its attendant companion cell(s) remain rather enigmatic. Zimmermann[746] has pointed out that the occasional absence of associated companion cells should be taken into account in any consideration of their functional role.

In early stages of differentiation, the sieve element is not distinguishable from the neighbouring companion cell, except in size. The sieve element and companion cell typically are formed by an unequal division of the type already discussed in previous chapters,[75] though occasionally procambial cells may differentiate into sieve elements without prior division.[511] At an early stage of development,, both cells have nuclei and dense cytoplasm, containing mitochondria, dictyosomes, endoplasmic reticulum and in some plants, though not in *Cucurbita*—on which many ultrastructural studies have been carried out—plastids. Discrete ***slime bodies*** are present in the cells; slime is a proteinaceous substance characteristically found in sieve elements (Fig. 9.1a), which is discussed in more detail below under the term ***P-protein***.

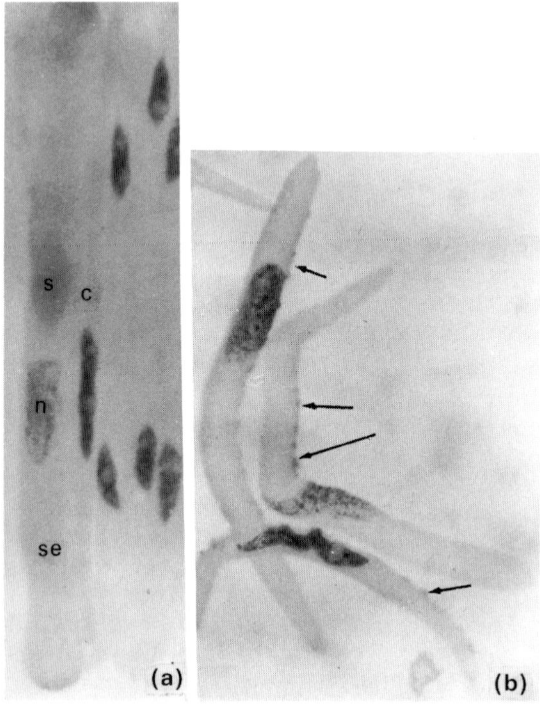

Fig. 9.1 Developing sieve tube elements and companion cells in squash preparations from young stems of *Vicia faba*. (**a**) The nucleus (n) of the sieve tube element (se) is beginning to disintegrate; slime (s) is present. c, companion cell. ×400. (b) Nuclei are becoming lobed. Note the nacreous cell wall (arrowed). ×300.

As in other plant cells, in young sieve elements the vacuole is delimited by a membrane, the tonoplast. In early stages of differentiation, too, numerous microtubules are present peripherally.[67, 130] Ribosomes and dictyosomes are present, and may be abundant.[385] Coated vesicles are sometimes associated with the dictyosomes.[495, 496] Both ribosomes and dictyosomes are absent from mature sieve elements.[189, 498] During differentiation of the sieve element the mitochondria undergo degenerative changes, and show disorganization of the inner membrane; thus eventually they have few or no cristae.[200] If present, chloroplasts also either fail to develop much internal structure or lose what they have.[67] Sieve elements may accumulate a carbohydrate, often called starch,[189] but of an atypical kind of lower molecular weight, which stains red with iodine.[728] The plastids of sieve elements may also be distinctive in other ways, for example in possessing crystalline inclusions.[445,495,496] It has been suggested[613] that the high levels of carbohydrates in sieve elements may be antagonistic to the formation of grana in plastids.

The breakdown of the tonoplast during development of sieve elements is another matter on which there has been some controversy. Although some workers have failed to observe rupture of the vacuolar membrane, many have now described its disappearance during differentiation, and it seems to be generally accepted that this usually occurs;[130,189,190,497] some authorities, however, are still inclined to reserve judgement.[498] During differentiation, changes also occur in the endoplasmic reticulum (ER), which in early stages is of the rough type, bearing ribosomes. Later, the ER tends to aggregate in stacks of parallel cisternae, which often occupy parietal positions. The cisternae may be parallel or perpendicular to the cell wall, or may be more convoluted or vesiculated. These modified types of ER, which have been observed in many different species, gradually lose their attached ribosomes.[189,192,207,496,498] These conformations of ER are sometimes called the 'sieve element reticulum', but this term is now considered inappropriate, since they are not restricted to sieve elements.[192,207] The occurrence of more vesiculate ER coincides with nuclear breakdown.[201] A recent study of the distribution of the enzyme acid phosphatase in sieve elements of French bean (*Phaseolus vulgaris*), in which the localization of a specific dye was observed with the electron microscope, revealed a frequent association of this enzyme with the ER.[196]

Delimitation of the sieve areas begins at an early stage of differentiation of the sieve element. Kollmann,[385] working with the gymnosperm *Metasequoia*, found that cell connections through the sieve areas were established early and made the interesting suggestion that the changes in the cell organelles which lead to degeneration of the sieve cells might be considered to be a consequence of, rather than a prerequisite for, the physiological functions of the conducting elements of the phloem. However this may be, it is clear that although recent work on ultrastructure now provides a fairly complete picture of the morphological changes occurring in differentiating sieve elements, the underlying causes of these have as yet scarcely been considered, far less understood.

Sieve elements sometimes have quite thick walls, and such walls have been termed nacreous, because of their glistening properties (Fig. 9.1b). The nacreous cell wall may develop very early in the ontogeny of the sieve element (Fig. 9.2), while it still possesses a nucleus.[56] Formation of the nacreous wall is followed by the

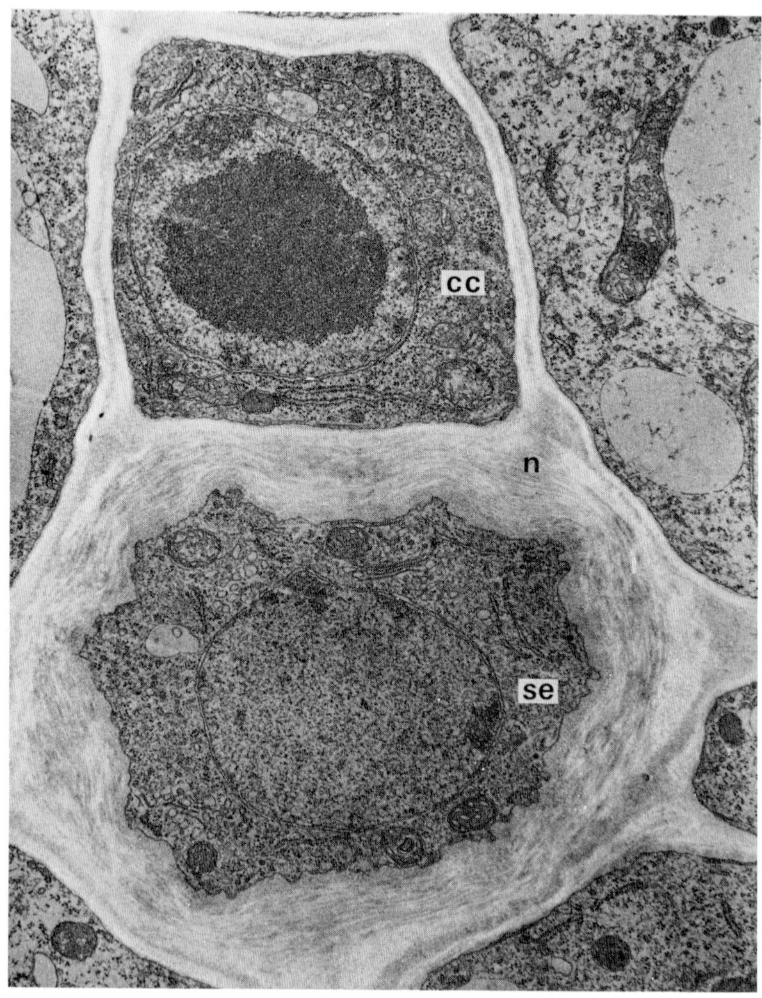

Fig. 9.2 T.S. phloem of *Mimosa pudica*, showing a sieve element (se) and companion cell (cc). Note the thick nacreous wall (n) of the sieve element, which at this early stage of development contains abundant cytoplasm with many organelles. ×16 000. (From Esau,[192] Plate 2A.)

disappearance of microtubules.[67] Sometimes the nacreous part of the wall may be distinguishable morphologically from that part of the wall formed still earlier. On occasion it may be very thick, up to half of the diameter of the cell, but it is absent from the region of the sieve plate.[197] The nacreous wall is lamellate, with close transversely oriented microfibrils,[159] which run perpendicular to the long axis of the sieve element.[498] The enzyme acid phosphatase is also associated with the nacreous wall in developing sieve elements.[196]

Some workers believe that it is the young sieve elements that function in translocation.[123] The question arises whether translocation occurs before or after the breakdown of the cell contents; however, it is considered that there is much evidence that it occurs in mature, enucleate cells.[190] Ways in which functioning sieve elements can be identified include their ability to translocate dye or radioisotopes, or the presence of the stylet of a feeding aphid.[85] Although in many plants the sieve elements may function for only a single season, in many perennial as well as arborescent monocotyledons there is evidence of much greater longevity.

In three perennial species, sieve elements were found to remain functional for at least two seasons, or in some cases up to eight years.[179] In the basal part of the stem of some palms the sieve elements may be up to a century old;[497] sieve elements are considered to be still functioning at well over 100 years old in a particular royal palm (*Roystonea* sp.) of known age.[748]

P-protein

In young sieve elements protein appears in the form of compact structures, formerly called slime bodies, these bodies consisting of aggregations of tubules, designated P_1-protein. During the differentiation of the sieve element, the tubules become dispersed in the cytoplasm (Fig. 9.3), and groups of striated fibrils, designated P_2-protein, become evident. It is believed that the P_1-protein gives rise to the P_2-protein. In *Cucurbita*, RNA has been found in slime bodies.[82] Observation of two further types of P-protein in the sieve elements of *Cucurbita* suggests that P-protein is rather labile and may be polymerized or depolymerized according to environmental conditions within the cell.[189] More than one type may occur in a single sieve element. It appears, indeed, that the form of P-protein may be related to conditions in the sieve element at the time, and that one form can give rise to another. In *Mimosa*, a transitional form was observed, interpolated between the fibrillar and tubular forms. This framework with pentagonal or hexagonal meshes was thought to constitute the assembly system for the tubular form.[191]

The ultrastructure of the P-protein tubules themselves has been examined. The tubules are 18–23 nm (180–230 Å) wide, and are composed of 6 more or less spherical sub-units of 6–7 nm (60–70 Å)

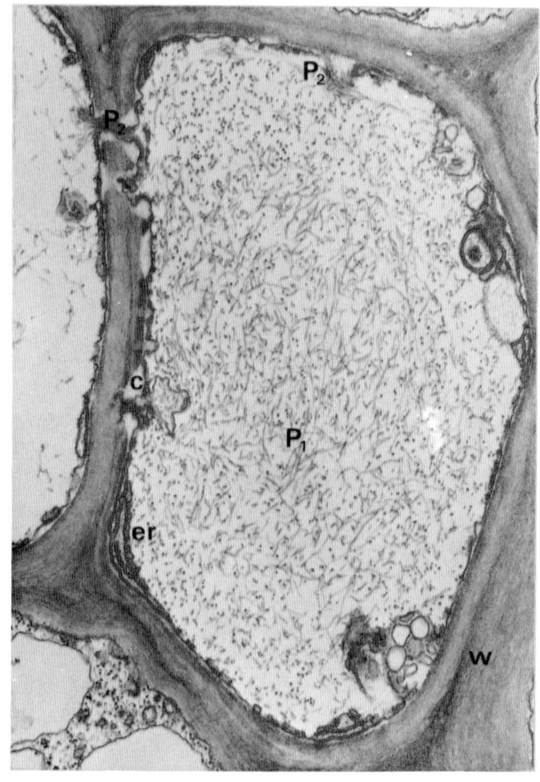

Fig. 9.3 Transverse section of a sieve element of *Nicotiana tabacum* at a late stage of differentiation. P-protein is present in the cell lumen. Both P_1 and P_2 protein are present but the former predominates. c, callose; er, endoplasmic reticulum; w, cell wall. ×15 500. (From Cronshaw and Esau,[133] Fig. 11, p. 813.)

diameter, arranged helically along the tubules. There may be a double helix of two strands of sub-units.[499] This proposed sub-structure has received considerable support. It is considered that the helices may become loosened and stretched; various images seen in the electron microscope could be interpreted in this way. The P-protein of *Saxifraga sarmentosa* indeed can be seen to have a helical structure, with occasional apparent double helices.[599]

Spiny vesicles, presumably formed by the dictyosomes, have been observed in close association with P-protein in a number of species, and it is thought that the spiny vesicles might play a role in the formation of the P-protein,[130] or in the synthesis of a precursor.[155]

P-protein has now been observed in the sieve elements of many dicotyledons, but is apparently sporadic in monocotyledons.[445,498] It is also absent from at least some gymnosperms,[212] and most ferns.[189,214] These facts are of some significance in considering possible functions of the P-protein.

The observation that sieve plate pores are frequently plugged with P-protein (Fig. 9.8)[202,599] has led to the suggestion that it may be involved in the sealing of the pores after wounding.[130,498] As an additional, perhaps more significant, function P-protein may bring about a situation which promotes movement of materials through the cell.[190] The P-protein might consist of contractile filaments which could actually generate motive force.[130] It has been speculated that the fibrils of the P-protein can be compared with actin in muscle; the force generated by such a system would be more than enough to result in the observed rates of flow, and such a view would be compatible with the observed elaboration of P-protein tubules at the stage of onset of translocation.[430] Histochemical methods have demonstrated that the enzyme adenosine triphosphatase (ATPase) is associated with the dispersed P-protein in mature, functional sieve elements.[268,269] This demonstration of ATPase activity may indicate that some P-protein may indeed be contractile or function in an activated mass-flow type of mechanism.[269] Although it is satisfying to envisage a possible functional contribution of the P-protein which is now regarded as such a basic constituent of many sieve elements, it must be emphasized that P-protein is apparently absent from the sieve cells of many lower vascular plants; if a mechanism of assimilate transport involving P-protein is proposed, it would be necessary to envisage at least one other mechanism for those plants which lack P-protein.[215]

Transcellular strands

Suggestions that protoplasmic streaming from one cell to another along transcellular strands[652] is the principal method of translocation have met with opposition. It is proposed that the transcellular strands, originally considered to be cytoplasmic, comprise several rings of protein filaments which are contractile, enclosing endoplasmic tubules. Flow through the strands is regarded as a peristaltic wave-like action.[653,654] Although the proponents of this view have presented micrographs of tissue treated in various ways and viewed in the light microscope with Nomarski optics and in the EM,[152,655] they appear unconvincing and, in general, opinion seems to be against the occurrence of transcellular strands of this type.[430,498,728] The lengthy controversy surrounding these proposals serves to emphasize the

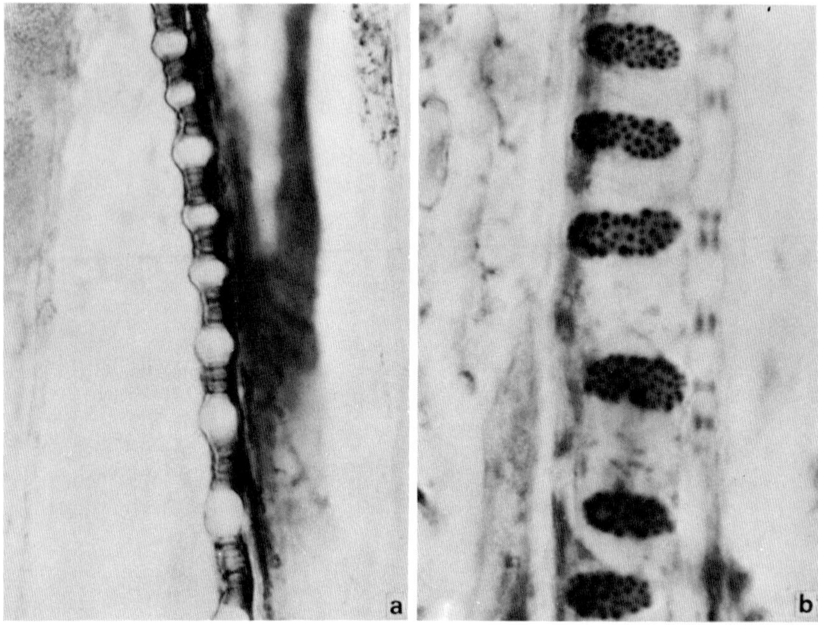

Fig. 9.4 Compound sieve plates in *Vitis vinifera*. (**a**) Sectional view of a sieve plate; darkly stained material is slime. (**b**) Face view of a sieve plate in a longitudinal wall. The connecting strands surrounded by callose appear as black dots; the whole sieve areas are covered with quantities of callose. ×750. (From Esau,[183] Pl. 12, B and D, p. 288.)

difficulties in interpreting observations on the structure and function of phloem, and has certainly stimulated further work.

Sieve plates

A sieve plate is the region of pores between one sieve tube member and another; thus sieve plates may be compared with perforation plates in the xylem. Sieve plates may be simple, consisting of one region of pores, or compound, consisting of several such regions separated by bars of wall thickening (Fig. 9.4).

In very early stages of development of the sieve tube member, when the nucleus is still present, the future sieve plate is smooth and there is no indication of primary pit fields. At a slightly later stage platelets of callose form isolated patches on the cell wall in the sites of the future pores.[67,204] Platelets are paired, occurring on both sides of the cell wall (Fig. 9.5a). Canny[85] has pointed out that the ability to form sieve areas belongs peculiarly to the sieve element, since in regions adjoining a

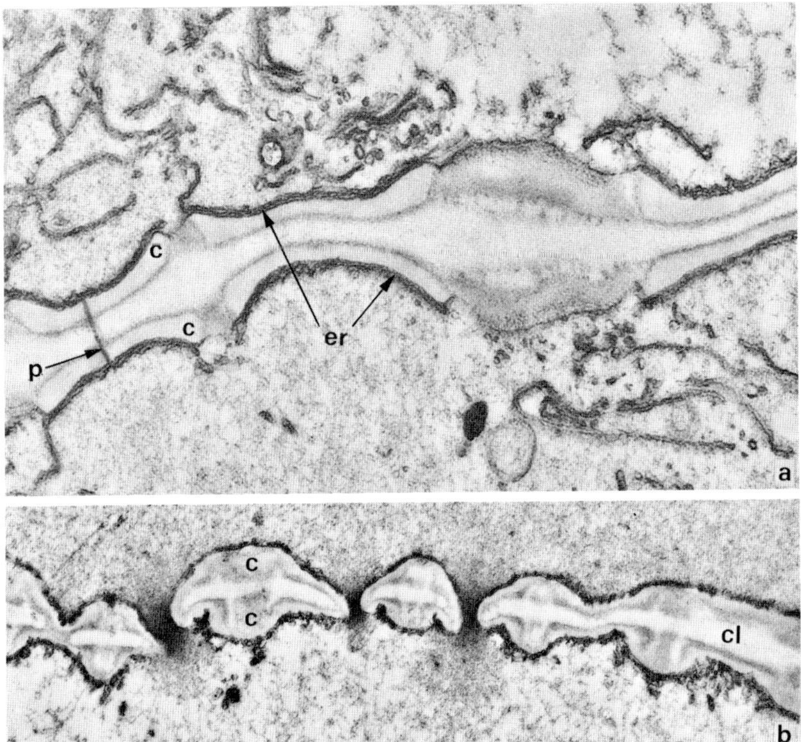

Fig. 9.5 Developing sieve plates of *Cucurbita*. (**a**) Platelets of callose (c) and endoplasmic reticulum (er) are present at the sites of each pore. One pore site is cut through the single plasmodesma (p). ×19 000. (**b**) Similar view of a sieve plate in which the pores have recently opened. cl, cellulosic part of sieve plate. ×14 000. (From Esau,[185] Figs. 2 and 5, pp. 52 and 54.)

companion cell a sieve area is formed only on one side, whereas a sieve plate represents the co-operative effect of two sieve elements. At least in some species, endoplasmic reticulum may be discernible applied to localized parts of the wall; the callose appears beneath it. A single plasmodesma is present in each pore site. The platelets of callose increase in thickness, and ultimately the wall material between paired platelets disappears, a break occurs and a pore is formed (Fig. 9.5b). As a consequence of this mode of formation, the pore is lined with callose; this substance thus surrounds the connecting strands (Fig. 9.6).

This description of the formation of sieve plate pores seems to be broadly applicable to many species,[154, 189, 496, 498] although in some recent work it is argued that the cell wall is not replaced by callose during pore formation, but that both the callose collars and the wall layer between are removed.[156] Enzymes are probably involved in the mechanism of perforation;[185] formation of the pores involves more than just the enlargement of existing plasmodesmata, although these may establish the pore sites.[204]

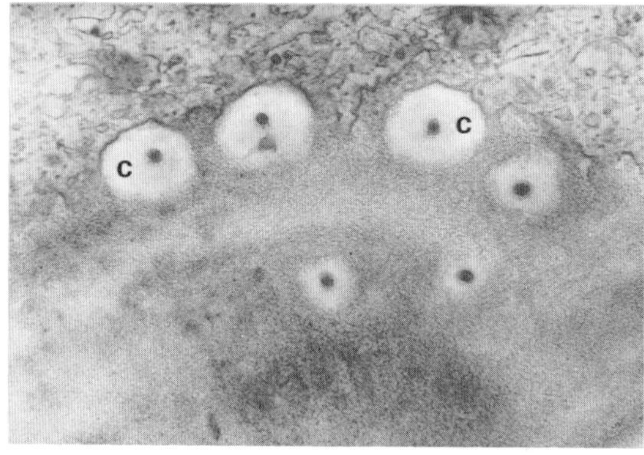

Fig. 9.6 Oblique longitudinal section of a developing sieve plate of *Pisum sativum*. Callose (c) has accumulated round the connecting strands. ×16 000. (From Bouck and Cronshaw,[67] Fig. 10, p. 91.)

Connecting strands pass through the sieve plate pores from one sieve tube member to another (Fig. 9.7). They may vary in size from the thickness of a plasmodesma to a diameter of about 10 μm. The appearance of connecting strands under the electron microscope is variable; they are usually seen as either solid or composed of fibrillar material.[199, 385] It is probable that slime is present in the connecting strands and indeed P_2-protein can be seen in this position (Fig. 9.8).[133] Aggregations of slime (P-protein) known as slime plugs may accumulate at the sieve plate in response to injury to the sieve tube. When the sieve tubes become non-functional the pores of the sieve plate become plugged with callose.

The question of whether the sieve plate pores in the living plant are open, or plugged as they are so often seen in micrographs, is an important matter in relation to theories of translocation, and one which has aroused much controversy. For example, plugged pores would be incompatible with the Münch mass-flow hypothesis. If the pores are indeed open and there is no tonoplast, the sieve tube

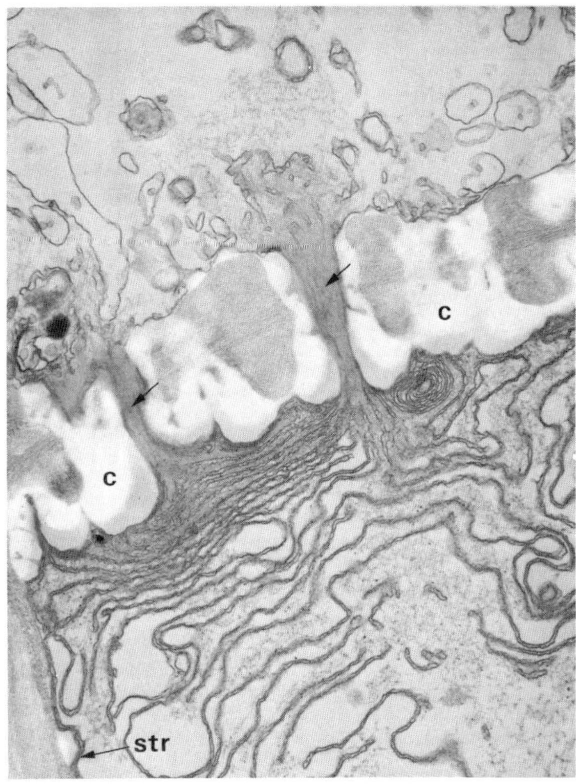

Fig. 9.7 Longitudinal section of mature sieve plate of *Pisum sativum*. The fibrous material filling the pores is arrowed. Many cisternae, which resemble the sieve tube reticulum (str) at the margin of the sieve tube element, are accumulated near the sieve plate. c, callose. ×26 500. (From Bouck and Cronshaw,[67] Fig. 12, p. 95.)

members would be in continuous contact and would behave as a single cell.[189]

Early electron micrographs showed the sieve plate pores plugged with slime. If rapid fixing or techniques designed to minimize the surge which results from wounding are used, however, some at least of the sieve plate pores are seen to be open (Fig. 9.9).[15,123,130,215] Evidence from material infected with mycoplasma-like bodies also suggests that the pores are normally open.[348,733] Other workers, however, still maintain that the pores are normally plugged with P-protein.[599] Certainly, it is noteworthy that even when precautions

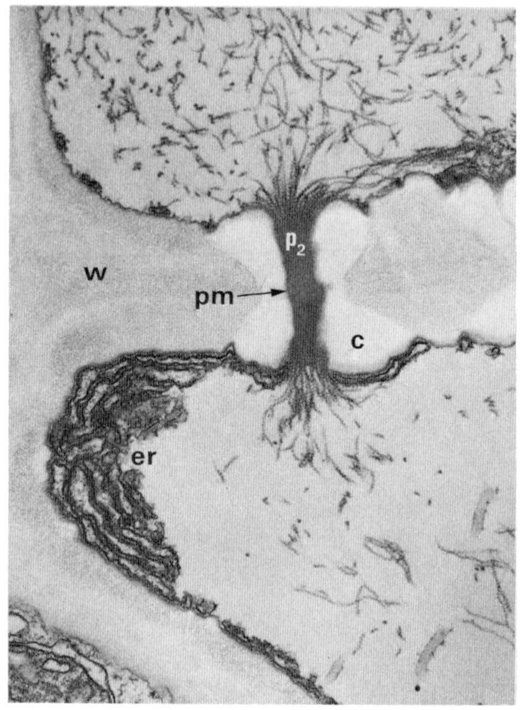

Fig. 9.8 Longitudinal section of sieve element of *Nicotiana tabacum* at a later stage of differentiation. The section passes medianly through a sieve plate pore. The plasma membrane (pm) is continuous through the pore, which is filled with P_2-protein fibrils. The endoplasmic reticulum (er) is closely applied to the plasma membrane of the sieve elements. c, callose; w, wall. ×31 000. (From Cronshaw and Esau,[133] Fig. 10, p. 810.)

are taken to prevent sudden changes in hydrostatic pressure some of the pores appear to be plugged.[15, 701] In this connection, it is interesting that when small slivers of palm stems were fixed, sieve elements from the central region had open pores, whereas those near the cut ends had filamentous or amorphous material in the pores.[497] Thus, on balance, it appears likely that in the living condition the pores are open; this is indeed suggested by the observation of even a few sieve plates with open pores.

Connection between the sieve elements and companion cells and also parenchyma is maintained by means of plasmodesmata; those situated in the lateral sieve areas are single strands on the side of the

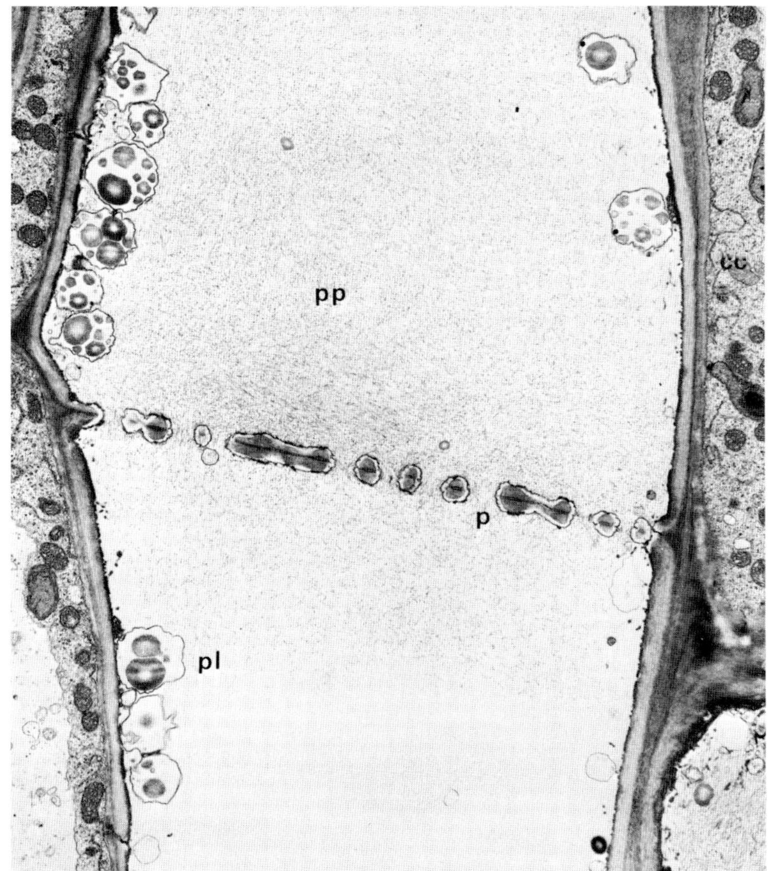

Fig. 9.9 L.S. sieve elements of *Nicotiana tabacum*, showing a sieve plate with open pores (p). cc, companion cell; pl, plastid; pp, P-protein. ×4400. (From Anderson and Cronshaw,[15] Fig. 1, p. 175.)

sieve element, but branched or multiple on the side of the companion cell.[185, 385]

Callose

Callose is a carbohydrate which is formed not only in sieve elements but also in some other types of cells. Apparently enzymes located in the plasmalemma are involved both in the synthesis and breakdown of callose in the sieve element;[124] since the timing of callose synthesis may be critical these enzymes assume some importance in the physiology of phloem.

Callose is found in sieve elements in association with the sieve plate and lateral sieve areas; it forms a sheath around the connecting strands which penetrate the pores in the sieve plate, and in the lateral sieve areas. It is indisputably formed in response to wounding, but much discussion is centred on whether callose is also normally present in uninjured sieve elements. This is obviously an important matter, but since killing material for microscopic study involves wounding it is also one which is very difficult to investigate. Experiments carried out by Engleman,[176] in which tissue was killed in an uncut condition by various gaseous or temperature treatments, showed that callose was present in sieve elements of intact plants killed within 4 sec after injury. It seems likely, therefore, that it is a normal component of living sieve elements, which is deposited in more massive amounts as a consequence of wounding, although in palms callose is apparently absent from the sieve elements[497] and may be absent from other species following certain fixation procedures.[152] It may even be present in some pores and absent from others.[192] Deposition of callose can be induced not only by wounding, but also by virus diseases and conditions of growth.[189] The question of the stage in ontogeny of the sieve element at which callose is formed should also be considered. Callose which accumulates while the sieve element is still functional is called definitive callose.[189]

In *Vitis*,[183] and no doubt also in other plants, there is a seasonal production of callose; in winter months there is an accumulation of dormancy callose, most of which disappears again on resumption of growth. Perennial monocotyledons may also show seasonal variation in the amount of callose.[179] The periodic removal of dormancy callose indicates that the amount of callose is related to physiological conditions in the cell.[189]

Structure and function of sieve tubes

Assimilates are known to move in the sieve tubes at rates which are estimated at from 2 to 7200 cm/h in different species, commonly 30–120 cm/h (see Table 11.3 of Crafts and Crisp[123]). Estimates of 60–100 cm/h,[728] or 200 cm/h,[130] seem to vary in part because of the doubt regarding the area of phloem occupied by sieve elements[430] already mentioned.

Diffusion is inadequate to account for rates of flow of this magnitude. An electro-osmotic mechanism would be favoured by plugged pores, a mass-flow mechanism by open pores. Thaine's[653] hypothesis of transcellular strands is the only one which could explain bidirectional movement within a single sieve element, but it is not certain that this has been unequivocally demonstrated in any case. Although labelling experiments seem to show that bidirectional movement is possible, it is pointed out that, although feasible, it may not actually occur under conditions of high rates of mass transfer.[430] Also, some of the evidence is subject to different interpretations.[694] The accelerated diffusion model[85] also depends on the existence and functional activity of strands that in fact appear not to be present. As already mentioned, an 'activated mass flow' mechanism[430] proposes a function for the P-protein; on such a view, a mass flow mechanism (dependent on source/sink relationships) would orientate the P-protein filaments, which in turn would provide the motive force. Despite the considerable amount of work on the structure and ultrastructure of phloem elements, it seems that there may be more than one mechanism for translocation, that there is some error in the accepted facts of structure or physiology,[498, 701] or that the right mechanism has not yet been envisaged.

Companion cells

The companion cell is formed from the same mother cell as the sieve tube element adjacent to it. Companion cells are of smaller diameter than sieve elements, and are usually somewhat angular in cross section. Observations with the electron microscope show that abundant organelles and membrane systems are present in the cytoplasm of the companion cells; both dictyosomes and endoplasmic reticulum occur and the mitochondria have well marked cristae.[201] Numerous ribosomes have been observed, contributing to the density of the cytoplasm and indicating the occurrence of active protein synthesis.[730] The nucleus, bounded by a normal double membrane, is usually rather elongated and may be lobed. The plastids show little internal structure, but in *Acer*, at least, are closely sheathed with endoplasmic reticulum, connected to the nuclear envelope and the cytoplasmic endoplasmic reticulum. It is suggested that the endoplasmic reticulum linking the plastids and the plasmodesmata may be a route for the transfer of sucrose from the sieve tubes to other neighbouring tissues, the plastids possibly providing temporary storage.[731]

The evidence seems to support the view that the companion cells and other nucleate cells probably form with the sieve elements a complex functional system for the transport of solutes. There are many plasmodesmata between the sieve element and its associated companion cell or cells; they are branched and have numerous branches on the companion cell side and a single trunk on the side of the sieve element.[194 495] Companion cells may develop wall labyrinths and form transfer cells (see Chapter 10).

Phylogeny

The evolutionary history of phloem is on the whole less well understood than that of xylem. Sieve elements are thought to have evolved from parenchyma cells, and it is evident that because of the specialized physiology of sieve elements there must have been profound changes in enzyme systems within the cell.[206]

Long sieve tube members with inclined end walls often have compound sieve plates, and these are probably the most primitive type of sieve element. The length of sieve tube members, however, is often deceptive, since transverse divisions of cambial derivatives may occur during the formation of the phloem, resulting in shorter elements. More specialized sieve elements are often characterized by relatively transverse end walls and fewer, thick connecting strands.[198, 318] There are statistical correlations between long end walls and possession of compound sieve plates, and between approximately horizontal end walls and simple sieve plates.[738]

Cheadle and Whitford[104] found that the least specialized types of phloem elements occurred in the roots of monocotyledons, and more highly specialized types in the aerial parts of the plant, the reverse of the findings relating to xylem elements. The prevalence of sieve elements with transverse sieve plates in the leaves of palms, and of elements with oblique sieve plates in the roots,[494] supports these observations. Perhaps physiological conditions in the various organs of the plant may be involved, although these reports remain puzzling.

Work with aphids

Some viruses pathogenic to plants are restricted to the phloem tissue. Often these are transported from plant to plant by insect vectors, e.g. aphids, which introduce the virus into part of the phloem, whence it is translocated through the plant. This offers a means of studying rates of transport in the phloem. Thus curly top virus was found to move 15 cm within 6 min in *Beta* (beet), and at least 30 cm in 3 h in *Cuscuta* (dodder).[184]

Rates of transport can also be calculated from the rate of flow of phloem sap up the severed stylets of aphids. These stylets penetrate the sieve elements very accurately (Fig. 9.10); if the body of the aphid is

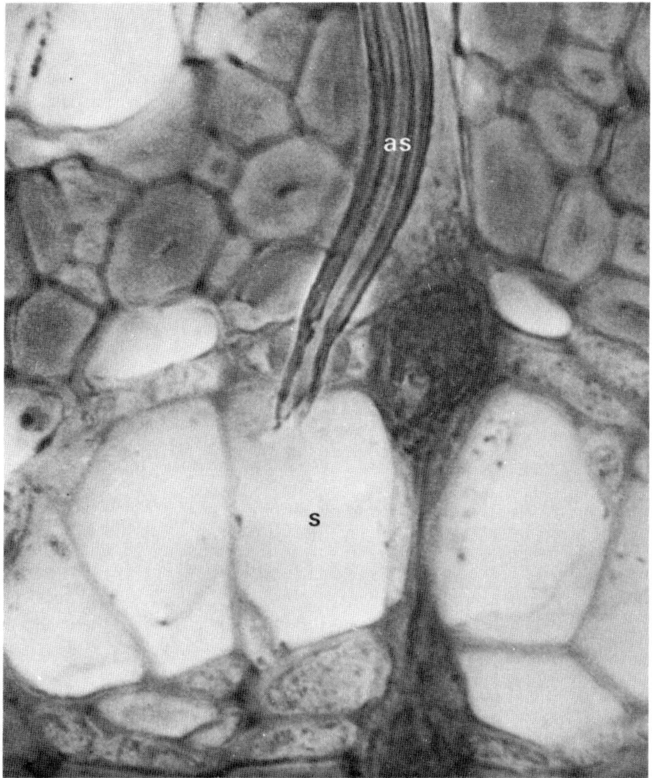

Fig. 9.10 Transverse section of stem of *Tilia* showing the penetration of an aphid stylet (as) into a single sieve tube element (s). The stylet has passed through the phloem fibres. ×1125. (From Zimmermann,[747] p. 132. Copyright © (1963) by Scientific American, Inc. All rights reserved.)

severed after it has achieved penetration of the sieve tube, leaving the stylet in the plant tissues, phloem sap continues to be exuded from the stylet for a prolonged period. The sap can be collected with a micropipette and its composition studied. The rate of exudation may be more than 5 mm^3 per h, which demands refilling of the sieve element 3–10 times per sec and rates of translocation of the order of 100 cm per h.[747] It is no wonder that sieve elements are structurally so specialized.

Studies of the path of aphid stylets inserted into plant tissues have indicated that this is mostly intercellular, until the sieve elements are approached.[64,216] Parenchyma cells may be punctured *en route*, but their protoplasts are pushed aside and remain more or less undamaged.[216] In one study of a species with internal phloem, the aphids appeared to feed preferentially on elements of the internal phloem, which were thought to contain more sap and be under greater pressure than external phloem elements. After an aphid has punctured a functional sieve element, it is essentially force-fed, and extracts amino acids and some sugars for its own nutrition, excreting the surplus.[64] Insects which feed on phloem often excrete some of the sugar in the form of a concentrated solution known as honeydew. This honeydew is believed to have been the manna used by the Israelites in the desert.[184]

Aphids have also been used in studies of the movement of hormones, or of the effect of hormones on translocation of assimilates.[68,410]

CONTROL OF PHLOEM DIFFERENTIATION

In attempts to investigate the factors controlling the differentiation of phloem a number of experiments have been carried out; most of these approaches to the problem parallel those employed in the study of xylem.

Using *Coleus* plants, La Motte and Jacobs[400] studied the regeneration of phloem in the vicinity of severed vascular bundles under various conditions (see Appendix). Incisions were made in the flat side of the internode below the fifth pair of expanded leaves, severing one or more vascular bundles. Under these conditions, there was regeneration of strands of phloem elements in the pith; many of these were formed in a plane diagonal to the long axis of existing cells (Fig. 9.11a).[399] In a series of experiments, either all leaves, all buds, both buds and leaves, buds and leaves distal to the wound, or buds and leaves proximal to the wound, were removed. A technique for clearing and mounting the injured internode was devised which made it possible to reach a quantitative estimate of the number of sieve tube elements regenerated[399] (see Appendix). Removal of all leaves and buds, and to a lesser extent removal of all organs distal to the wound, greatly reduced the regeneration of phloem.

More recently, Aloni[9] has shown that mature leaves induce the differentiation of primary phloem fibres in internodes below but not above them. In wounding experiments on *Coleus* internodes[10] phloem fibres were found to differentiate in the regenerated strand of phloem after about three weeks. Fibres develop by redifferentiation of parenchyma cells, and elongate at both ends by intrusive growth (Fig. 9.12). They can be observed by means of a lacmoid clearing

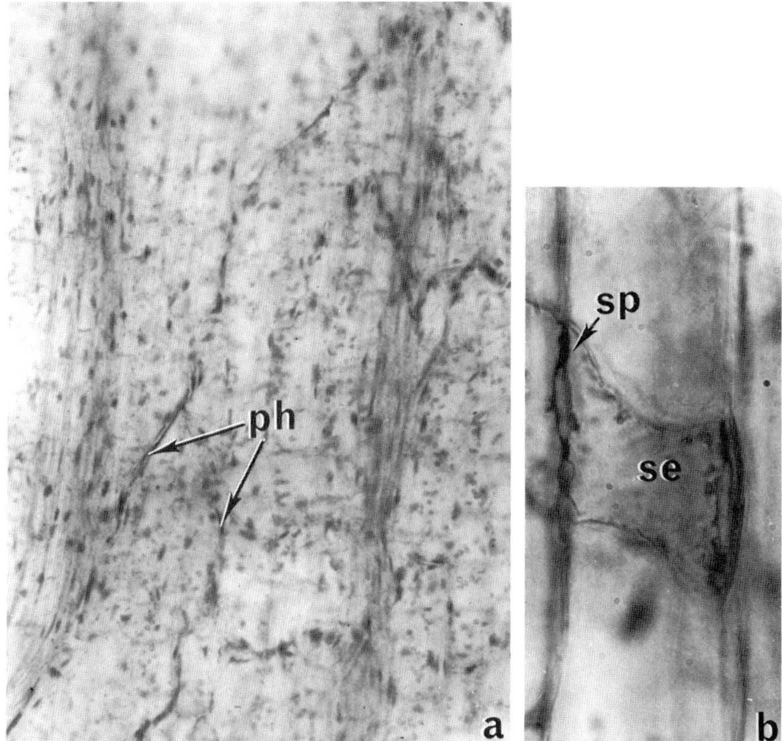

Fig. 9.11 (**a**) Cleared, stained preparation of a wounded fifth internode of *Coleus* showing regeneration of a strand of phloem (ph) from parenchyma cells between vascular strands. *See* Appendix. ×150. (**b**) Part of a strand of regenerated phloem, passing obliquely across the stem, showing the formation of a sieve element (se) by division of a parenchyma cell. sp, sieve plate. ×800.

method.[11] Sieve elements always differentiate in a strand before the fibres.[10]

In another series of experiments, indoleacetic acid (IAA) in lanolin or in aqueous solution was applied to the cut stump of wounded stems from which all organs had been removed; sucrose was also applied.[400] IAA applied in this way induced more regeneration of sieve elements than the organs of the intact shoot; sucrose had little effect (Table 9.1). Regeneration of phloem in excised internodes was also stimulated by IAA, although in this instance some regeneration took place in controls.

That regenerated sieve elements of this kind are able to function

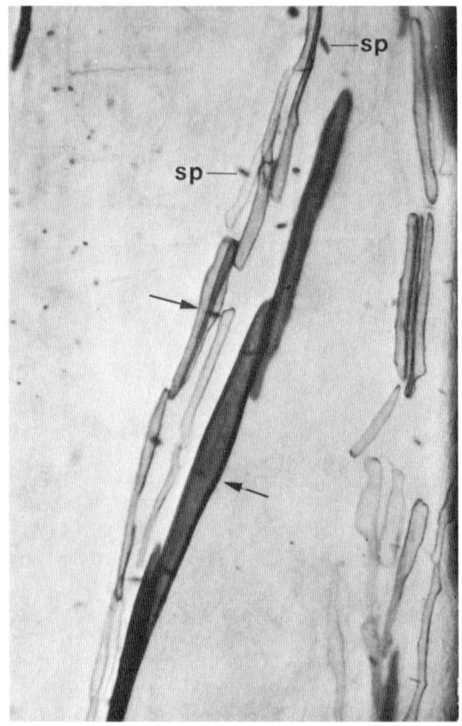

Fig. 9.12 Part of a strand of phloem which has regenerated following wounding of a young internode of *Coleus*, 60 days after wounding. Phloem fibres (arrowed) have differentiated from parenchyma cells; sieve plates (sp) in adjacent sieve tubes can also be seen. ×125. (From Aloni,[10] Plate 1B.)

normally is shown clearly by the experiments of Eschrich[209] on wounded stems of *Impatiens*. He found that a dye could be transported even in the undifferentiated strand of regenerated phloem before the formation of sieve plates; after this stage the sieve tubes formed as a result of wounding were fully functional.

LaMotte and Jacobs[400] concluded that auxin normally derived from the lateral organs of the shoot was probably a limiting factor in phloem regeneration. In subsequent experiments with intact plants of *Coleus* it was shown that the regeneration of sieve tubes preceded that of xylem cells by one day.[657] Xylem differentiated adjacent to existing sieve elements. These observations suggest that the primary action of the IAA may be on the regeneration of phloem. The importance attributed[548] to severing the primary phloem may support this view. It is pointed out[657] that this pattern of regeneration—sieve elements, then xylem—resembles that of normal differentiation in a primary stem. In many species the differentiation of secondary phloem also often precedes that of secondary xylem by several weeks.[217] The effect of gibberellic acid (GA) on cambial activity and the formation of secondary phloem[150] is discussed more fully in Part 2,[136] Chapter 4. It has been suggested that the effect of GA is on cambial division, the actual differentiation of sieve elements being controlled by IAA.[343]

Table 9.1 The effects of IAA and sucrose on phloem regeneration in plants of *Coleus* without shoot organs. (From LaMotte and Jacobs.[400])

Treatment	Mean number of strands regenerated in 5 days
Controls (intact plants)	34·6 ± 2·8
All shoot organs off	
Plain lanolin	9·1 ± 1·7
0·1% IAA in lanolin	42·8 ± 1·4
1% IAA in lanolin	64·1 ± 11·4
Water	13·2 ± 5·2
Water*	11·2 ± 0·9
20 g/l. sucrose*	14·6 ± 3·7

* 0·025% sulphanilamide present in the aqueous solution.

There was evidence from some of the experiments discussed above, also, that both sugar and unidentified substances leaking from the cut strands of phloem might sometimes be limiting. Previous observations on normal differentiation in *Coleus* suggested that differentiation of sieve tube elements is normally limited by a factor which moves upwards from the more mature parts of the shoot.[345]

The experiments of Wetmore and Rier[707] on callus, already discussed (see p. 166), indicate that the balance between sucrose and auxin may be critical in controlling the differentiation of cells as xylem or phloem. It is interesting that well developed sieve elements can occur at random in callus cultures, grown on a medium containing an auxin and sucrose but not otherwise treated,[613] in addition to the xylem cells which have been frequently observed. The situation in the wounding experiments with *Coleus* differs from that in undifferentiated callus in that various gradients must already have been set up in the stem; it is noteworthy that xylem regenerates in proximity to the xylem of the severed vascular bundle, and phloem near existing phloem. This is an example of homoeogenetic induction—the induction by a tissue of the differentiation of adjacent tissue in a similar manner. Several earlier experiments also yield evidence of this phenomenon.[74]

In callus tissue from carrot, sieve elements were observed, and an experiment was carried out to test whether translocation took place. The callus was fed with labelled CO_2, and half of each piece was kept shaded. Translocation from the illuminated side to the shaded did not occur, indicating that the phloem was discontinuous.[304] In tobacco pith cultures, phloem consisting of sieve elements, companion cells and parenchyma was induced by sucrose and an auxin, either IAA or NAA. The phloem differentiated in nodules round the periphery of the pith cylinders, but only after the development of tracheids; this is the reverse of the order of differentiation in wounded internodes. Sieve elements were formed in short rows with sieve plates between them, and they showed normal ultrastructure, including tubules of P-protein.[131] As suggested in Chapter 8, a constant source of stimulatory substances emanating from the shoot apex and young leaves may be necessary for the formation of continuous vascular strands.

From a consideration of these experiments it is clear that several attempts to understand the causes underlying the differentiation of phloem have already been made. However, one needs only to recall the extremely complex changes which occur at the sub-cellular level during the differentiation of sieve elements to conclude that scarcely a beginning has been made in attempting to understand the underlying causes of such changes or the inter-relationships between neighbouring cells in the phloem.

10

Transfer Cells

Although we have a fair understanding of the transport of materials from one part of the plant to another by means of the conducting elements of the xylem and phloem, until recently little was known about short distance transport of metabolites, and, indeed, much still remains to be discovered. Such short distance transport includes the loading and unloading of the conducting elements themselves. As long ago as 1884 certain specialized companion cells were believed to be involved in this process in minor veins of the leaf,[236] and more recently Gunning and Pate[284, 502] have conducted a series of investigations into the nature, distribution and probable function of these and other similar cells.

Because of their probable function, these cells have been given the name ***transfer cells***. Although variable in form in different species, and even in different regions of the same plant, these cells are always characterized by ***wall ingrowths*** (Fig. 10.1). These consist of a specialized form of unlignified secondary cell wall, deposited on the inner side of an ordinary, unspecialized primary wall.[502] The plasma membrane of the cell follows the outline of these wall ingrowths (Fig. 10.1), which may occur all round the periphery of the cell or may be restricted to localized regions of the primary wall (Fig. 10.2). An important consequence of the close association of the plasma membrane with the contours of these projections of wall material is that the area of the former is greatly increased, presumably facilitating increased absorption or secretion of solutes by the protoplast. In the field pea, *Pisum arvense*, the area of the plasma membrane in a transfer cell associated with a minor vein may be increased by more

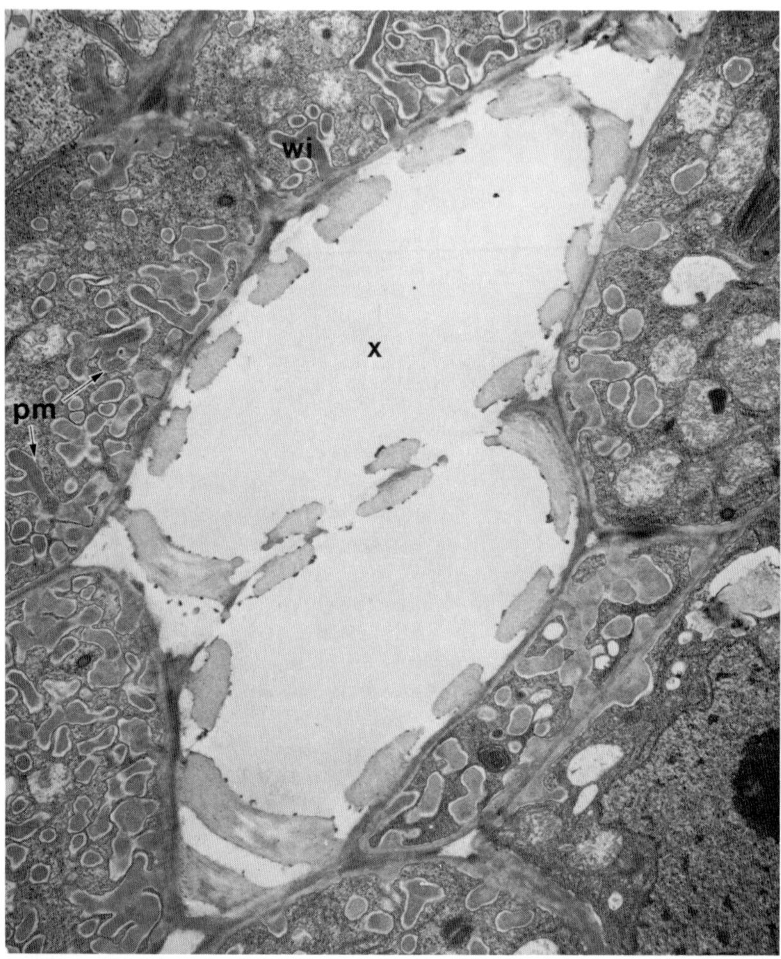

Fig. 10.1 Transmission electron micrograph of xylem transfer cells surrounding two tracheary elements (x) in a leaf trace of a seedling of goosegrass, *Galium aparine*. Wall ingrowths (wi) are present on the walls of these xylem parenchyma cells that abut on the tracheary elements. The plasma membrane (pm) follows the contours of the wall ingrowths. ×5140. (From Gunning and Steer,[286] Plate 13b, p. 209.)

than ten-fold over that of a cell of similar size with a smooth wall.[282] According to estimates on other species, the surface area of the plasma membrane may be more than 20 times that of an ordinary parenchyma cell, even in transfer cells in which the wall ingrowths are restricted to only portions of the periphery of the cell.[502] The increase in potential absorptive surface in such cells is thus by no means negligible. This association of plasma membrane and ingrowths of secondary wall is sometimes referred to as the wall membrane apparatus.[500,502]

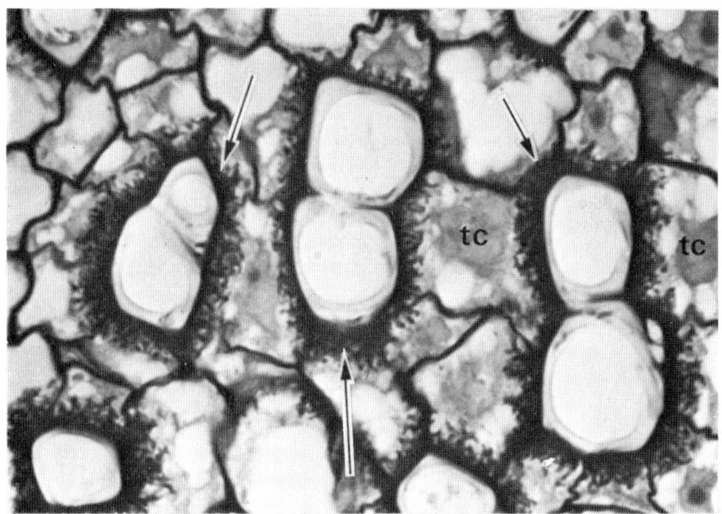

Fig. 10.2 Transverse section of the rhizome of *Hieracium floribundum* as seen with the light microscope, showing xylem parenchyma cells with wall ingrowths (arrowed). The transfer cells (tc) have this labyrinth of wall ingrowths only in regions adjacent to tracheary elements. ×1100. (From Yeung and Peterson,[736] Fig. 12, p. 161.)

Another feature of transfer cell walls is their possession of plasmodesmata (Fig. 10.3).[502] These may occur between two adjacent transfer cells.[737] Cytoplasmic contents are dense; there is a large, sometimes lobed nucleus, numerous mitochondria and abundant endoplasmic reticulum (Figs. 10.1, 10.3).[282,737] The mitochondria are often closely associated with the wall ingrowths.[279,737] Other organelles, such as dictyosomes and ribosomes, and in a few instances microbodies with crystalloid cores, are also present. It is thought that these microbodies may have a role in wall metabolism.[737]

Transfer cells can usually only be recognized with the light microscope in very thin sections of material embedded in resin, though those in the nodal region of *Tradescantia* shoots and *Hieracium* rhizomes can be detected in stained hand sections.[502,735] From such

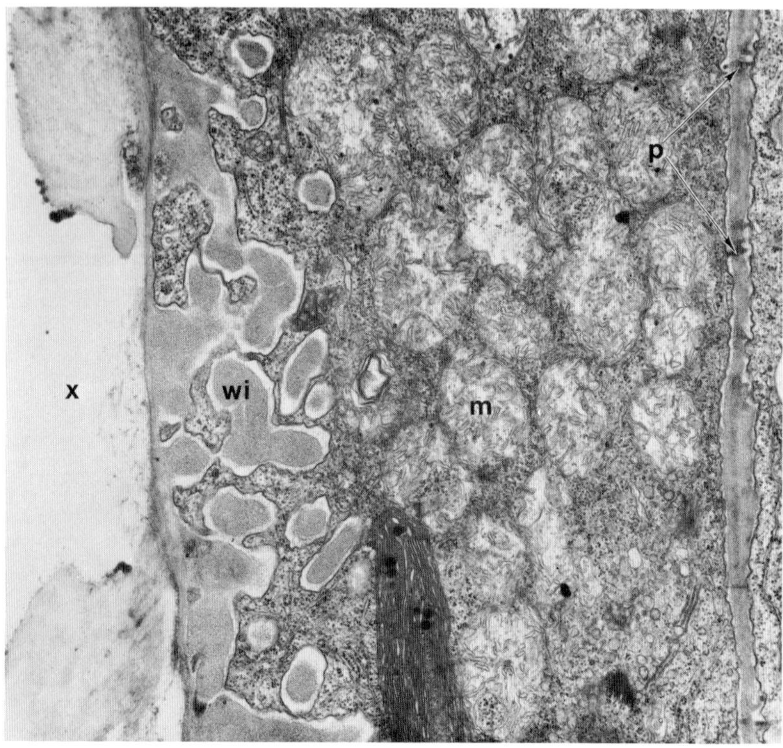

Fig. 10.3 Part of a xylem transfer cell in the shoot of *Galium aparine*, showing wall ingrowths (wi) adjacent to a tracheary element (x), and plasmodesmata (p) in the opposite wall. Many mitochondria (m) are present. (From 'Transfer cell, plant', by B. E. S. Gunning and J. S. Pate, Fig. 1, p. 420, McGraw-Hill Yearbook of Science and Technology, 1971. Copyright © 1971 by McGraw-Hill, Inc. Used with permission of McGraw-Hill Book Company.)

thin sections, and the still thinner ones required for transmission electron microscopy, it is quite difficult to build up a three-dimensional picture of the wall ingrowths, which may be further obscured by the characteristic dense cytoplasm. Recent techniques involving the fracturing of embedded material, or the removal of cytoplasm prior to the usual preparative techniques, have permitted observation of the walls of transfer cells with the scanning electron microscope.[70, 360] This three-dimensional view reveals a veritable labyrinth of wall ingrowths (Figs. 10.4 and 10.5) which, as will be seen below, can be quite variable in form.

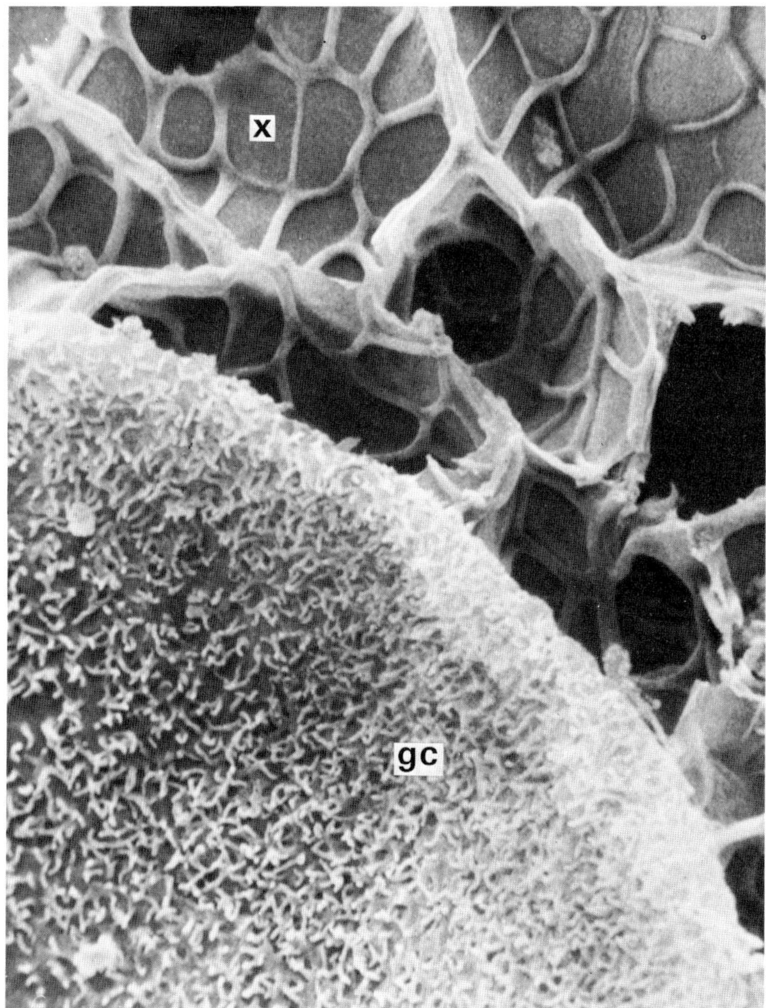

Fig. 10.4 Scanning electron micrograph of the wall of a giant transfer cell (gc) induced in the root of *Impatiens balsaminea* by the root-knot nematode, *Meloidogyne incognita*, following removal of the cell contents. Numerous wall ingrowths are present on the giant transfer cell, in contrast to the patterned walls of adjacent abnormal xylem elements (x). ×1670. (From Jones and Dropkin,[360] Fig. 7, p. 153.)

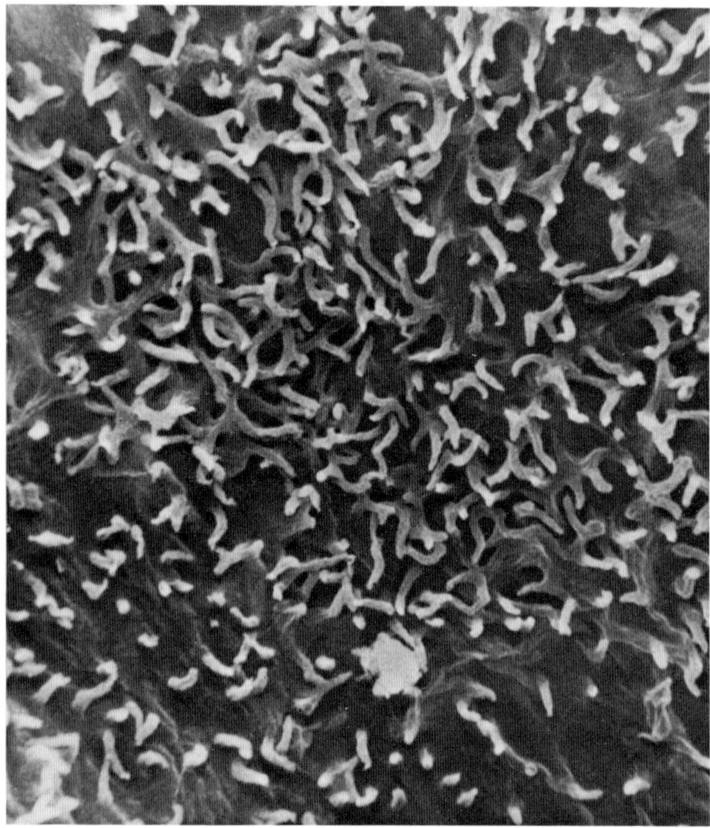

Fig. 10.5 SEM of part of the wall of a giant transfer cell of *Impatiens balsaminea*, as in Fig. 10.4, showing the labyrinth of wall ingrowths. ×3250. (By courtesy of Dr. M. G. K. Jones.)

Occurrence of transfer cells

Transfer cells as defined above, that is, cells with invaginations of secondary wall material, have now been described in many parts of the plant. Some of these sites are illustrated in Fig. 10.6. Transfer cells are associated, *par excellence*, with xylem and phloem (Fig. 10.7a), especially in the minor veins of leaves and at nodal regions of shoots. These have received the greatest amount of study to date, and are discussed in more detail below.

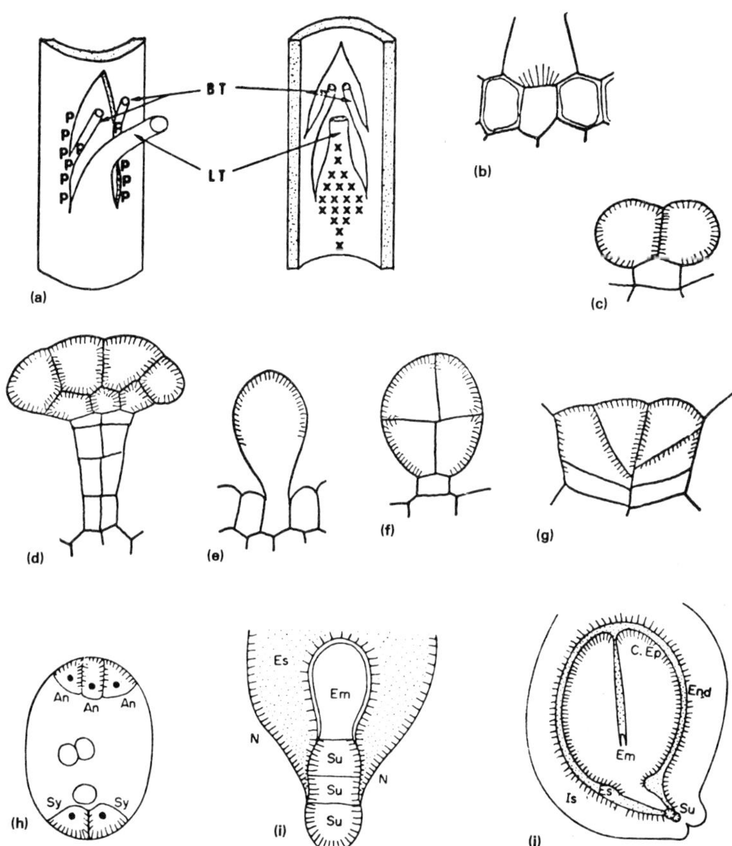

Fig. 10.6 Diagrams showing some of the locations in which transfer cells are found in plants. (**a**) Distribution of xylem transfer cells (x) and phloem transfer cells (p) at a node. BT, branch trace; LT, leaf trace. (**b**) Root hypodermis. (**c**) and (**d**) Glands of insectivorous plants. (**e**) Nectariferous trichome. (**f**) Glandular hair with wall ingrowths on exterior walls. (**g**) Salt gland. (**h**) Embryo sac of an angiosperm with wall ingrowths in synergids (sy) and antipodal cells (An). (**i**) Young embryo (Em) with wall ingrowths on suspensor cells (Su) and endosperm (Es); N, nucellus. (**j**) Older embryo (Em) with wall ingrowths on cotyledon epidermis (C. Ep), endosperm (Es), suspensor (Su) and endothelium (End), on inner side of integuments (Is). In (**b**)–(**j**) lines indicate wall ingrowths. (From Gunning and Pate,[281] Fig. 13.9 B–D, p. 453, and Pate and Gunning,[502] Figs. 3 and 20–28, pp. 181 and 186; reproduced with permission from *Annual Review of Plant Physiology*, **23**. Copyright © 1972 by Annual Reviews Inc. All rights reserved.)

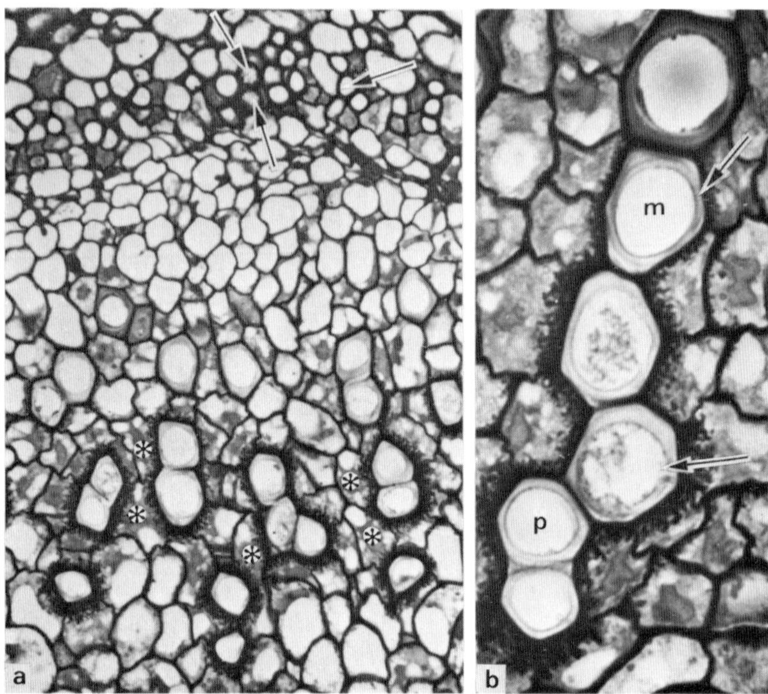

Fig. 10.7 Transfer cells in the rhizome of *Hieracium floribundum*. (**a**) Part of a vascular bundle showing both xylem (asterisks) and phloem transfer cells (arrows). ×400. (By courtesy of Dr. R. L. Peterson.) (**b**) Part of the protoxylem (p) and metaxylem (m) of a vascular bundle, showing the much more elaborate wall ingrowths of xylem parenchyma transfer cells adjacent to protoxylem, as compared to metaxylem, elements. ×900. (From Yeung and Peterson,[736] Fig. 15, p. 163.)

Other structures of the vegetative shoot in which transfer cells are found are glandular hairs of various kinds, including the secretory trichomes of insectivorous plants, hydathodes, nectaries and salt glands (Fig. 10.6c–g);[426, 502, 579] see also Chapter 11. Clearly, these are all structures involved in the secretion and transfer of water and solutes. Wall ingrowths also occur in a localized band in the leaf epidermal cells of water plants such as *Ranunculus fluitans* (Fig. 10.8).[279] They have been reported in other rather specialized situations, such as the haustoria of the parasitic angiosperm *Cuscuta*,[279] and in multicellular hairs in the leaf cavities of the aquatic fern *Azolla*.[169] The location of these hairs suggests that in this instance the wall ingrowths may have a role in the well-known symbiotic relationship between *Azolla* and the blue-green alga *Anabaena*, which fixes atmospheric nitrogen, but since the hairs have wall ingrowths even in the absence of the phycobiont, this is by no means certain. Transfer cells are also present at the junction between gametophyte and sporophyte in bryophytes and ferns.[218,280]

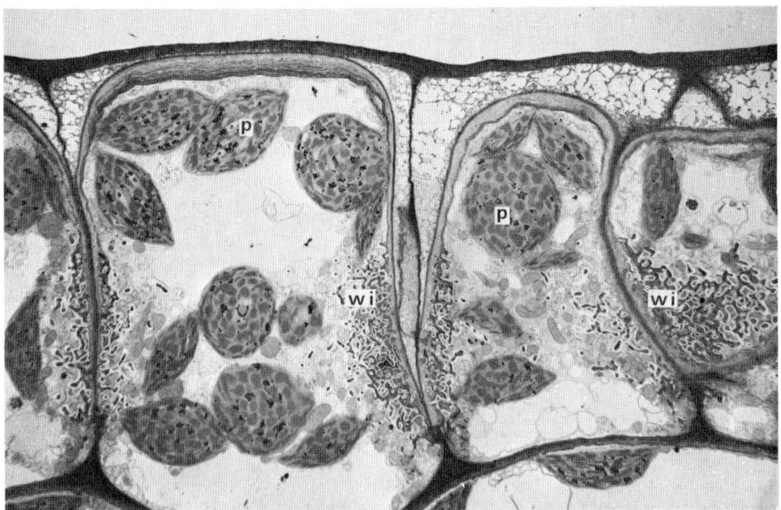

Fig. 10.8 Electron micrograph of the epidermal cells of a submerged leaf of *Ranunculus fluitans*. Note the wall ingrowths (wi) in the anticlinal cell walls, and the abundant chloroplasts (p). $\times 2520$. (By courtesy of Prof. B. E. S. Gunning.)

Transfer cells are also well differentiated in reproductive parts of angiosperms. For example, the synergids in the embryo sac have wall ingrowths extending from the outer walls; in some species, the antipodal cells are similarly endowed, and wall projections may occur on the outer faces of the basal cell of the suspensor, or of all suspensor cells (Fig. 10.6h–j; see also Part 2,[136] Chapter 8). The cells of the tapetum of the anthers of *Paeonia* also have wall ingrowths.[279,281] In alfalfa, transfer cells are present throughout the reproductive system; in particular, they line the stylar channel and may help to guide the pollen tube to the embryo sac.[358] In *Lathyrus* and *Vicia* (the latter genus is particularly well endowed with transfer cells), the outer and inner layers of the endosperm have wall ingrowths early in development, and at a later stage, following enlargement of the embryo, a wall membrane apparatus occurs in the epidermal cells of the cotyledons.[281] In the seeds of some members of the Gramineae, some of the cells of the aleurone layer are specialized and possess irregular wall thickenings on the outer tangential wall and the outer regions of the radial walls. These aleurone cells are situated adjacent to the region of attachment of the seed, and have been called 'transfer aleurone cells'.[553]

Transfer cells occur only rarely in roots, and then often in rather unusual situations. However, the recent report[412] of the occurrence of both xylem and phloem transfer cells in *Hieracium* at the junction of lateral roots with the main stele may indicate that these cells are commoner in roots than has been supposed. In roots grown in mineral nutrient solution, xylem transfer cells occurred adjacent to protoxylem poles associated with a developing lateral root, and phloem transfer cells were observed in adjoining strands of phloem. Both xylem and phloem transfer cells also occurred in the stele of the lateral root itself.[412] Apart from these observations, however, reports of transfer cells in roots are infrequent. In *Azolla* and a few other ferns, the pericycle cells adjacent to tracheids develop sparse wall ingrowths on walls abutting on the xylem elements.[281] The root hairs of a few aquatic plants have bulbous bases which bear wall ingrowths.[279] Wall ingrowths are not, however, common in root hairs or other root epidermal cells, though they do occur in certain epidermal cells of the onion root, which lacks root hairs.[510] In some leguminous root nodules, that is, regions of the root in which symbiotic bacteria fix nitrogen, the cells which surround the vascular tissue, comprising the pericycle, have wall ingrowths that are irregularly distributed over the surface of the primary wall.[503] Since such cells are not associated with root nodules in all species, they are apparently not essential for the functioning of the root nodules, which again are regions where active transport of materials must be occurring. Another situation in roots where wall ingrowths occur is in the giant cells that result from infection by nematodes (Figs. 10.4, 10.5). These cells develop as a result of repeated mitoses without accompanying cytokinesis, and are therefore multinucleate. Branched depositions of secondary wall were observed in giant cells of nematode-infected roots of broad bean, *Vicia faba*, and cucumber, *Cucumis sativus*.[332] The giant cells (formed by repeated endomitosis) induced in *Coleus* and *Impatiens* by the root-knot nematode, and the syncytia (formed by dissolution of cell walls) induced in potato roots by the potato cyst-nematode, are also characterized by wall ingrowths, especially where they abut upon xylem vessels.[360–362] These giant structures have been called multinucleate transfer cells.

In leaves infected with potato virus M, wall thickening observed in cells round the virus lesion was compared with that found in the giant cells,[675] but it is not altogether similar. A very interesting site of transfer cell development was the outer walls of peripheral cells in cultured cell aggregates of *Acer pseudoplatanus*.[145] These walls, in contact with the culture medium, would no doubt be sites of solute uptake. Similarly the wall protuberances of the nematode-induced giant cells may be formed in response to the flow of solutes from the xylem vessels of the root across the transfer cells to the nematode, which acts as a nutrient sink.[361, 362]

As already mentioned, vascular transfer cells are the most widely distributed form of these specialized cells. They are commonly present in the nodal regions of stems. Here there is a strong association between xylem transfer cells and the leaf traces, whether in uni-, tri- or multi-lacunar nodes.[283] Well developed transfer cells are also associated with leaf traces departing from rhizomes.[737] Phloem transfer cells, by contrast, border the leaf gap and occur in association with branch traces, if present (Fig. 10.6a).[504, 741] Phloem transfer cells occur as a kind of 'collar' at the cotyledonary node of seedlings at a very early stage of development, and are thought to function in the early nutrition of the plumule before it has any fully developed vascular connections.[502, 504] This is a particularly interesting suggestion, since very little is known about the transport of materials to actively growing meristematic regions that are devoid of vascular tissue. Phloem transfer cells seem to be most fully developed in regions of the stem vascular network which have only inconspicuous xylem transfer cells, or where these are absent. This complementary pattern of distribution of xylem and phloem transfer cells in stem vascular tissues may facilitate a cross transport of materials.[281]

Vascular transfer cells were first recognized in the minor veins of leaves of angiosperms, where they occur very commonly.[202, 501] A definite association between vein transfer cells and the herbaceous habit was noted among dicotyledons, in which vein transfer cells were identified in 22 of 200 families investigated; they were found in only one out of 42 families of monocotyledons studied. Four types of transfer cell have been described in minor veins (see below).[501]

Taxonomically, transfer cells are widespread, occurring in some situation in representatives of most groups of vascular plants and in bryophytes. Transfer cells occur in leaf traces of angiosperms, ferns, and *Equisetum*, but not in the microphyllous genera *Lycopodium* or *Selaginella*; apparently they are associated predominantly with the

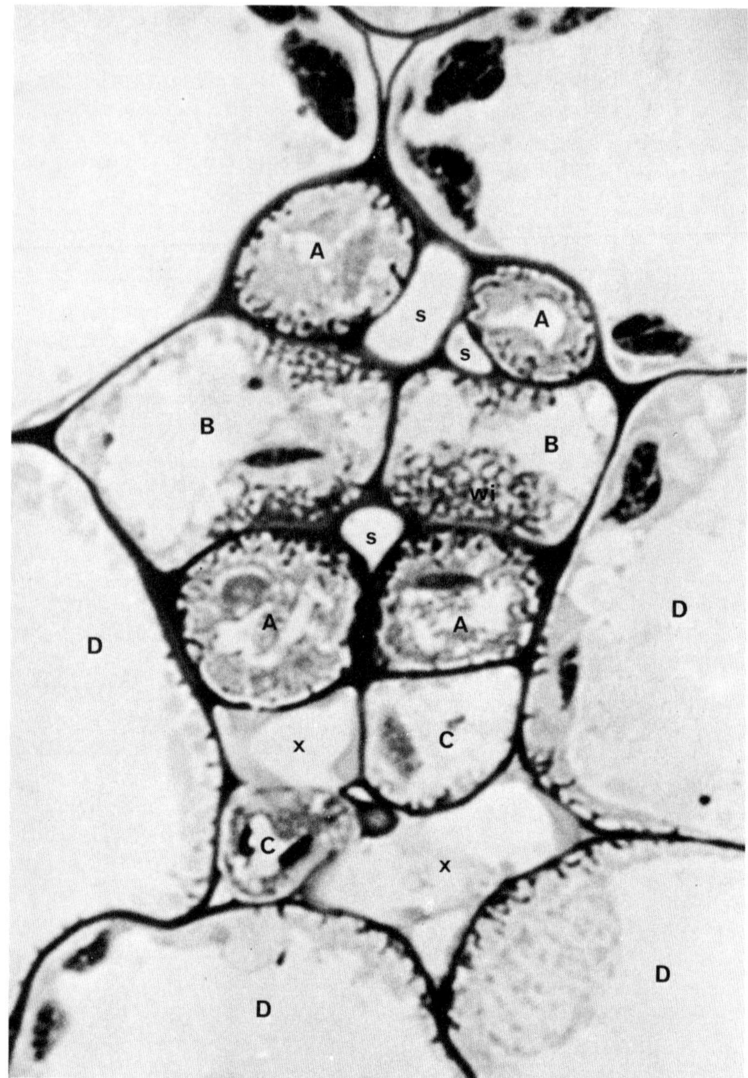

Fig. 10.9 Vascular transfer cells in a minor vein of the leaf of *Anacyclus pyrethrum*. Types A, B, C and D are present; note the ingrowths (wi) of the cell wall. s, sieve element. x, xylem. ×2530. (From Pate and Gunning,[501] Fig. 2, p. 139.)

megaphyllous condition.[283] It is thought that vascular transfer cells probably exist in the stems of most monocotyledons and dicotyledons.[281]

Form and position of wall ingrowths

As mentioned above, four types of transfer cells were defined in a study of minor veins in leaves.[501] Types A and B are associated with phloem. Cells of type A are considered to be specialized companion cells, with dense contents and wall ingrowths over the whole periphery of the cell, though sometimes less prolific on one wall adjacent to a sieve element. In type B cells, which are specialized phloem parenchyma, the wall ingrowths are best developed on walls contiguous with sieve elements, or with type A transfer cells.[279, 501] Transfer cells of types C and D are associated with xylem. In these, as also in xylem transfer cells located elsewhere in the plant, the wall ingrowths are restricted to portions of the wall that abut directly upon tracheary elements of the xylem (Fig. 10.2). Type C comprises cells of the xylem parenchyma (Fig. 10.1); type D are bundle sheath cells which border xylem elements.[501] Representative cells of all four types are illustrated in Fig. 10.9. The main point, perhaps, is that in xylem transfer cells the wall ingrowths are usually restricted to regions abutting upon xylem elements, whereas in phloem transfer cells they usually occur all round the periphery of the wall, though they may not be uniformly distributed and may, indeed, be less well developed on walls abutting upon sieve elements.

The form of the wall labyrinth itself may be species-specific. Wall ingrowths may be long, filiform structures, or Y-shaped flanges, or they may be quite short and papillate.[281, 479] However, the wall ingrowths of xylem and phloem transfer cells in the same plant may be totally different. This is true, for example, in the shoot of the rock rose, *Helianthemum*,[281] and in the rhizome of the rosette plant *Hieracium*.[511]

Ontogeny of transfer cells

As yet, relatively little is known about the development of transfer cells, but their ontogeny has been followed in various seedlings and in the shoot of *Hieracium*. In seedlings, phloem parenchyma transfer cells develop at an early stage relative to the differentiation of vascular tissue.[502] Xylem transfer cells are discernible one day after the tracheary elements differentiate, and phloem transfer cells can have fully developed wall ingrowths at least 4 days before the neighbouring xylem elements become lignified.[504]

In the rhizome of *Hieracium*, a Composite, differentiation of the xylem transfer cells occurs after that of the tracheary elements. The procambial cells adjacent to protoxylem elements divide; one daughter cell becomes a xylem parenchyma cell, the other, which is more densely cytoplasmic, becomes a transfer cell. It is not certain whether the preliminary mitosis is essential for transfer cell development. Wall ingrowths only form in cells adjacent to protoxylem elements after these have achieved connection with the xylem of a more mature vascular bundle. Transfer cells that are associated with protoxylem elements always have the greatest development of wall ingrowths (Fig. 10.7a, b), and the degree of development of these decreases centrifugally in any single vascular bundle.[736] Developing xylem transfer cells are very densely cytoplasmic (Fig. 10.10), and transfer cells from vascular bundles in older parts of the rhizome are more vacuolated and

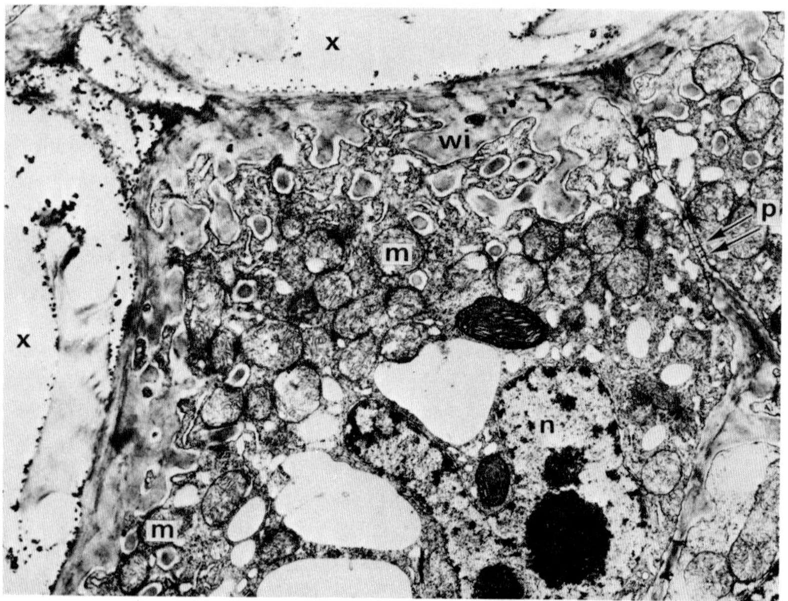

Fig. 10.10 Transfer cell showing wall ingrowths (wi) adjacent to two tracheary elements (x) in the rhizome of *Hieracium floribundum*. A lobed nucleus (n) is shown. Many mitochondria (m) are present in the vicinity of the wall ingrowths. Plasmodesmata (p) occur in the wall between two transfer cells. ×7300. (From Yeung and Peterson,[737] Fig. 1, p. 435; reproduced by permission of the National Research Council of Canada from the *Canadian Journal of Botany*, **53**, 432–438, 1975.)

characterized by the possession of microbodies with crystalloid cores. During development new wall materials are probably laid down between existing projections, and those regions of wall not characterized by wall ingrowths also become thickened.[737] In later stages of the ontogeny of xylem transfer cells, the cytoplasm begins to degenerate, and the wall ingrowths also seem to degenerate and disappear, though the whole wall may become somewhat thickened (Fig. 10.11).

During the differentiation of the primary phloem in *Hieracium*, some of the procambial cells probably differentiate directly as sieve elements. Others divide twice: one daughter cell of the first division becomes a phloem parenchyma cell and the other divides again to form a sieve element and a companion cell. In the metaphloem and less often in the protophloem, companion cells develop wall ingrowths and become transfer cells. During later development, more wall ingrowths are formed and there is a considerable increase in mitochondria in these companion cells (Fig. 10.12). Many parenchyma cells in the primary phloem also develop as transfer cells, becoming highly

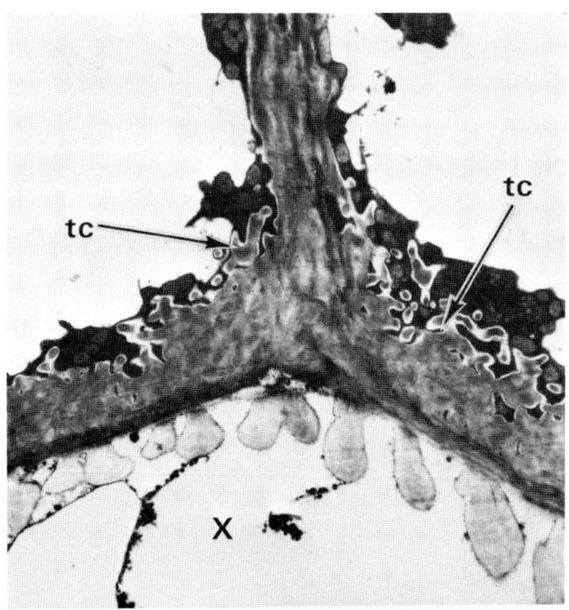

Fig. 10.11 Old transfer cells (tc) adjacent to a protoxylem element (x), showing thickening of the wall and the wall projections (arrowed). ×3950. (From Yeung and Peterson,[737] Fig. 6, p. 437; reproduced by permission of the National Research Council of Canada from the *Canadian Journal of Botany*, **53**, 432–438, 1975.)

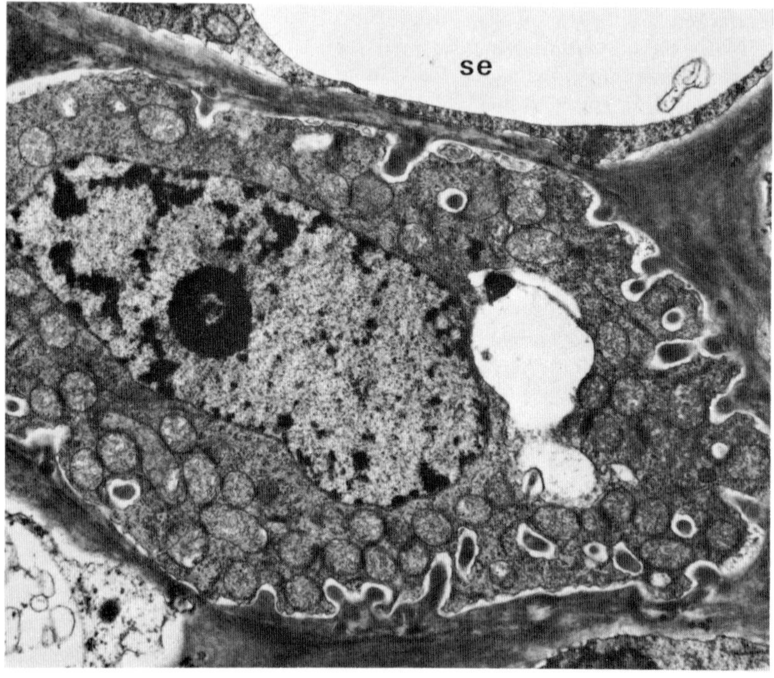

Fig. 10.12 Electron micrograph of a phloem transfer cell which has developed from a companion cell in *Hieracium floribundum*. The cell has wall ingrowths, a large nucleus, and many mitochondria, and is adjacent to a sieve element (se). ×10 100. (From Peterson and Yeung,[511] Fig. 19, p. 2755; reproduced by permission of the National Research Council of Canada from the *Canadian Journal of Botany*, **53**, 2745–2758, 1975.)

cytoplasmic in the process. Wall ingrowths are less well developed in regions of wall adjacent to a sieve element (Fig. 10.13). Parenchyma cells of the secondary phloem develop few wall ingrowths. During secondary growth, also, those companion cells which had elaborated wall ingrowths begin to degenerate.[511]

These studies indicate that there is a build-up of both xylem and phloem transfer cells during primary growth, and progressive degeneration during secondary growth. More work is required, however, before it can be certain that this situation is a general one.

Function

Transfer cells are apparently involved in the intensive transport of solutes,[279, 281] although the evidence for this is largely circumstantial.

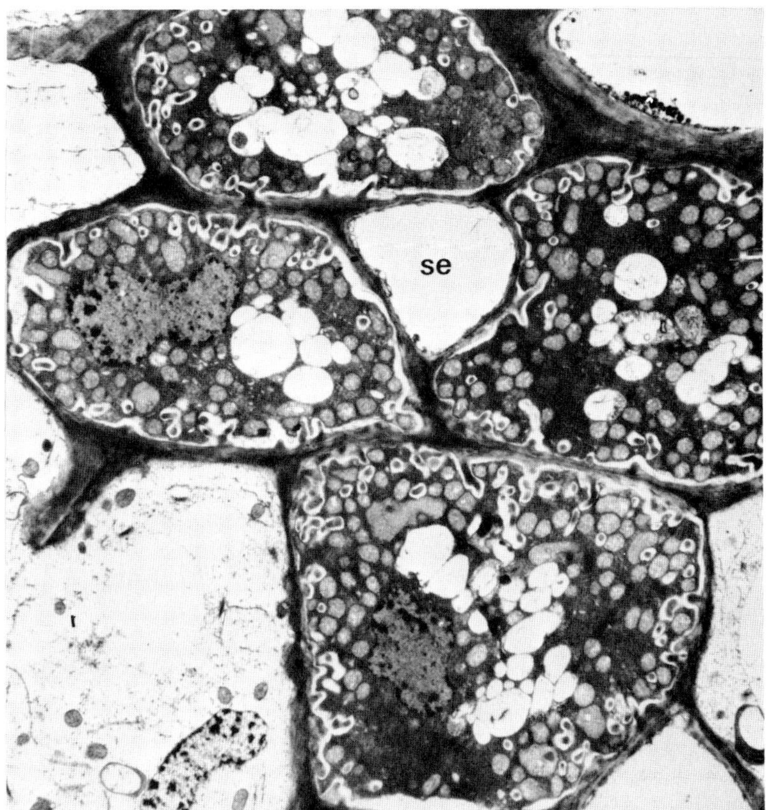

Fig. 10.13 Densely cytoplasmic phloem parenchyma cells which have developed wall ingrowths in *Hieracium floribundum*. Wall ingrowths are least developed where the cells abut upon a sieve element (se). × 3800. (From Peterson and Yeung,[511] Fig. 7, p. 2748; reproduced by permission of the National Research Council of Canada from the *Canadian Journal of Botany*, **53**, 2745–2758, 1975.)

For example, in minor veins the formation of wall ingrowths slightly precedes or coincides with the beginning of export of substances from the leaf, and further outgrowth accompanies the build-up of export from the leaf.[502] This was demonstrated with experiments using carbon-labelled CO_2.[281,282]

The position of transfer cells within plants, outlined above, suggests that they are involved in 4 kinds of trans-membrane flux, namely: absorption of solutes from the external environment; secretion of solutes to the external environment; absorption from internal extracytoplasmic compartments; and secretion of solutes into external extracytoplasmic compartments.[279]

Studies of the giant transfer cells induced in roots by nematodes suggest that all transfer cells may be characterized by the occurrence of action potentials.[363] Recent measurements of rates of flux across the plasma membrane of transfer cells suggest a very high relative rate on a tissue-volume basis.[500] Thus the wall ingrowths and the associated plasma membrane are probably well equipped to participate in the exchange of solutes between the apoplastic (i.e. cell wall, air spaces) and symplastic (i.e. cytoplasmic) compartments of a tissue.

Factors controlling the differentiation of transfer cells

There is unfortunately little positive knowledge about the factors which induce the development of wall ingrowths, although there is undoubtedly a strong correlation between their development and the onset of intensive solute transport. This has led to the suggestion that the existence of solutes to be transported may be enough to induce the formation of transfer cells, in a fashion resembling the effect of available substrate on the formation of an adaptive enzyme.[504]

In experiments with seedlings, it was found that if one cotyledon was removed or darkened, transfer cells failed to develop in association with the foliar trace to the mid vein of that cotyledon. In lettuce seedlings grown in the absence of CO_2, wall ingrowths in the xylem transfer cells were greatly diminished. If seedlings so grown were subsequently returned to an atmosphere containing CO_2, wall ingrowths developed in the xylem transfer cells progressively with time.[504] Wall ingrowths also fail to develop normally in leaves grown in darkness, or in non-green regions of variegated leaves or those of albino mutants.[281] In pea plants, the presence of a developing pod in the axil of a node results in greatly increased development of wall ingrowths in the nodal xylem parenchyma, as compared with nodes subtending only an unfertilized flower.[502] In *Hieracium*, the development of wall ingrowths was much greater and more elaborate in plants grown in higher concentrations of nutrient solution; plants grown in water showed very few wall ingrowths.[736]

All of these observations (and also those on nematode-induced transfer cells) support the view that transfer cells develop, both temporally and topographically, in association with a need for solute transport. It is, however, difficult to imagine how this might bring about a patterned deposition of secondary wall material, as well as the cytoplasmic changes mentioned above. The localized development of wall ingrowths in xylem transfer cells on walls adjacent to xylem elements suggests that some stimulus comes from the xylem.[281] The causal factors involved in the differentiation of transfer cells remain, however, something of a mystery, and much more work is needed in this field.

Genetic and evolutionary aspects

The form and distribution of wall ingrowths may be genetically determined, since they are relatively constant within a species but vary between species.[281] Since, however, considerable differences may exist between xylem and phloem transfer cells of the same plant, physiological factors must also be important during differentiation. The presence of transfer cells at the junction of gametophyte and sporophyte generations of bryophytes and pteridophytes, and the presence in *Equisetum* of all types of transfer cell found in the leaves of higher plants,[283] suggest that they may have had an ancient origin phylogenetically. This has led to the idea that the potentiality to develop transfer cells may be present in all flowering plants, though it may remain latent.[281] This in turn implies that if an appropriate selection pressure relating to solute transport were exerted, any higher plant could form transfer cells.[502] Further work may throw light on the truth of these speculations. It is, however, already clear that transfer cells are widely distributed in plants, and are of considerable functional importance.

11

Secretory Cells and Tissues

In many species throughout the plant kingdom, cells or whole tissue regions are involved in the processes of secretion or excretion. Secretory cells occur in different parts of the plant and particular kinds of secretory structures are sometimes characteristic of certain families.

Secretion implies the release of substances produced in the cytoplasm and moved outside the cell. These substances may be of no further use to the plant, in which case the process may more properly be defined as excretion, or they may be functionally important, as in the case of hormones or enzymes. However, the borderline between secretion and excretion remains ill-defined, largely because the function of many plant secretions is often still obscure. Many of the products secreted by plants are of considerable economic importance; among these may be mentioned rubber, gutta-percha and opium.

It is difficult to classify structures so diverse as those involved in secretion, since individual structures sometimes transcend particular categories. Some secretory structures, such as the glandular trichomes, are of epidermal origin and are thus completely external. Others comprise both epidermal and more deep-seated tissues, and still others, such as resin ducts and laticifers, occur in internal tissues.

Recent work on secretion in plants has been concerned with the fine structure of secretory cells and the mode of transport of the secretion. Glandular cells often contain large numbers of mitochondria, and may have numerous dictyosomes or well developed endoplasmic reticulum. They are also often characterized by ingrowth of cell wall protuberances (Fig. 11.1), thus resembling, and presumably function-

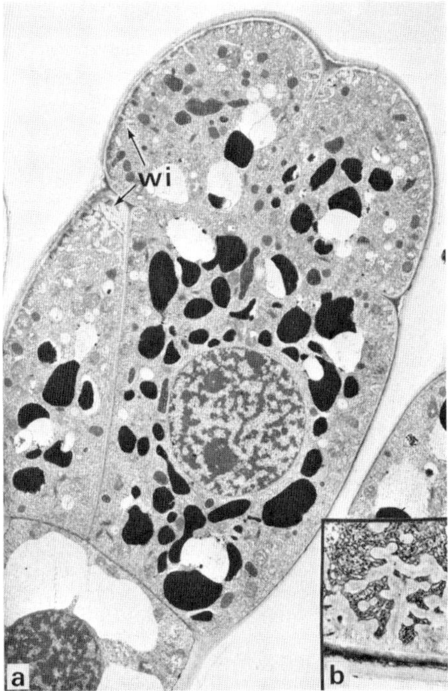

Fig. 11.1 Wall ingrowths (wi) in the secretory cells of an extra-floral nectary trichome of broad bean, *Vicia faba*. (**a**) Head of the trichome. ×1850. (**b**) Enlarged portion of cell wall with ingrowths. ×5000. (By courtesy of Dr. A. J. Browning.)

ing as, transfer cells (see Chapter 10).[426, 579, 580] Examples of various structures involved in secretion are discussed below.

Glandular trichomes

These plant hairs have already been briefly described in Chapter 7. They have a stalk and a head region; the stalk may be unicellular or multicellular, and may even have several rows of cells. The head, which is the secretory part, may also be unicellular (Figs. 7.14d, 11.2) or multicellular (Fig. 11.3). The formation of glandular trichomes on the leaves of *Callitriche* is illustrated in Fig. 11.3. The first step is outgrowth of the wall of an epidermal cell, and unequal distribution of the cytoplasm, which accumulates at the outer end of the cell. Cell division follows, giving a densely cytoplasmic, unicellular head and a vacuolated supporting cell. This is thus another example of the

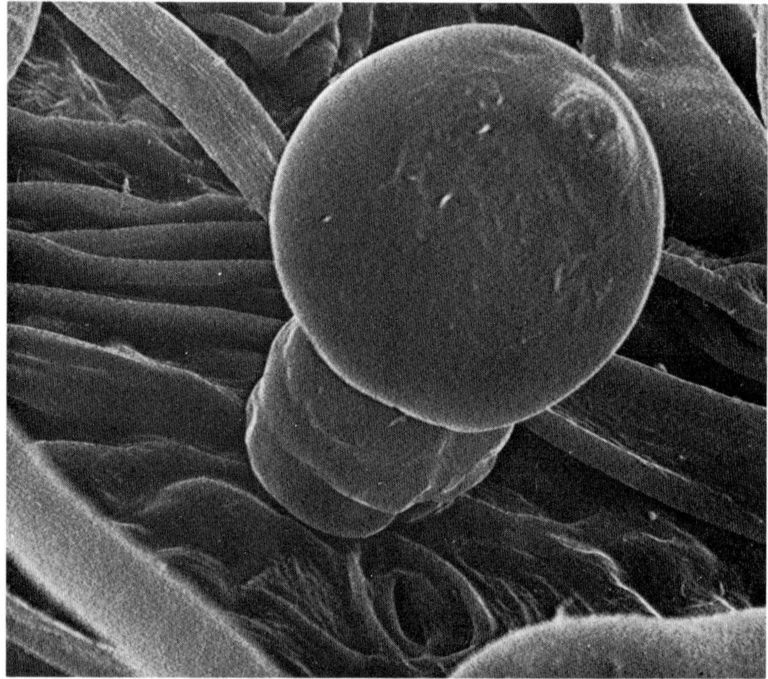

Fig. 11.2 Scanning electron micrograph of a glandular trichome with a single-celled head, from the leaf of *Balsamorhiza sagittata*. ×1500. (From Lott,[423a] *A Scanning Electron Microscope Study of Green Plants*, St. Louis, 1976, The C. V. Mosby Co., Fig. 18, p. 18.)

importance of unequal cell divisions in differentiation (see Chapter 2). Further division of the head cell follows, but the cytoplasm remains more or less uniformly distributed. The cells of the head of glandular trichomes are covered with a cuticle, beneath which the secretion accumulates. The cuticle is apparently permeable. Trichomes of this type secrete volatile oils, e.g. peppermint oil, balsams, resins and camphors. Some variation in form occurs; for example, in mint the head usually has eight cells and has a rather scale-like appearance in surface view. In the trichomes of *Cannabis* the head is rather similar, but the stalk is multicellular; a type of resin is secreted. The hallucinogen tetrahydrocannabinol is in fact thought to be secreted by the glandular trichomes of *Cannabis*, which have been examined recently by several workers using the scanning electron microscope.[149,303,407] There is some disagreement over the number of different morphological types of hair that can be recognized. The

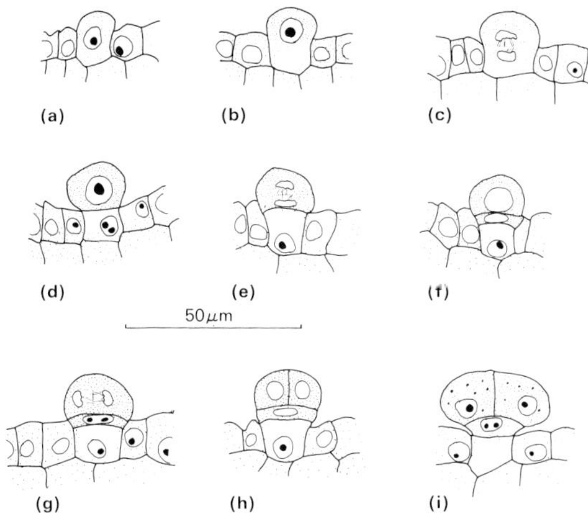

Fig. 11.3 Stages in the development of glandular trichomes on the leaf of *Callitriche*. (**a**) Elongation of an epidermal cell; cytoplasm uniformly distributed. (**b**) Cytoplasm now denser towards the outer region of the cell. (**c**) Telophase in the epidermal cell; cytoplasm unequally distributed. (**d**) Formation of the glandular, densely cytoplasmic, head cell. (**e**) Telophase in this cell. (**f**) Formation of the supporting cell; cytoplasm less dense than that of the head cell, but much denser than that of the mother epidermal cell. (**g**) Telophase in the head cell (prior divisions may have occurred in other planes). (**h**) Formation of 2 head cells in this plane. (**i**) Head region now multicellular. Cytoplasm contains larger particles, perhaps related to the secretory function of the cells. ×430.

massive stalks of one type are thought to originate partly in the hypodermis.[149]

The glandular trichomes of *Pharbitis* originate as enlarged epidermal cells on the leaf or stem, and in less than 36 hours develop a 4-celled head, a foot cell and an epidermal cell which at maturity differs from other, adjacent epidermal cells. Many plasmodesmata connect the foot cell with both the subjacent epidermal cell and the head cells. During the period of secretion the head or cap cells contain many mitochondria and dictyosomes, and an elaborate network of rough endoplasmic reticulum (ER). Contact between both the ER and coated vesicles, produced by the dictyosomes, and the plasmalemma are commonly observed, and it is thought that both the ER and dictyosomes may be involved in transport of the secretion.[680] In these trichomes, and also in those of some other species, histochemical tests show that the secretion contains both protein and carbohydrate.[235, 330, 679, 680] Trichomes with lipophilic secretions, such as oil or flavonoids, have extensive tubular ER.[581]

Some trichomes may be regarded as nectaries or hydathodes (see below), since they secrete nectar or water respectively. The extra-floral nectaries on the stipules of *Vicia sepium* consist of groups of trichomes.[293] Trichomes functioning as hydathodes occur on the leaves of the aquatic plant *Hygrophila*,[540] as well as in other species.

Many glandular trichomes have a basal cell or cells with cutinized or suberized anticlinal walls (Figs. 11.4 and 11.5). It is thought that this prevents apoplastic transport in the cell wall, and ensures passage of the secretion through the cytoplasm, functioning in much the same way as the Casparian strip in root endodermal cells. The presence of many plasmodesmata connecting the secretory cells may facilitate symplastic transport.[426] These suberized walls may block water outflow rather than back-flow of secretion.[580]

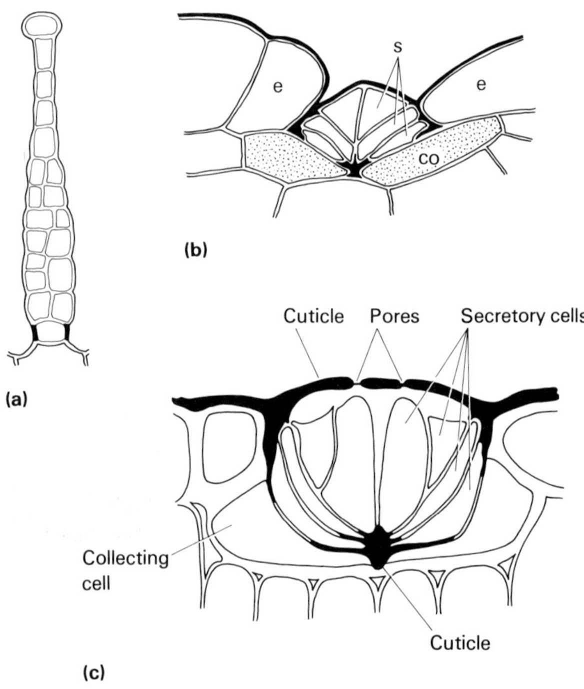

Fig. 11.4 Secretory glands. (**a**) Nectar-secreting trichome from the calyx of *Abutilon* sp.; wall thickenings resembling a Casparian strip shown black. (**b**) Salt gland of *Tamarix aphylla* as seen in cross section, with secretory (s) and collecting (co) cells; e, epidermal cells. (**c**) Salt gland of *Limonium latifolium*. (**a** and **c** from Schnepf,[580] Fig. 9.7G, p. 351 (copyright © 1974 McGraw-Hill Book Co. (UK) Ltd. From Robards: *Dynamic Aspects of Plant Ultrastructure*. Reproduced by permission); **b** from Shimony and Fahn,[595] Fig. 1.)

Several of the secretions of glandular trichomes are repellent to insects, some of which have contact chemoreceptors in their legs by means of which they can detect unpleasant secretions before actually feeding on the plant.[414] In other instances, insects such as aphids may dislodge the heads of glandular trichomes and become trapped by the sticky exudate thus released. This phenomenon is observed both in the tomato and in wild potato species,[265, 357] and attempts have been made to breed cultivated potatoes which possess these useful trichomes.[266]

Insectivorous plants have an enhanced ability to trap insects and possess numerous specialized trichomes of great interest. These contain proteolytic enzymes, and may have evolved from hydathodes. Two kinds of glandular hairs, stalked and sessile, are present on the leaves of *Pinguicula* (Figs. 11.6, 11.7). The stalked glands have a large basal cell, stalk and columellar cell with a head of 8–32, usually 16, cells. Drops of secretion are evident on the head (Fig. 11.7). This mucilage achieves the capture of the insect.[323] Sessile glands have basal and columellar cells supporting a head of 2–8 cells. The columellar cell resembles endodermal cells in that its lateral walls are cutinized (Fig. 11.8).[326]

The mature secretory cells of all insectivorous plants examined have wall ingrowths, resembling transfer cells. In *Pinguicula* these are best developed in the radial walls of both sessile and stalked glands.[324] The heads of the trichomes in *Pinguicula* have no well developed cuticle, and experiments have shown that colloidal lanthanum nitrate penetrates readily;[324] consequently the secretion can also pass out.

Histochemical staining of the glandular trichomes of *Pinguicula* showed that activity of the enzymes acid phosphatase, esterase and ribonuclease was essentially restricted to the glandular trichomes. The main site of activity of these enzymes was in the anticlinal, or radial, walls of the head cells (Fig. 11.9), which are characterized by wall ingrowths.[326] Indeed, the principal sites of stored enzymes are these regions of wall ingrowths in all leaf-trapping species studied.[324] It is thought that the enzymes are moved into these walls and then through a poorly developed cuticle, as in *Pinguicula*, or in other species through pores in the cuticle. When the sessile glands of *Pinguicula* are stimulated, e.g. after the capture of an insect, secretion begins within an hour and the products accumulate in a pool on the leaf surface in 2–3 hours. The fluid has detergent properties and is able to wet the water-resistant exoskeleton of the insect. Other glands which fall within the boundaries of the pool are stimulated to secrete, too. When stained for acid phosphatase, these trichomes stain less densely than the still unstimulated ones beyond the pool of secretion (Fig. 11.10). In experiments with labelled protein it was found that digestion had

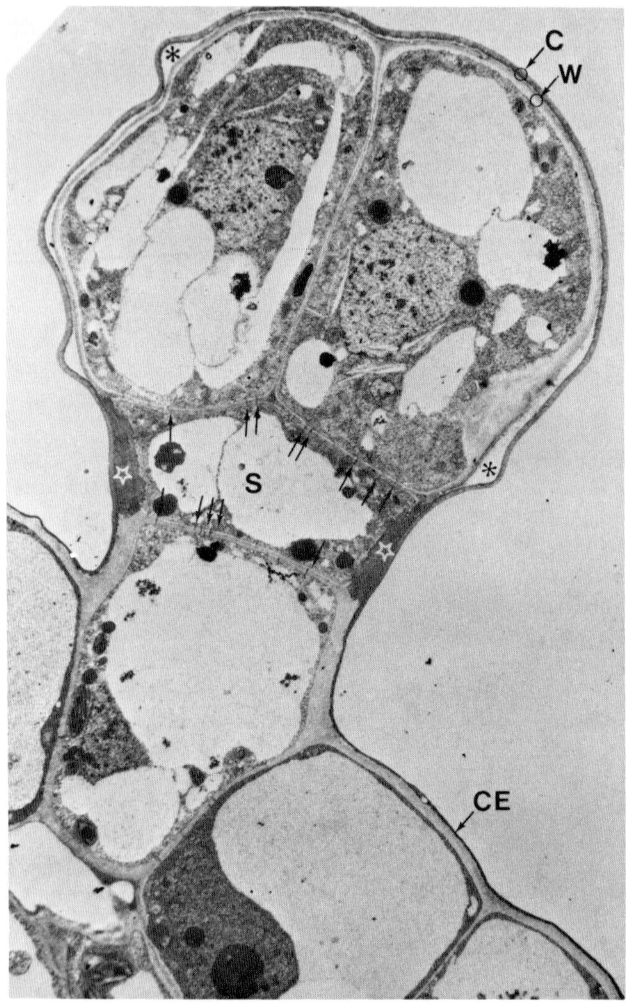

Fig. 11.5 Section of a glandular trichome from a young leaf of *Lamium*, showing the endodermis-like region in the wall of the stalk cell (S). This region (white asterisks) is impregnated with suberin or cutin, and ensures that materials move through the plasmodesmata of the transverse walls (arrows) rather than through the lateral walls. The secretion from the head cells passes through the cell wall (W) and accumulates under the cuticle (C), which becomes raised in places (black asterisks). The cuticle of the epidermal cells (CE) is continuous with that of the glandular trichome. ×3200. (From Gunning and Steer,[286] Plate 10a.)

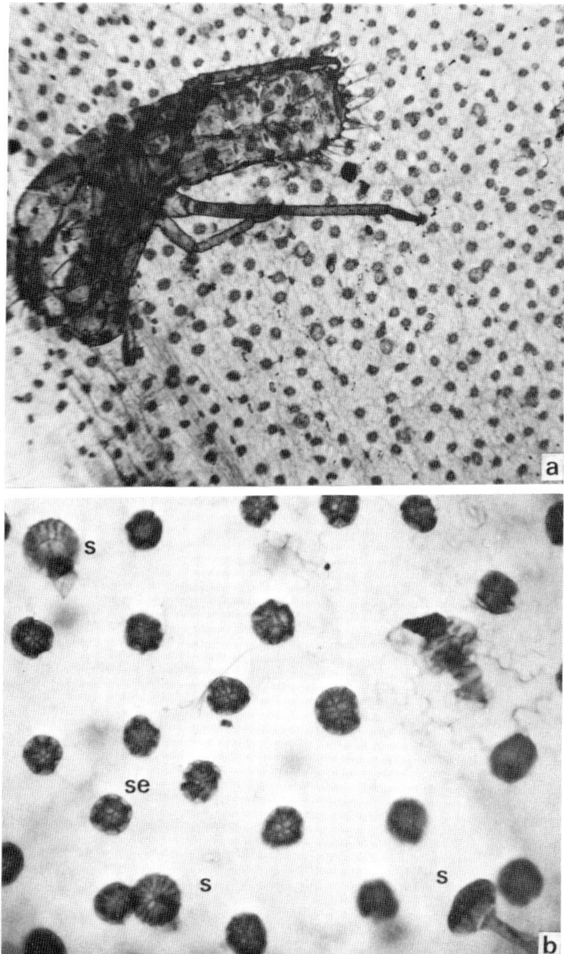

Fig. 11.6 Portions of cleared leaf of *Pinguicula vulgaris*. (**a**) Part of a leaf showing an insect victim that has been trapped by the hairs. ×40. (**b**) Enlarged view of part of the leaf surface, showing the sessile (se) and stalked (s) glandular hairs. The head of the stalked hairs is larger, and has 16 cells. ×120.

begun within 2 hours and the products entered the leaf, apparently by way of the head cells of the sessile glands.[324, 326]

The glandular trichomes of insectivorous plants are particularly interesting functionally because not only does the secretion move out, but also the products of digestion move into the leaf by the same route.

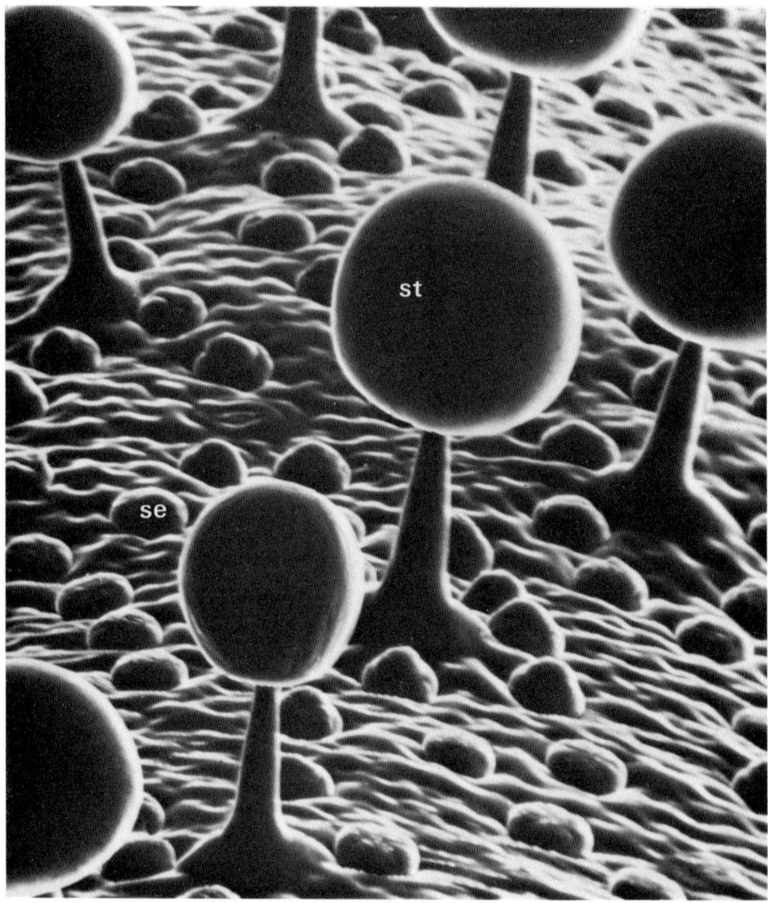

Fig. 11.7 Scanning electron micrograph of part of the surface of a fresh leaf of *Pinguicula grandiflora* with large stalked (st) and smaller sessile (se) glands. The stalked glands have a large mucilage globule over the heads. (After Heslop-Harrison,[323] Fig. 1A. Copyright 1970 by the American Association for the Advancement of Science.)

It has been pointed out that for this reversal to occur loss of control of fluid movement by the cytoplasm of the basal endodermal cell would be necessary.[324]

In *Drosophyllum*, both stalked glands that secrete a mucilage which is effective in capturing insects and sessile ones secreting digestive enzymes are again present. The mucilage is secreted by the outer cells of the stalked glands or tentacles, in which many dictyosomes are

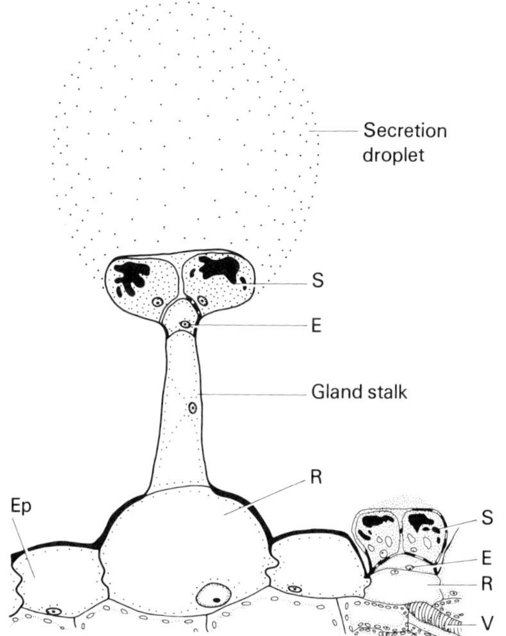

Fig. 11.8 Part of a transverse section of the leaf of *Pinguicula grandiflora*, showing (left) a stalked gland, and (right) a sessile gland. The cuticle (black) is thin over the secretory cells. E, endodermal cell; Ep, epidermis; R, reservoir cell; S, secretory cell; V, vessel connecting with the vascular system of the leaf. (From Heslop-Harrison,[324] Fig. 6.)

present; it is believed to be secreted by these organelles.[573, 574] Studies of the fine structure of the glands of *Drosophyllum*, *Pinguicula* and *Drosera* show that larger and more numerous Golgi vesicles are present in the cells during the period of secretion.[578] Experiments indicate that secretion of the capturing-mucilage in *Drosophyllum* is respiration-dependent,[575] and that the rate of secretion is affected by temperature.[578] The surface area of the secretory cells of *Drosophyllum* is greatly increased by plate-like wall ingrowths which themselves bear protuberances. The cuticle possesses pores through which the mucilage passes.[324, 579, 580] The proteolytic enzymes of *Drosophyllum* glands will digest the gelatin layer of processed colour film, the first holes occurring over the stalked glands.[325]

In the Venus's fly-trap, *Dionaea*, also, the effect of the secreted proteases can be shown by feeding the leaf with small pieces of exposed colour film. The trap closes on the film, the glands are stimulated and

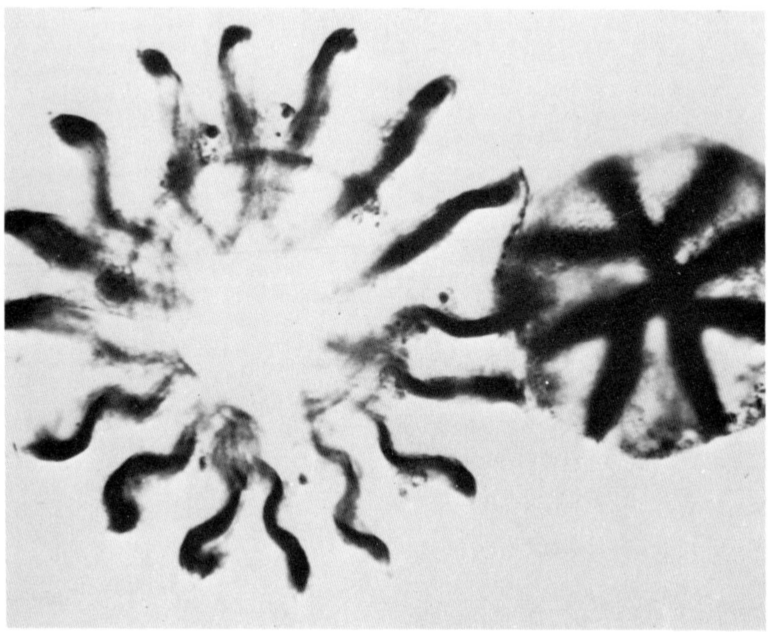

Fig. 11.9 Stalked (left) and sessile (right) glands from an unstimulated leaf of *Pinguicula grandiflora*, stained for the enzyme ribonuclease. Enzyme activity is associated with the radial, anticlinal walls, which have wall ingrowths, and is greater in the sessile glands. ×c. 1000. (After Heslop-Harrison and Knox,[326] Fig. 7.)

the gelatin is digested before the leaf opens again.[325] Closure of the leaf is brought about by electrical signals originating in the multicellular sensory hairs, each of which consists of distinct basal and distal regions.[347] In *Dionaea* digestion begins within 36 hours of closure of the trap. The cell walls of the resting digestive gland have many protuberances, which decrease in size during later stages of digestion.[584]

The familiar tentacles of the leaves of sundew, *Drosera* spp., may be considered as large stalked glands which are not entirely of epidermal origin. These structures have a multicellular stalk, containing a bundle of tracheids and a head with three or four layers of cells covered by a cuticle.[423] The cuticle is perforated by a number of pores over the gland cells.[532] An endodermal layer, the cells having a Casparian strip, surrounds the central tracheid mass. The inner stalk cells have the characteristics of transfer cells.[366] Many plasmodesmata are present between cells in the head region.[532] Transport of calcium from applied

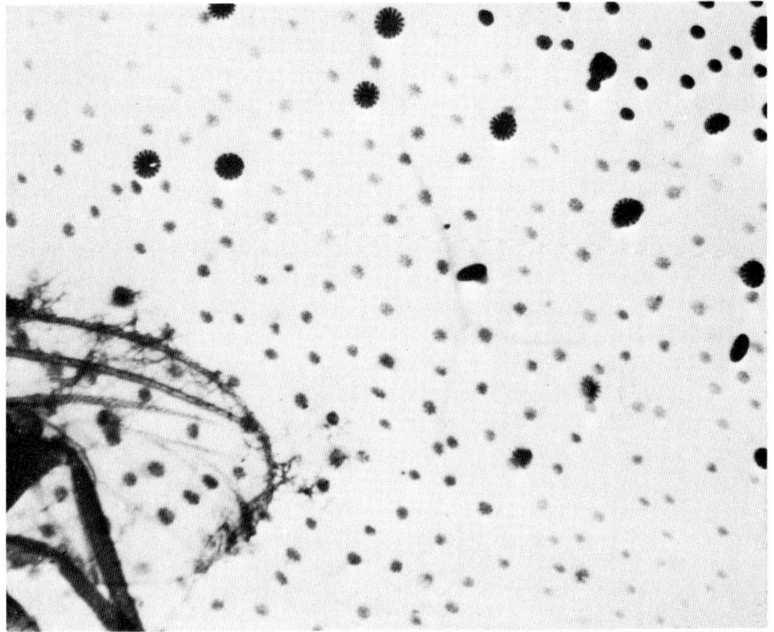

Fig. 11.10 Part of a leaf of *Pinguicula grandiflora*, with a captured insect (left), stained for acid phosphatase. Glands adjacent to the fly have been stimulated; those beyond the area of the secretion pool (top right) still show strong enzyme activity and have not yet been stimulated. ×50. (From Heslop-Harrison and Knox,[326] Fig. 16.)

cow's milk has been studied using a special analysing microscope. Calcium is detectable within one minute in the gland cells, and appears to be absorbed by them and transported in the apoplast as far as the endodermis, which it traverses in the symplast. On the basis of these observations it is believed that resorption of the products of digestion can occur against the direction of flow of mucilage and digestive enzymes.[366] It is thus not surprising that the mechanisms are quite complex.

The pitchers of *Nepenthes* possess several different kinds of glands, which seem to differ more in position and in their supposed function than in structure. Multicellular 'alluring' glands, of epidermal origin, are present on the under surface of the lid of the pitcher and elsewhere; they secrete nectar. Numerous digestive glands having the same structure are present on the inner surface of the pitcher wall, and glands with a possible digestive and absorptive function are present towards the base of the pitcher.[423]

The glands serve as short-distance transport between the pitcher fluid and the tissues of the pitcher wall.[426] The multicellular glands below the level of the fluid again secrete acid phosphatase and esterase. The pitchers of one species of *Nepenthes* are said to contain as much as one litre of fluid.[324]

The mechanism of the bladder-like traps of species of *Utricularia* (bladderwort) also depends on glandular structures. Stalked and sessile glands on the outside of the trap secrete mucilage and sugar, which may attract animals. Short-stalked glands on the inside of the bladders extract water from the contents, which creates a tension. If an animal then happens to touch the four stiff bristles on the outside of the door of the trap, it opens slightly, releasing the tension and causing an inrush of water. The victim passes in with the water and the door then closes again.[587] In addition to three types of mucilage-secreting trichomes near the entrance to the bladder, bifid and quadrifid secretory hairs occur on the lower lip and the inner wall of the utricle.[664]

It is thus evident that glandular trichomes attain their greatest degree of specialization of both form and function in carnivorous plants.

Hydathodes

Hydathodes occur on leaves, and are usually associated with the marginal teeth or serrations, or the tip. They secrete water and discharge it in liquid form by a process known as guttation. In the case of the so-called active hydathodes, the energy involved in secretion is supplied by the glandular cells themselves. If the surface of the leaf is painted with an alcoholic solution of mercuric chloride, which kills the living cells of the hydathode, no liquid emerges from the treated hydathodes even if water is forced into the leaf under pressure.[293]

During the development of a hydathode procambium differentiates towards the lobes or serrations of the leaf, which may undergo some swelling. The cells adjacent to the procambium may proliferate to give rise to the ***epithem***, a tissue of small, thin-walled, rather densely cytoplasmic cells with an extensive system of intercellular spaces. The cells of the epithem usually contain no chloroplasts,[642] unlike the adjacent mesophyll cells of the leaf. In the epidermis over-lying the epithem the mother cells of the water pores appear; these divide to give two guard cells, sometimes undergoing earlier divisions to give rise to subsidiary cells.[540] The guard cells of the water pores associated with hydathodes lose the power of controlling the aperture of the stoma.

New water pores may form later between those already present. In some species, only one water pore is associated with each hydathode, but usually there are several and the number may be fairly constant within a particular species.

The cells of the epithem are in close contact with the terminal tracheary elements, usually tracheids, of the vein ending, and water moves through the numerous intercellular spaces of the epithem and is discharged through the open water pores. The epithem may be bounded by cells with Casparian strips.[220] Hydathodes, perhaps especially those of aquatic plants, seem to function during only part of the life of the leaf.[464,642]

Salt and chalk glands

There is no clear distinction between hydathodes and salt glands.[580] The latter usually occur in halophytes, and presumably function in the removal of excess salt. The submerged marine angiosperm *Thalassia*, turtle grass, does not have salt glands, but it is thought that all the leaf epidermal cells, which have a highly invaginated plasmalemma, may function in osmoregulation.[350]

The simplest salt glands are those of the maritime grasses *Spartina* and *Chloris*, which consist of two cells, a large basal cell and a smaller cap cell.[413,419] Salt glands of *Limonium*, *Tamarix* and *Avicennia* are characterized by collecting and secretory cells, as well as a stalk cell in some instances (Fig. 11.4b, c).[595,596,660] The stalk cell or basal cell has suberized lateral walls (Fig. 11.4), again apparently functioning as a Casparian strip. The walls of the secretory cells, and sometimes adjacent cells, are characterized by protuberances of the transfer cell type.[413,660,745] The cuticle over salt glands usually has one or more pores over each cell (Fig. 11.4c).[413,580,595,596,745]

In a study of the localization of sodium and potassium using antimonate precipitation and electron microprobe analysis, a gradient from a high concentration in cells near the xylem decreasing progressively to the outermost cells of the complex glands of *Avicennia* was observed. This would be consistent with diffusion of ions through the symplast from the leaf mesophyll to the secretory cells, perhaps from a pump in the region of the xylem parenchyma.[596]

In the Plumbaginaceae, including the sea thrift, *Armeria maritima*, chalk-secreting glands occur on the leaves. In *Plumbago* these comprise 4 secretory cells encircled by 4 subsidiary cells. Each secretory cell has a surface depression which apparently corresponds to the pores of salt glands (Fig. 11.11). The secretion accumulates over these pores.

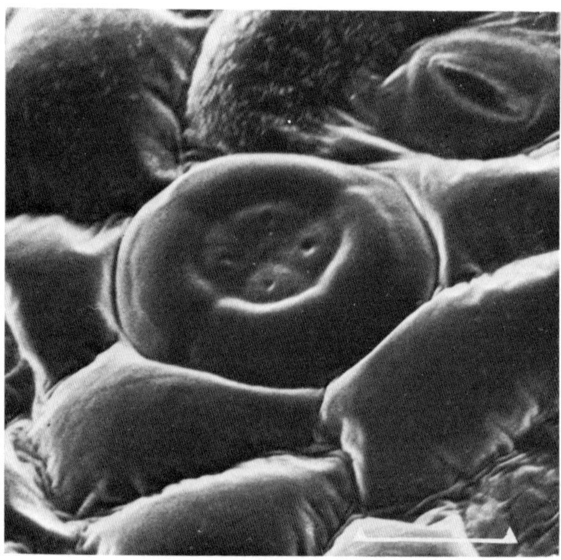

Fig. 11.11 Scanning electron micrograph of a young chalk gland of *Plumbago capensis*. Pores in the surface of the secretory cells are evident, but no secretion is as yet present. ×2000. (From Sakai,[566] Fig. 2, p. 95.)

Analysis of the secretion by X-ray analysis indicated the presence of calcite ($CaCO_3$) and nesquehonite ($MgCO_3.3H_2O$). Chalk glands are probably not completely distinct from salt glands.[566]

Nectaries

Nectaries are most commonly associated with parts of the flower, though the so-called 'extra-floral' nectaries may occur on stems or parts of leaves. In cotton, a nectary consisting of external multicellular papillae and internal subglandular tissue is situated on the abaxial side of the leaf.[704] Nectaries secrete nectar, a sugary fluid attractive to many insects. As in the case of hydathodes, tissue lying below the epidermis may be associated with the epidermis in forming the nectary. The cells of the nectary are usually densely cytoplasmic. Apparently passage of the nectar to the exterior may occur by diffusion through the cell walls, by rupture of the cuticle, or through stomata present in the epidermis.[219] In nectar-secreting trichomes, the nectar accumulates between the secretory cells and the cuticle.[225,447] In *Abutilon*, patches of nectar-secreting trichomes occur on the inner side of the sepals (Figs. 11.4a and 11.12a, b).[278] The stalk cell of these trichomes has a

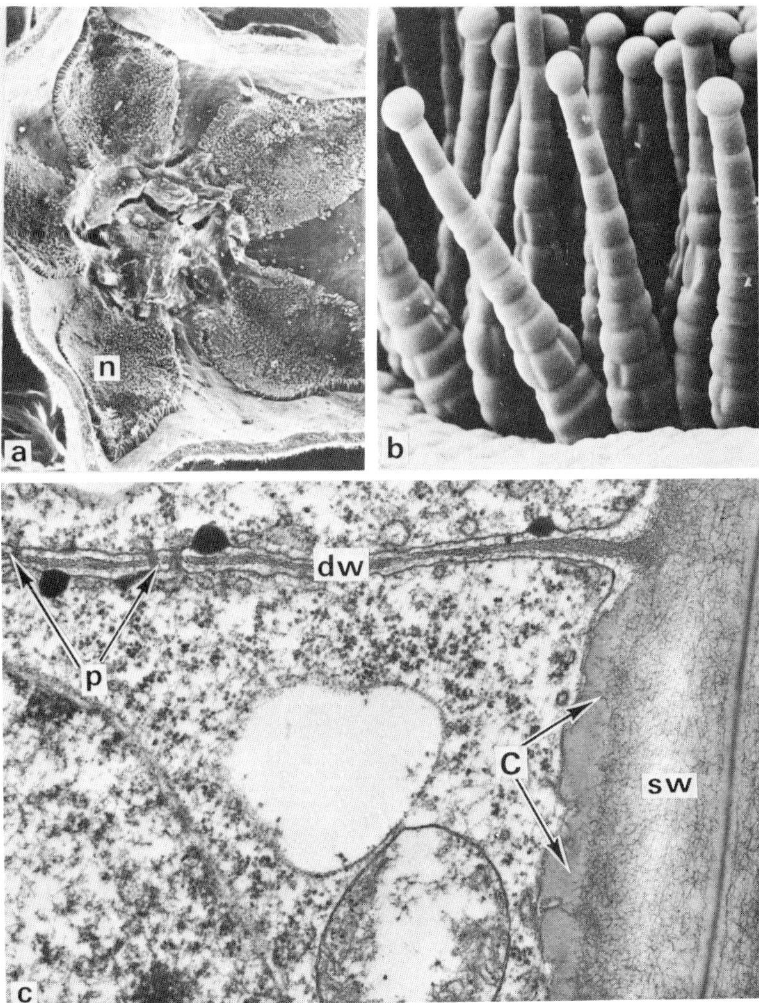

Fig. 11.12 Nectary trichomes of *Abutilon*. (**a**) Scanning electron micrograph of the nectaries (n), which consist of groups of trichomes, on the inner side of the bases of the sepals. (**b**) SEM of some of the nectar-secreting trichomes in side view. Note the hemispherical apical cells, from which the nectar emerges. ×600. (**c**) Transmission electron micrograph of part of the side wall (sw) of the stalk cell of a trichome, showing the 'Casparian strip equivalent' (C, arrowed). Part of the distal wall (dw) of the stalk cell is included and shows plasmodesmata (p). ×37 000. (See also Figs. 11.4**a** and 11.5.) (By courtesy of Professor B. E. S. Gunning; **b** from Gunning and Hughes,[278] Fig. 2, p. 622.)

Casparian strip-like region in the side wall (Fig. 11.12c), which constitutes an apoplastic barrier, so that the pre-nectar must pass by symplastic transport through the plasmodesmata of the distal wall (Fig. 11.12c; see also Chapter 3).[278] The nectar is eventually ejected to the exterior through the cuticle of the apical cells of the multicellular hairs every few seconds or minutes, and the cuticle snaps back on to the head cells. Pores in the cuticle open when a positive hydrostatic pressure builds up.[447]

Nectaries secrete sugar supplied by the phloem. The secretion differs from sieve-tube sap to a variable extent according to the degree of anatomical specialization of the nectary.[580] Nectar secretion often occurs for a very limited period, perhaps a single day at a particular stage of floral development. The glandular cells are usually characterized by wall protuberances.[225, 531] In the septal nectaries of *Gasteria*, occurring on the ovary, the wall protuberances extend some distance into the cells (Fig. 11.13); they are formed just before the secretory phase and are removed after it is over.[580]

In studies of nectar-secreting cells with the electron microscope,

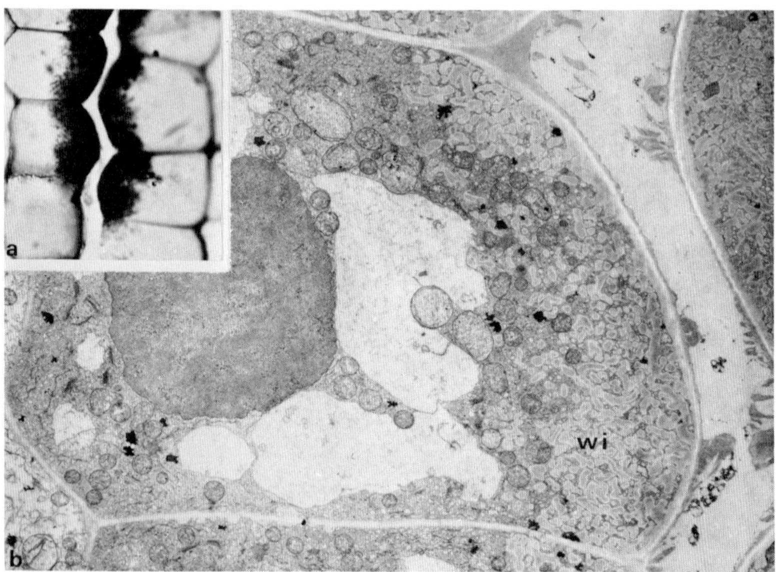

Fig. 11.13 Sections of nectaries with pronounced wall ingrowths at the secretory surface. (**a**) Light micrograph of a section of the nectary of *Eucharis grandiflora*, showing the extensive region of wall thickenings (darkly stained). ×735. (**b**) Electron micrograph of the septal nectary of *Gasteria trigona*, showing wall ingrowths (wi). ×3330. (**a**, from Schnepf,[576] Fig. 2, p. 141; **b**, from Schnepf,[577] Pl. 13, Fig. 7.)

changes in cellular components can be observed during the secretory phase. Dictyosomes are thought not to be the principal organelles involved in nectar secretion.[459] In the nectaries of several different species the amount of endoplasmic reticulum (ER) increases markedly before the period of secretion, so that it becomes the dominant cell component. In later stages of secretion the ER is mainly represented by distended vesicular elements; in nectaries which have ceased secretion none of these is present. It is thought that these vesicles may contain a sugar solution and secrete it by fusion with the plasmalemma.[225, 530, 531]

This view is supported to some extent by experiments in which excised flowers were fed with sucrose labelled with tritium. The nectar of flowers fed with tritiated sucrose for 15 minutes was analysed in a scintillation counter and found to be radioactive, and autoradiographs indicated that most of the label was in the ER.[226]

The amount and type of secretion of floral nectaries of *Antirrhinum* were affected by supplying the auxin indoleacetic acid (IAA) to excised flowers. If supplied in water or dilute sucrose solutions, IAA led to an increase in the volume of fluid secreted. When supplied in 10% or 20% sucrose solutions, IAA treatment effected a reduction of 25–30% in the nectar sugar. It was not clear whether IAA affected the secretory mechanism itself, or the transport of sugar to the nectary.[598]

Analysis of the sugar content of nectar has shown that the major components are sucrose, glucose and fructose.[69, 175] Recent work has also shown that nectar contains amino acids. In a survey of nectar from 266 species, all the essential amino acids were found in the nectar of one species or another, at least twelve different amino acids occurring in the nectar of *Dianthus barbatus*. Flowers which had nectar that was richest in amino acids were those that attract carrion and dung-flies by their smell or aspect, or those visited by butterflies, and it is suggested that the nectar may be a significant source of amino acids for these insects.[38]

Studies with radioactive isotopes indicate that the cells of nectaries can not only secrete, but also absorb, a sugary fluid;[506] whether such a process occurs in nature appears doubtful. However, this indicates that these secretory cells, like those of carnivorous plants, can undertake bidirectional transport. In the experiments the fluid applied was readily translocated to other parts of the plant; this ease of movement is undoubtedly linked with the fact that the vascular tissue associated with nectaries secreting a concentrated sugar solution may consist only, or predominantly, of phloem,[187] contrasting with the strands consisting entirely of xylem which are associated with hydathodes.

INTERNAL SECRETORY STRUCTURES

Secretion may be carried out internally by single cells, by small groups of cells, or by a whole tissue. Cells which secrete oils or enzymes may be distributed throughout a tissue, as in the endosperm of *Ricinus* and the cotyledons of *Arachis*, the sources of castor oil and ground nut (peanut) oil respectively. Other important oils are palm oil, extracted from the fleshy mesocarp of the fruit of *Elaeis guineensis*, and safflower oil, an unsaturated oil derived from the seeds of *Carthamus tinctorius*. In the shoot of *Ricinus*, secretory cells may differentiate in meristematic tissue at the apex, or much later, associated with secondary tissues.[59]

In some tissues, oil- or resin-secreting idioblasts occur. Examples of these are the oil cells in the phloem of *Cinnamomum zeylanicum*, which secrete the aromatic cinnamon oil, and the oleo-resin cells in the ground tissue of the rhizome of ginger, *Zingiber officinale*. Crystal-containing cells are also sometimes considered to have a secretory or excretory function. Occasionally, a whole layer of specialized secretory cells is present. For example, a layer of cells just within the testa secretes an aromatic oil in the seeds of cardamom, *Elettaria cardamomum*. The importance of the aleurone layer in the grains of cereals in secreting the enzyme α-amylase has already been discussed in Chapter 3.

In the phloem of *Mimosa pudica*, secretory cells form longitudinal files. Study with the EM shows that during ontogeny the protoplast becomes disorganized, and the end walls become perforated. The secretion is interpreted as a carbohydrate, and osmiophilic particles are also present.[195]

Glands and ducts

In many instances small groups of thin-walled, densely protoplasmic cells which are a part of another tissue have a secretory function. Their secretions collect in an internal cavity, which may be more or less isodiametric (gland) or considerably elongated in one plane (duct). These cavities may originate either by a splitting apart of the cells at the middle lamella (schizogenous glands) or by actual breakdown or lysis of some of the cells (lysigenous glands). In some instances, schizogeny may be followed by lysis (schizolysigenous glands). In the case of schizogenous glands, a ring or lining of intact cells, the epithelium, surrounds the cavity, the boundary of which is well defined by their cell walls; lysigenous or schizolysigenous glands have no such clear-cut boundary.

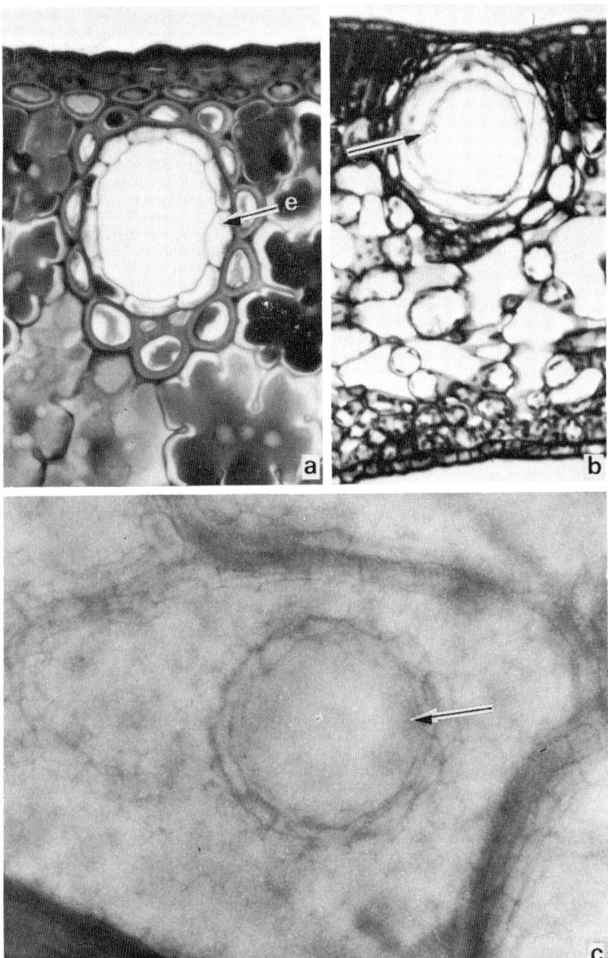

Fig. 11.14 Schizogenous and lysigenous secretory cavities. (**a**) Schizogenous resin duct from transverse section of leaf of *Pinus resinosa*. e, epithelium. ×300. (**b**) Transverse section of leaf of *Citrus* sp. with lysigenous oil gland. ×200. (**c**) Cleared leaf of *Citrus* sp., showing an oil gland in optical section. The views shown in **b** and **c** together demonstrate that the gland is isodiametric. ×200. In (**b**) and (**c**) the cells around the cavity which are undergoing lysis are arrowed.

Lysigenous or schizolysigenous glands are present in the leaves and fruits of *Citrus* spp. (Fig. 11.14b, c). The glands occur, for example, in orange and lemon peel, in the outermost part of the pericarp.

Schizolysigenous oil glands are also found in the floral parts of clove, *Eugenia caryophyllata*, and are the source of clove oil.

The schizogenous oil glands of *Eucalyptus* illustrate the difficulty of classifying glands as internal or external, since they originate from a single epidermal cell, though a subepidermal cell contributes to the 'casing', or surround, of the gland. The epidermal cell divides to give three tiers of cells (Fig. 11.15), the innermost of which gives rise to the epithelial cells and some of the casing cells (Fig. 11.15a, b). It has long been a controversial matter whether these glands were schizogenous or lysigenous, but newer techniques of embedding material in resin have preserved the epithelial cells better and shown that the space into which the oil is secreted is formed schizogenously, i.e. arises as an intercellular space (Fig. 11.15c, d).[92] In *Eucalyptus* the oil glands are formed in the embryo at about the stage of cotyledon formation. During their formation they are centres of meristematic activity, and indeed are distributed rather regularly, like meristemoids (see Chapter 2).[92] In more mature material of *Eucalyptus*, it was found that oil glands or ducts occurred in the pith of some species, and in the secondary phloem or bark of others, and that this could prove a useful taxonomic feature.[93]

The secretory ducts of *Rhus glabra* also develop schizogenously. It is thought that lipophilic material may be synthesized or transported (or both) by all organelles. The gum-resin is formed from secreted osmiophilic droplets and the disintegration of cell walls facing the duct lumen.[223]

Resin ducts

The resin ducts of conifers and other genera are examples of schizogenous ducts (Fig. 11.14a). A study of the resin ducts of *Pinus* with the electron microscope[732] showed that the epithelial cells have many more relatively undifferentiated plastids than adjoining cortical cells. In the epithelial cells the plastids are sheathed with endoplasmic reticulum, and it is believed that this ER may play a role in transporting the resin to the duct. In conifers the ducts usually form a rather complex vertical and horizontal system within the plant, and may be branched. In *Pinus halepensis* the inner end of each radial duct is connected to a vertical duct in the secondary xylem, and the lumina of the ducts are continuous. Vertical ducts develop from the fusiform initials of the cambium, and radial ducts from the ray initials.[705] It is interesting that isolated portions of the resin ducts of *P. halepensis* dissected out of the wood are still capable of secreting resin.[740] Resin ducts occur in species from many families, but only a few species have

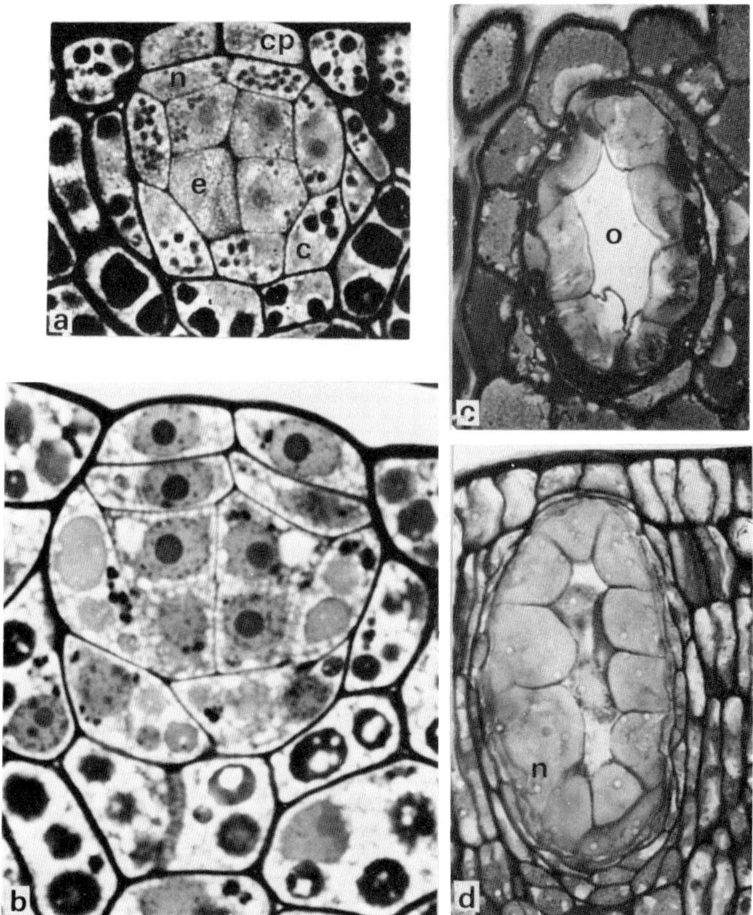

Fig. 11.15 Schizogenous oil glands in the cotyledons and young leaf of *Eucalyptus* spp. (**a**) Complete gland in cotyledon of the mature seed of *E. phoenicea*. Epithelial cells (e) are present centrally, with casing (c), neck (n) and cap (cp) cells peripherally. ×900. (**b**) Oil gland in T.S. cotyledon of *E. St. Johnii*. ×1500. (**c**) More mature gland (*E. nesophila*) with central oil cavity (o). The epithelial cells are intact. ×550. (**d**) Mature oil gland from T.S. seedling leaf of *E. brachyphylla*. n, nucleus of an epithelial cell. ×750. (By courtesy of Professor D. J. Carr; (**a**) and (**c**) from Carr and Carr,[92] Figs. 25 and 40, pp. 198 and 204.)

resins of commercial importance. Frankincense and myrrh are soft resins produced by members of the Burseraceae. In some plants, resin ducts apparently develop in response to injury, and wounding is consequently sometimes practised in the commercial collection of resins. In *Pinus halepensis* experimental wounding led to a localized increase in the number of vertical ducts. Several auxins, applied to stems from which part of the periderm had been removed, also stimulated the formation of vertical ducts.[229] More work is required on the factors which control duct formation.

Resin is sometimes regarded as a chemical defence mechanism, since it promotes wound healing and is repellent to animal marauders.[174] An interesting interaction between animals and plants has been reported, in which the sawfly larvae which feed on conifers such as *Pinus sylvestris* have developed the ability to sequester and store the plant's resin and use it in their own defence. When disturbed or poked, the larva rears up, emits a drop of fluid and dabs it on the intruder. The fluid contains all the components of *Pinus* resin, and proves an effective defence against ants and spiders and a deterrent to birds preying on the larvae.[174]

Laticifers

Perhaps the most important of all plant secretions is latex, a fluid secreted by specialized cells or groups of cells known as laticifers. Laticifers are unique among plant cells, and their structure and growth have many features of interest. Latex is found in representatives of about twenty plant families.[449]

Latex is most commonly milky white in colour, but may be clear and colourless, or yellow to orange. It may contain carbohydrates, organic acids, alkaloids, etc., in solution, and also various dispersed particles, including terpenes, oils, resins and rubber. Starch is also sometimes present. In *Euphorbia milii* the starch grains in the laticifers are elongated or dumb-bell shaped, whereas those in adjacent parenchyma cells are oval (Fig. 11.16). This raises the problem of how the starch is synthesized in both types of cell. Studies of the starch grains with the scanning electron microscope have shown that several forms of starch grain occur in the laticifers of species of *Euphorbia*, and that their morphology is species specific and controlled by the laticifer.[435,436] It is not understood what controls the pattern of starch deposition and the resulting form of the grain.

The latex of *Carica papaya* is rich in the proteolytic enzyme papain,[709] and many other enzymes may be present in latex of other species.[20] The medically important alkaloid opium is present in the latex of *Papaver somniferum*. Small particles found in the latex of *Hevea* after centrifugation, considered to be possibly a kind of plastid, suggest that latex is composed of cytoplasm rather than vacuolar

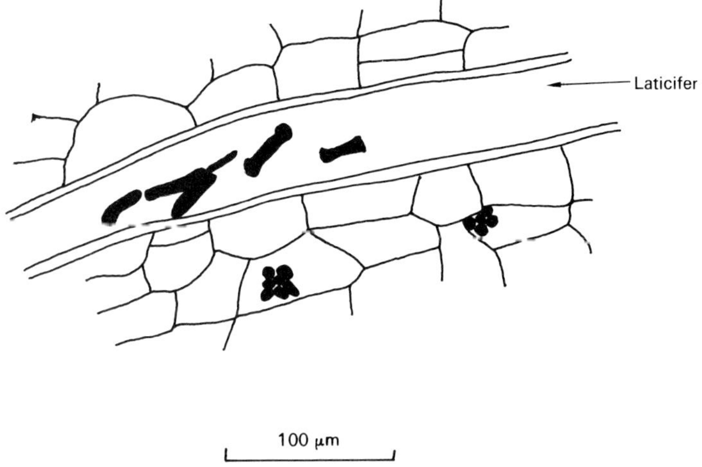

Fig. 11.16 Longitudinal section of stem of *Euphorbia milii*, showing the dumb-bell shaped starch grains in the laticifers, and the oval starch grains in adjacent parenchyma cells. Starch grains black. ×230.

sap.[621] Preliminary studies of the fine structure of the laticifers of *Achras sapota* reveal both electron-transparent organelles, which later degenerate, and electron-dense organelles, which constitute the main part of the latex.[569] In *Papaver somniferum*, many vesicles are present in the laticifers, possibly derived from the ER. It is thought that the dense material which lines the vesicle membrane may be alkaloids. Cytoplasm persists at the periphery of the laticifer and contains the usual organelles.[662] In *Nelumbo nucifera*, also, a thin layer of cytoplasm remains, although the protoplast of the laticifer partially disintegrates.[208] It is thought that the vacuoles in mature laticifers may be autophagic, and probably digest a considerable portion of the cytoplasm.[444]

Centrifuged latex separates into heavy vesicles, with associated alkaloids, and the supernatant.[161,231,444]

Analysis of the alkaloids morphine, codeine and thebaine in the latex of *Papaver somniferum* showed variation in amount, suggesting that the alkaloids are not just waste products but are actively metabolized.[232] In poppy fruits fed with ^{14}C-labelled tyrosine, an amino acid, a rapid turnover of alkaloids and the periodic disappearance of morphine were noted.[233] When labelled tyrosine was mixed *in vitro* with latex isolated from poppy capsules, radioactivity was incorporated into the three alkaloids mentioned above within 10 minutes, and again the 'disappearance' of morphine was observed, suggesting that it was incorporated into a non-alkaloidal substance in the latex.[234] Expelled latex is capable of complex alkaloid biosynthesis, perhaps because of the presence of viable mitochondria and other organelles.[161,231] Rubber particles are synthesized in the cytoplasm of the laticifers of *Ficus elastica*.[311] These observations support the view of electron microscopists that latex is cytoplasmic and possesses viable organelles.

Laticifers may be divided into two main types, articulated and non-articulated. ***Non-articulated laticifers*** originate from single cells, which are apparently capable of potentially unlimited growth. By their continued growth these cells develop into long tube-like structures, which may branch but usually do not undergo anastomosis (Fig. 11.17). In *Nerium oleander*, the cells which will differentiate as laticifers become evident during the late globular stage of embryogeny. They appear in an irregular ring just below the corpus of the shoot apical meristem and differentiate from cells on the periphery of the future procambium.[433] Usually 28 such cells differentiate (Fig. 11.18). The laticifer initial cells grow more rapidly than neighbouring cells, and their nuclei also enlarge and subsequently divide without accompanying wall formation. Development of the laticifers in *Euphorbia marginata* is broadly similar, except that only 12 initials are formed in the embryo of this species.[438]

The developing non-articulated laticifer thus becomes a multinucleate cell; it is also characterized by extremely rapid growth. All mitotic figures are oriented in a plane approximately parallel to the long axis of the laticifer. There is a total absence of cell plate formation.[431] Nuclei often occur in groups of two or more,[94] and many nuclei are usually present in the youngest part of the cell.[23] Apparently a wave of mitoses passes through the laticifer, so that nuclei are at successive stages of division along a region of the cell. This suggests that a mitotic stimulus may pass along the cell axis;[437] the nature of this stimulus is not known.

The non-articulated laticifers of *Nerium* branch repeatedly, and the laticifers of the shoot system are continuous with branches formed in the cotyledonary node of the embryo. These cells thus ramify through the tissues, by means of rapid and predominantly apical growth; the rate of growth of the various branches may not be uniform.[434] Laticifers grow intrusively, occupying the intercellular spaces, and apparently general cell elongation is involved as well as tip growth.[432] In some species, non-articulated laticifers are unbranched.

It has recently been shown that the enzyme pectinase is present in abundance in the latex of milkweed, *Asclepias syriaca*. It is thought that the intrusive growth of the ramifying laticifers of this species is

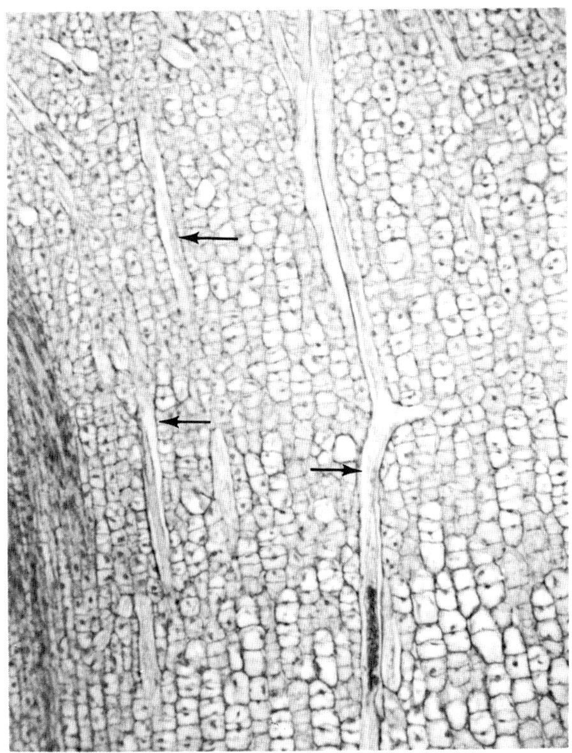

Fig. 11.17 Part of a longitudinal section of a shoot tip of *Nerium oleander*, slightly below the shoot apex, showing the non-articulated, branched laticifers (arrowed) ramifying through the tissue. ×150.

facilitated by secretion of pectinase by the growing tip of the laticifer, resulting in pectolytic dissolution of the middle lamella of adjacent cells.[705] Intrusive growth of fibres has been attributed to a somewhat similar phenomenon (see Chapter 6).

Laticifers usually have fairly thick walls, except at the extreme tip, thicker than those of neighbouring cells. The wall is formed by apposition, by a process of multi-net growth[460] (see Chapter 4). Experiments indicate that laticifers may have some mechanism for the closure of a wound—thus retaining the latex—but this may merely be coagulation of latex particles.[622]

Articulated laticifers are compound in origin, consisting of longitudinal files of cells, the end walls of which break down wholly or in part. In *Hevea brasiliensis*, the major source of rubber, laticiferous tubes or vessels occur associated with the phloem of the vascular bundles of the embryo in late development. Perforation of the lateral walls is apparently more advanced than that of the end walls, and a complex anastomosing system becomes established early in ontogeny.[585] In *Achras sapota*, disappearance of the end walls is also gradual.[370] Observations of the laticifers of *Nelumbo nucifera* with the electron microscope confirm the presence of single perforations in the end walls of the laticifers.[208]

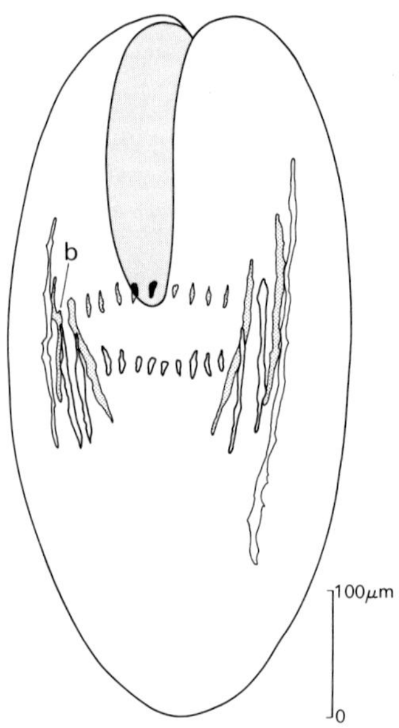

Fig. 11.18 Reconstruction of the laticifer tissue in an immature embryo of *Nerium oleander*. Twenty-eight laticifer initials of variable length are distributed along the periphery of the vascular tissue in the cotyledonary node. b indicates a branch. (From Mahlberg,[433] Fig. 1, p. 91.)

In *Papaver somniferum*, the opium poppy, laticifers are absent from the embryo and young seedling;[230] but differentiate within a day or two of germination.[662] In *Taraxacum kok-saghyz*, also, laticifers are not present in the embryo, but differentiate in the pericycle of the primary root soon after germination. Resorption of the end walls occurs, and also lateral anastomosis.[22]

Laticifers may be fairly generally distributed in the tissues, but seem to be most commonly associated with the phloem. In *Taraxacum* root, the laticifers are located in the secondary phloem. Concentric cylinders of laticifers and sieve tube elements alternate with cylinders of parenchyma; no anastomoses take place between laticifers from different cylinders.[22] In *Hevea*, also, the main laticiferous system is in the secondary phloem.

ECONOMIC IMPORTANCE OF PLANT SECRETIONS

Many plant secretions are of commercial importance, but none more so than latex, because of the substances that it contains. Of these, rubber is the most important. The Indians of the Amazon region utilized rubber for making containers, balls for playing, etc., in quite early times; the earliest record is about 1510, but probably it was used considerably earlier.[709] The principal source of rubber is *Hevea brasiliensis*, which can yield about 2000 lb rubber per acre (2240 kg per hectare) per year; yields from *Taraxacum kok-saghyz* are only about 100 lb per acre (112 kg per hectare) per year, though yields of 400 lb per acre (448 kg per hectare) have been reported.[20] 175 lb per acre (196 kg per hectare) can be obtained from *Cryptostegia grandiflora*.[449] The production of rubber in *Hevea* can be stimulated by application of hormones, and can also be affected by various minerals; in guayule (*Parthenium*), in which rubber occurs principally in the cells of the vascular rays, not in laticifers,[21] low night temperatures increased rubber production.[20] How these factors affect the physiology of the laticifers remains to be determined.

Gutta-percha is obtained by coagulating the latex of species of *Palaquium*. It has been used by the peoples of Borneo and neighbouring regions since early times to make containers and tools. At the present time it is employed in the manufacture, among other things, of dentures, golf balls, and underground and underwater cables; up to 1914, some 300 000 miles (480 000 km) of submarine cable had been covered with gutta-percha.[722] The latex of another member of the same family (Sapotaceae), *Achras sapota*, is the original source of chicle, from which chewing gum is made. The importance of opium and its derivatives in medicine needs no emphasis.

Many oils secreted by plants are also of considerable economic value. Olive oil is widely used in a variety of ways. Safflower oil is becoming increasingly important, because of its unsaturated nature; it is utilized in the production of margarine and cooking oil. Palm oil is also used in margarine, as well as in the manufacture of soap and candles.

12

Vascular Cambium and Periderm

VASCULAR CAMBIUM

During the process of secondary growth a lateral meristem, the vascular cambium, comes to form a thin cylinder of cells surrounding the primary xylem in both the stems and roots of most dicotyledons and gymnosperms. The cylinder of cambial cells, once complete, divides predominantly periclinally, forming a cylinder of secondary phloem towards the outside of the organ, and a cylinder of secondary xylem towards the inside. The process of secondary growth in roots and stems is discussed in Part 2,[136] Chapters 2 and 4. Most workers consider that the true cambium consists of a single layer of initials, the adjacent undifferentiated xylem and phloem mother cells combining with it to form the cambial zone. If there were several initiating layers, it is difficult to imagine how exact radial continuity of xylem and phloem mother cells could occur.[41,71] On the other hand, it has not proved possible to distinguish definite initials, distinctively different from adjacent cells, with the electron microscope.[95]

Some workers believe that the cambium is a distinct tissue with a characteristic ontogeny, arising by differentiation of procambial cells.[95] The results of wounding the stem of castor bean, *Ricinus communis*, showed continuity between the procambium and cambium,[222] which are sometimes considered to represent two stages of a vascular meristem; these observations are supported by some experiments discussed later in this chapter.

The thin-walled cells of the vascular cambium are highly vacuolate and in this respect are unlike most other meristematic cells. Examination of cambium cells with the electron microscope confirms their highly vacuolate nature. Many ribosomes and dictyosomes, and well developed endoplasmic reticulum, are present.[624] The cells of the resting cambium of beech, *Fagus sylvatica*, contain many lipid droplets and protein bodies and abundant smooth ER, as in other species. When the cambium resumes activity, it is thought that vacuoles may form by dissolution of the stored lipids and protein.[379] Seasonal changes occur in the structure and number of organelles in the cambium. The plastids of cambial cells of arborescent species do not form grana, although those of herbaceous species may do so.[95]

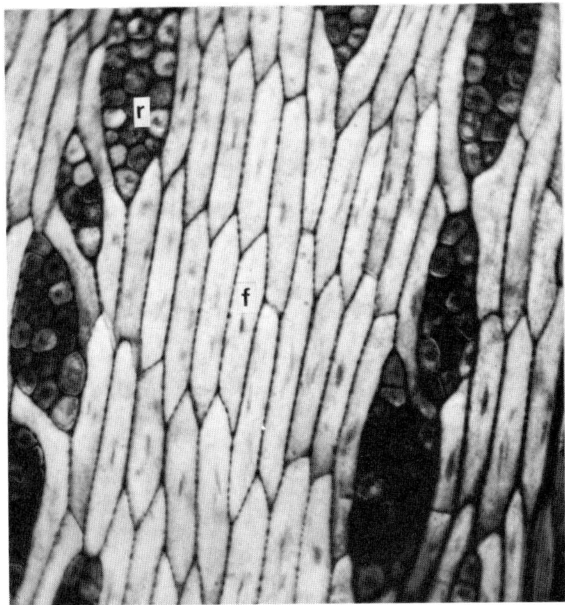

Fig. 12.1 Tangential longitudinal section through the cambial region of *Robinia pseudacacia*. f, fusiform initials; r, ray initials. ×200.

The cambium is made up of two kinds of cell, the ***fusiform*** and ***ray initials***. The latter are almost isodiametric and constitute the radial system of the vascular cambium, their products differentiating as parenchymatous rays. The fusiform initials, the axial system of the cambium, are considerably elongated in the longitudinal plane of the axis and are approximately prism-shaped (Fig. 12.1). In some species the fusiform initials are arranged in regular rows, having a stratified structure; in others the cambium is non-stratified. The former would be a storied, and the latter a non-storied cambium; storied types (Fig. 12.1) are considered to be more highly evolved.[71]

Development of fusiform initials

The fusiform initials of the cambium do not obey the usual laws of cell division. They usually divide vertically, in the longitudinal plane, thus contravening Errera's law, for example, which asserts that a cell will divide by a wall of minimal area. In the fusiform initial, a wall of minimal area would be transverse (horizontal). Divisions of this kind do occasionally occur, during the formation of additional ray initials, but are much less frequent than divisions resulting in a vertically oriented wall. Such divisions are mainly periclinal, but some anticlinal divisions also occur; the latter keep pace with the growth in girth of the stem or root. In longitudinal divisions the wall forms first in the region of the nucleus and grows towards the ends of the cell, which it may not reach for some time after mitosis.

Divisions leading to formation of additional fusiform initials, accommodating the increase in girth of the stem, are radial (i.e. anticlinal) in storied cambia, and obliquely radial, often called pseudo-transverse, in non-storied cambia.[71,512] Detailed studies of these pseudo-transverse divisions, which result in daughter cells of approximately half the length of the mother cell and are followed by longitudinal growth of the former, have been made in conifers.[39–42] Such divisions are usually restricted to a narrow band of cells of the cambial zone.[40,41] Often only one of the daughter cells survives in that form.[40,42] More new fusiform initials are formed in this way than are required; in conifers an initial may divide in this way 3 or 4 times in a year.[512] The rate of division is quite variable, and may be three times as rapid in some trees as others.[40] The majority of pseudo-transverse divisions occur late in the season; this is a contributory cause of the gradual decrease in size of the fusiform initials at this time.[39]

By periclinal divisions of the fusiform initials, radially oriented files of cells are produced. Usually more xylem than phloem is formed. Towards the inside of the stem or root these cells differentiate to form the axial system of the secondary xylem; towards the outside they differentiate into the axial system of the secondary phloem. At certain seasons a fairly wide zone of undifferentiated cells is present between the secondary xylem and phloem; these cells constitute the cambial zone. Division and differentiation of xylem mother cells may occur before those of the phloem, or *vice versa*, in different species.[71,153,461] Indeed, at any one time there is usually a predominance of one tissue or the other being formed.[512] Little is known about the factors controlling the differentiation of some cambial derivatives as xylem and others, not far removed in space, as phloem elements. After an analysis of the amount of auxin in secondary xylem, cambium and secondary phloem of several woody species, Sheldrake[590] suggested that since there was a gradient of auxin from xylem (highest) to phloem the cambial derivatives would be in different environments with respect to growth substances. The proximity of functional sieve elements would ensure the presence of relatively high concentrations of sucrose on the outer side of the cambium. As we have seen in Chapter 9, a high sucrose:auxin ratio promotes the differentiation of phloem. Sheldrake points out that this situation would be self-perpetuating.

The length of fusiform initials within a tree varies, the average length increasing acropetally within a single growth ring but falling off again towards the top.[512] Cell length is also related to the width of the growth ring.[39,40] Fusiform initials show apical intrusive growth, and there is a tendency for elongation to occur in a particular direction among groups of cells.[42]

Development of ray initials

Cells produced by the ray initials differentiate as the parenchymatous vascular rays. The fine structure of the fusiform and ray initials of the cambium is similar,[624] and the basis for the differences in size and shape and in the fate of their products is not yet understood. Ray initials may be relatively few, or may represent more than 50 per cent of the cambium.[512] Primary rays originate from primary tissues, but secondary rays are formed during secondary growth, the ray initials which give rise to them originating by division of fusiform initials. The ray initial may be formed at the end of a fusiform initial, or laterally from a short segment of the radial face, or sometimes the whole fusiform initial may divide transversely to form ray initials.[71,512]

The arrangement and spacing of the vascular rays, especially as seen in tangential longitudinal section (Fig. 12.1), suggest that they may be mutually inhibitory. Some ingenious experiments in which specimens of the tree *Ailanthus altissima* were partially girdled indicate that the distance between ray initials is not the only regulating factor.[89] The girdling left a narrow vertical strip of bark and cambium which formed the only connection between shoot and root at that level in the tree. Growth of these strips was very rapid, and occurred in both radial and tangential planes. Half of the bridges were confined laterally by metal barriers, so that only growth along the radius of the tree trunk could take place. Sections showed that in this fast-growing tissue a greater proportion of the area, as seen in tangential section, was occupied by rays. The number of rays per unit area increased by up to 100%. New rays were formed between existing ones. The experimenters concluded that rays may differentiate along channels of a stimulus moving between the phloem and xylem. The spacing of rays would thus be controlled by factors in the differentiating vascular tissues rather than within the cambium itself.[89]

Seasonal effects on cambial activity

Since some specimens of *Sequoiadendron*, for example, are found to be 3000 or 4000 years old, the cambial initials are evidently capable of periods of intermittent but considerable activity more or less indefinitely. Bristlecone pines (*Pinus aristata*) attain even greater age; the oldest known living specimen is 4600 years old, but wood some 9000 years old has been examined.[461] This virtual immortality of the cambial cells is one of their most interesting features. In perennial plants of temperate regions cambial activity is a seasonal phenomenon and occurs during the period of active growth, beginning in the spring. Considerable research has been directed towards elucidating the factors that stimulate the seasonal activity of the vascular cambium. These factors, which include photoperiod and temperature, are further discussed in Part 2,[136] Chapter 4. It may be mentioned here, however, that plants of tropical origin also have some seasonal activity of the cambium. In plants of *Acacia raddiana* growing in Israel the stem cambium became active in patches in January, was very active in April and May and inactive in October–December.[228] *Ziziphus* was greatly affected by the availability of water, two flushes of cambial activity occurring in a wet locality and only one short period of activity in a dry locality.[420] In these experiments plants were supplied with $^{14}CO_2$, and active cambial cells were defined as those with radioactivity in the walls. In *Zygophyllum*, also, availability of water controls cambial activity in young plants; adult plants show a period of true dormancy.[689] There is often a relationship between rainfall and cambial activity, even in temperate regions.[512]

One interesting study compared cambial activity in pairs of shrub species from comparable habitats in Chile and California. The shrubs were paired on the basis of morphological resemblances, but were not closely related. Considerable similarities were found in the seasonal activity of the cambium in four such pairs, though no direct correlations could be found with temperature or rainfall.[34] It seems possible that growth substances of some kind may have been the intermediary between the environment and stimulation or otherwise of the cambium.

Factors affecting cambial activity

It was first shown many years ago that cambial activity was inhibited in disbudded shoots.[364] Later it was found that auxin applied to decapitated sunflower seedlings stimulated the cambium.[615] The conclusion that auxin synthesized by young leaves in growing buds promoted cambial activity has been supported by much subsequent work. Some of these studies,[679,700] in which interactions between indoleacetic acid (IAA) and gibberellic acid (GA) in stimulating cambial activity were demonstrated, are discussed in Part 2, Chapter 4. Relationships can also be shown between cambial activity and seasonal levels of endogenous growth substances.[162] Even in large trees decapitation or ring barking led both to a severe reduction in auxin concentration in the cambial region, and to cessation of division of the cambium.[726] The fact that under normal conditions seasonal cessation of cambial activity occurs first in upper twigs, and proceeds basipetally, also suggests that a continuous source of diffusible auxin is required to maintain growth of the cambium. Alternatively, inhibitors produced by the leaves late in the season could inhibit the cambium.[71] Abscisic acid might well be one such substance. In a species in which activity of the cambium was affected by the availability of water, the former was correlated with active development of young leaves.[420] Thus seasonal effects may well be mediated by growth substances.

A number of effective experiments have been carried out on excised tissues. For example, the cambium in isolated stem segments of *Pinus* responded to IAA and sucrose in the culture medium. Much less auxin was required in spring and summer to stimulate a degree of cambial activity comparable to that occurring in autumn.[739] This suggests the possible presence of inhibitors late in the season.

Siebers[601–603] has carried out a series of ingenious experiments on the interfascicular cambium of the hypocotyl of castor bean, *Ricinus communis*. When metal blades were inserted radially into young hypocotyls, so that they intervened between the vascular bundles and the region of the future interfascicular cambium, the latter developed normally, indicating that it was not dependent on stimulatory effects from the vascular tissue, as has often been suggested. Isolated blocks of interfascicular tissue grown in culture also gave rise to normal cambium.[601] Later Siebers[602] excised and cultured small blocks of fascicular or interfascicular tissue from the prospective hypocotyl region of mature embryos. In the presence of kinetin, cambium differentiated in many interfascicular blocks of tissue. In fascicular blocks, little vascular tissue differentiated from the procambium, but cambium often developed. Since the cambium was formed in the expected normal position in interfascicular blocks, Siebers[602] concluded that this tissue had essentially already been determined (or pre-programmed) in the embryonic phase. He also pointed out that both fascicular and interfascicular cambium originated from the residual meristem, a cylinder of cells usually recognizable just below the apex (see Part 2,[136] Chapter 3).

By use of seedlings of *Ricinus*, it was found that although removal of the plumule had no effect on cambium development, removal of the cotyledons totally inhibited its formation.[603] Sucrose stimulated cambial development; IAA had little effect, but GA strongly promoted it in the presence of sucrose. Cultured hypocotyls supplied with only mineral nutrients and sucrose showed about half the rate of development of the cambium of intact seedlings. Cambial development could be stimulated to the level found in the intact plant by GA, suggesting that the cotyledons have a role in gibberellin synthesis.[603]

These various experiments show that different growth substances can stimulate the cambium in different species, or even in the same species at different stages of ontogeny or of the seasonal cycle of development. This is probably explicable in terms of changes in levels of endogenous growth substances, about which there is all too little information.

The position of the cambium with respect to gravity affects both its functioning and the differentiation of its products. Reaction wood is formed in horizontal or inclined stems of dicotyledons and gymnosperms (see Part 2,[136] Chapter 4).

Polarity of the cambium

In other skilful experiments with *Ricinus*, Siebers[600] excised a rectangle of interfascicular tissue from the hypocotyl, cut out the piece of tissue in which the cambium would be expected to develop several days later, inverted it through 180° and then re-inserted the whole rectangle into the hypocotyl (Fig. 12.2). About half such grafts became necrotic. In successful grafts, however, cambium formed in the expected position, and gave rise to secondary xylem and phloem in the positions normal for the original block of tissue. Thus these vascular tissues of the inverted scion were situated at 180° to those of the surrounding stock (Fig. 12.3). This spectacular finding shows that radial polarity is determined in the cambial region at least several days before the cambium is discernible. The cambial region does not adapt at all to its new, imposed situation. Siebers[600] concludes again that there is direct ontogenetic continuity between the residual meristem and the interfascicular cambium.

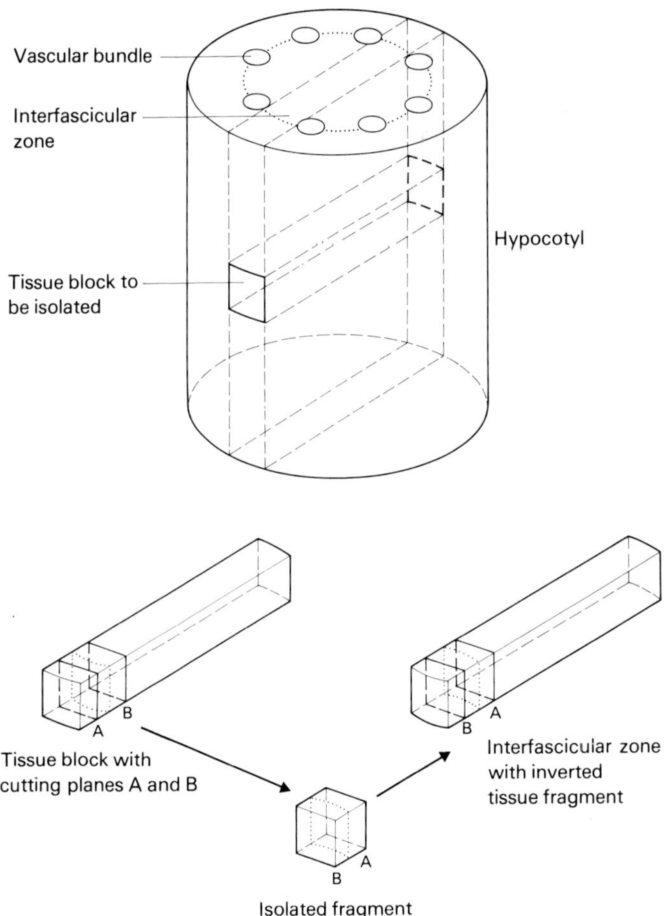

Fig. 12.2 Diagram showing experimental inversion of a region of interfascicular tissue of the hypocotyl of *Ricinus communis*. After excision of a rectangular piece of tissue, a small block of tissue with side A–B, containing the region of the prospective interfascicular cambium, was turned through 180° and re-inserted in the rectangular block. This was then grafted back into the hypocotyl. (From Siebers,[600] Fig. 1, p. 213.)

Experiments[656] in which square pieces of bark (of 5 mm side) were grafted back into the trunks of five woody species bear out this remarkably strong polarity of the cambium, even in later stages of its activity. The pieces were grafted back in the normal position, turned through 90°, or turned through 180°, and the main experiment was carried out on crabapple cultivars, other species giving confirmatory results. The cambium did not become reorientated, but again behaved as it would have done if left *in situ*. Grafts turned through 180° showed rather slow growth. In transverse sections of the trunk of the stock cut in the first or third years after grafting, pieces of bark turned through 90° showed the production from the cambium of relatively normal new

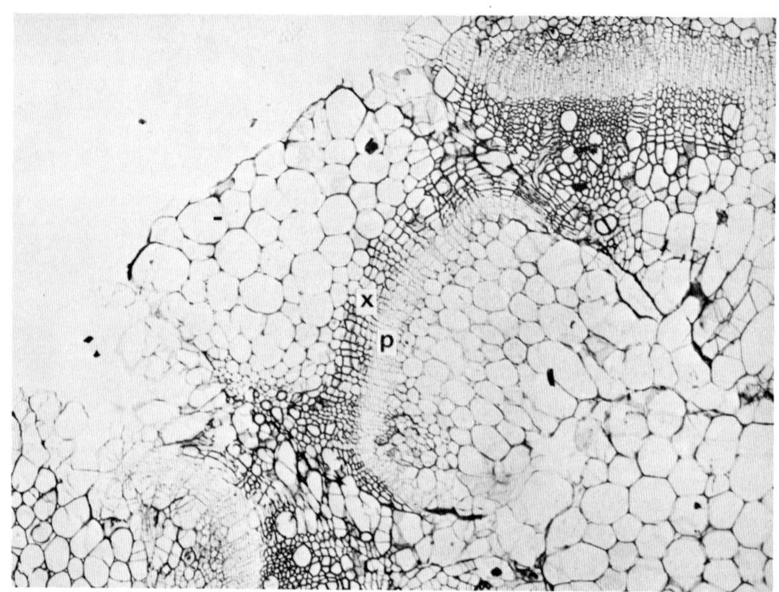

Fig. 12.3 Part of a transverse section of a hypocotyl of *Ricinus communis*, with grafted tissue as indicated in Fig. 12.2. The inverted block of tissue and adjacent tissues of the stock are shown. The secondary xylem (x) and phloem (p) of the small block of grafted tissue are reversed with respect to those of the stock. See text. ×30. (From Siebers,[600] Fig. 3, p. 215.)

secondary xylem, which was seen in radial longitudinal section (Fig. 12.4). Thus the long axis of the grafted tissue did not correspond with that of the stem into which it was grafted. These results suggest that, as some of the older workers proposed, the cambium has an inherent polarity.

If this is indeed so, the experiments on *Ricinus* suggest that it is imposed very early in ontogeny, indeed probably in the embryo. How this is brought about, however, remains a fascinating problem.

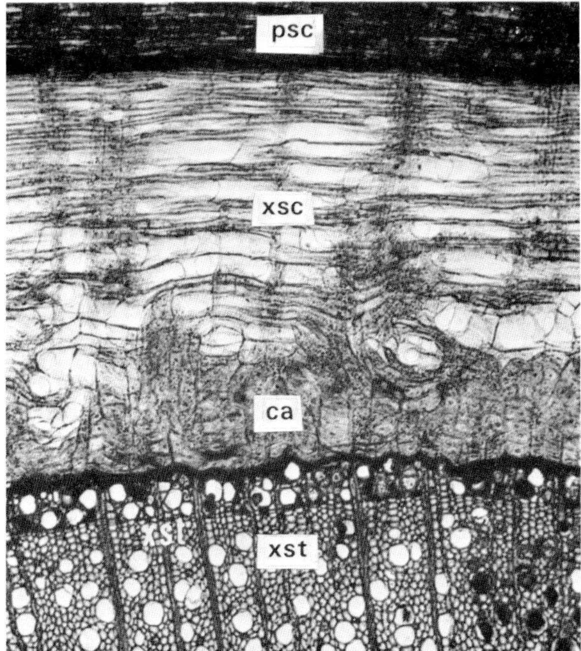

Fig. 12.4 Part of a section of the trunk of Siberian crabapple, *Malus baccata*, at the end of its first season of growth after a graft of cambial tissue rotated through 90° was made. The secondary xylem of the stock (**xst**) is seen in transverse section, that of the scion (**xsc**) in radial longitudinal section. **ca**, callus at the interface between stock and scion; **psc**, secondary phloem of the scion. ×92. (After Thair and Steeves,[656] Fig. 5, p. 367; reproduced by permission of the National Research Council of Canada from the *Canadian Journal of Botany*, **54**, 361–373, 1976.)

Vascular cambium in monocotyledons

In some monocotyledons, a type of vascular cambium is present in outer regions of the stem; it gives rise on the inner side to entire vascular bundles comprising both xylem and phloem, embedded in ground parenchyma or conjunctive tissue, and on the outer side to parenchyma.[99] The cells of the cambial region may be fusiform, rectangular or polygonal and may vary even in a single plant.[99] The occurrence of two kinds of cells in this tissue is illustrated in Fig. 12.5.

Until recently, very little was known about the structure and functioning of this type of cambium, or of the factors controlling its initiation and activity. The cambial zone is thought not to have a single initial layer at first, though one is present later.

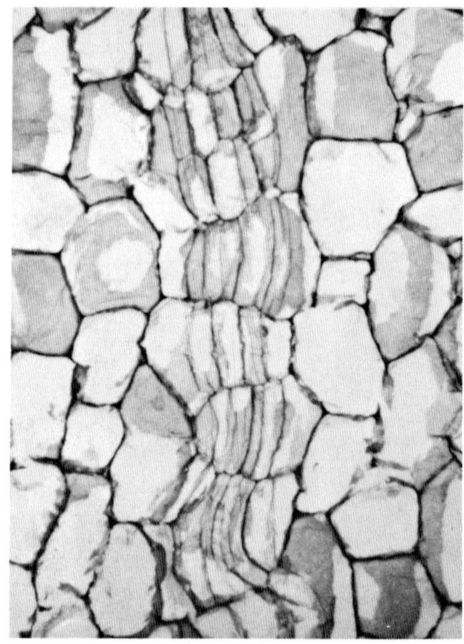

Fig. 12.5 Tangential longitudinal section through a stem of *Agave* in the cambial region. Two kinds of cells are present: squat, parenchymatous cells; and narrow, elongated cells, formed by division of the former, that will give rise to a vascular bundle. ×150.

Fisher[238] has recently carried out some interesting experiments on *Cordyline terminalis* and other members of the Agavaceae. One-year-old seedlings were grown upright or horizontally, the shoot of horizontal plants projecting through a slit in the side of the pot. The lower side of horizontal shoots showed increased secondary growth, decreased bundle density and decreased lignification of the ground parenchyma on the lower side as compared with the upper. This was even more evident in some other species, such as *Dracaena reflexa*, which had 13 times as much growth on the lower side. Thus gravity affected the mitotic activity of the cambium, as it does also in dicotyledons and gymnosperms, but no typical reaction wood was formed.

It seems likely that these effects were due to gravity-controlled accumulation of auxin on the lower side of the stem. In other experiments it was found that up to seven times as much labelled auxin accumulated in the cambial region on the lower side of inclined stems.[239] When applied to shoots, auxin promoted cambial activity. Thus it appears that, as in many dicotyledons, auxin stimulates the activity of the vascular cambium of several monocotyledons, even though the products of the activity of the two kinds of cambia are so differently organized. Further work is undoubtedly needed to elucidate what controls this organization.

PERIDERM

The periderm is a protective tissue which usually replaces the outer tissues of stems and roots that undergo secondary thickening. It is formed by another lateral meristem, the ***phellogen*** or cork cambium. The cells of the phellogen are meristematic, but like those of the vascular cambium are highly vacuolate; however, unlike the vascular cambium, with its fusiform and ray initials, the cells of the phellogen are all of one kind. The phellogen divides periclinally to give radially seriate files of cells; those towards the outside differentiate as ***phellem*** or cork, those towards the inside as ***phelloderm*** or secondary cortex. When the phellogen is highly active, it gives rise to large, thin-walled phellem cells, in contrast to the flat cells with thicker walls formed by a less active meristem.[688]

The cells of the phellem or cork are dead at maturity, and have suberized cell walls. Suberin is a fatty substance, and it is this which makes the cork essentially impervious and confers upon it its protective properties. Cork cells of the potato contain numerous mitochondria and ribosomes. It is thought that breakdown of the cytoplasm may yield wall thickening materials.[533] The cells of the phelloderm, formed towards the inside of the phellogen, are living and are often only distinguishable from the cortical cells by their radial alignment with the phellogen and phellem.

Like the vascular cambium, the phellogen gives rise to an unequal number of derivatives on either side during a growing season. The situation is the inverse of that in the vascular cambium, however, greater numbers of cells differentiating on the outer (phellem) side of the phellogen. Up to 20 rows of cork cells may be produced in a single season.

In some stems the inception and development of one phellogen may be followed by others. The later-formed phellogens are successively more deep-seated. Each phellogen functions normally and produces phellem and phelloderm. Pockets of tissue, usually secondary phloem, become isolated between the periderms, and these cells die. All of this tissue is sometimes referred to as ***rhytidome*** (Fig. 12.6). If the various

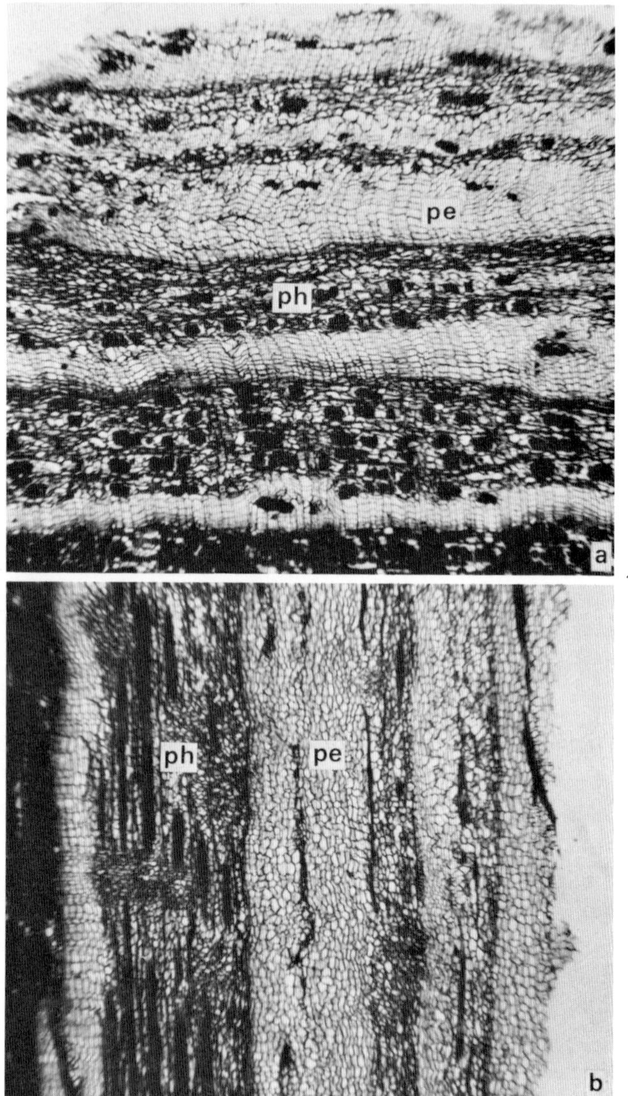

Fig. 12.6 (**a**) Transverse section and (**b**) longitudinal section of rhytidome in *Robinia pseudacacia*. Successive phellogens originate more and more deeply in the stem, giving rise to bands of suberized periderm (pe) which cut off and isolate regions of the secondary phloem (ph). These sections also show the isodiametric nature of the cells of the periderm. ×40.

phellogens form complete cylinders around the stem, a ring bark is formed; if they form separate arcs, a scale bark is formed.

Bark is a more inclusive term than periderm; it includes all the tissues outside the vascular cambium, viz., secondary phloem, primary phloem, cortex, periderm and any tissues outside the periderm. Barks usually contain a quantity of sclerenchyma, including phloem fibres and often also sclereids. Indeed, barks are made up of a considerable variety of tissues, and their structure is very varied in different species.[186]

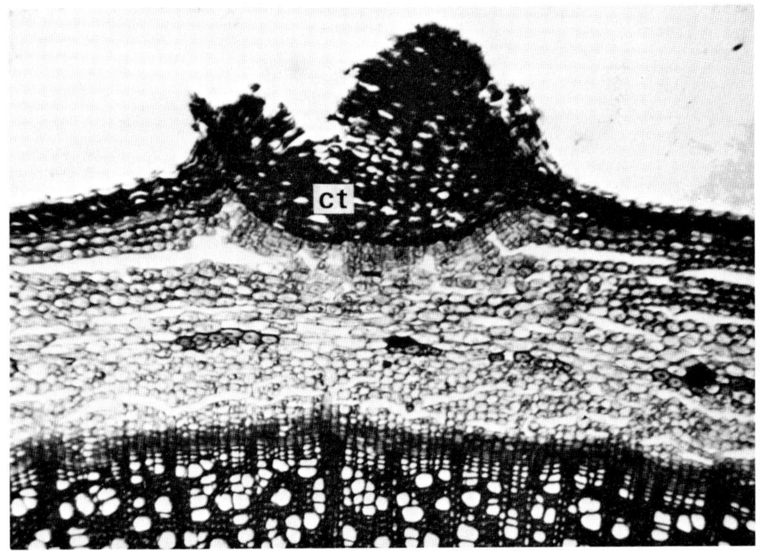

Fig. 12.7 Transverse section of outer part of the stem of *Sambucus*, showing a lenticel. ct, complementary tissue. ×60.

Certain regions of the periderm are differentiated as ***lenticels*** (Fig. 12.7). Some parts of the phellogen, usually below a stoma, and continuous with the rest of the phellogen, function rather differently and form a mass of unsuberized and loosely arranged cells called complementary tissue. In contrast to cork, in which no air spaces occur, many intercellular spaces are present in complementary tissue; lenticels are believed to function in gaseous exchange. The complementary tissue eventually breaks through the epidermis and may protrude. These properties of lenticels enabled their number to be estimated in potato tubers by placing the latter in a solution of methylene blue under vacuum. This gave a pattern of blue spots, indicating a variable number of lenticels.[719] With the scanning electron microscope, pores can be seen between the cells which form a dome in the centre of lenticels of potato tubers. Gaseous exchange is believed to occur here. Thread-like waxy outgrowths were present on cells from the sides of the pores; these may function in the regulation of water loss.[309]

Cork cells are dead and filled with air and function more or less as air cushions. Cork is resilient, resistant to pressure, impermeable to liquids and resistant to acids, organic solvents and many other chemicals.[118] All of these properties render it of considerable commercial importance. The chief source of cork is *Quercus suber*, the cork oak. The first cork formed is useless for commercial purposes, and this is stripped off. This wounding process stimulates the formation of another phellogen, and the whole process is repeated periodically. Corks used as stoppers for vintage wines are cut in a plane tangential to the surface, so that leakage does not result from vertically orientated lenticels.[461]

In some monocotyledons a type of cork known as storied cork is formed. This does not originate from a true phellogen, but by successive periclinal divisions of parenchyma cells. Their derivatives are suberized and arranged in radial files. This occurs only in species with secondary growth. In many monocotyledons, such as bamboo, the epidermis is retained throughout ontogeny, and may become sclerified.[71]

Whitmore[717] has studied the strains that may be set up in phloem and in mature bark by the formation of secondary xylem and phloem. Differences in rates of growth of the phloem, calculated from measurements of total tangential growth in the bark and the xylem growth rate, seemed to be related causally to the surface pattern and structure of the bark in different species.[718]

Factors affecting formation and activity of phellogen

Little is known about the factors leading to the formation of phellogen. A type of phellogen layer was induced to form in willow stems treated with naphthaleneacetic acid in lanolin,[411] but it is perhaps doubtful if this tissue was very normal. Combinations of daylength and temperature also had an effect on phellogen activity.[688] As already mentioned, the formation of a phellogen can sometimes be stimulated by wounding. Perhaps it might be worth considering whether the strains resulting from formation of secondary xylem and phloem and consequent increase in girth of the stem, discussed above, might be of some causal significance in the initiation or stimulation of phellogen. In this connection it is interesting to note that stems of *Robinia pseudacacia* show only one annual period of cambial activity, from about March to September (in Israel), whereas the phellogen shows two discontinuous periods of activity, approximately in April and in July–August (Fig. 12.8).[688] In *Acacia*, also, activity of the phellogen was periodic, and apparently independent of cambial activity.[26] It seems possible, at least, to interpret these interesting observations in the following way: After the vascular cambium becomes active, the stem increases in girth, placing a strain on the tissues outside the vascular cylinder; this activates the phellogen, which begins to function and, by the production of periderm with accompanying radial divisions, relieves the strain and consequently ceases activity. Continued activity of the cambium, however, leads to renewed stimulation of the phellogen some months later, and the same sequence of events occurs. However, in comparable work with *Robinia*[25] no relationship was found between the xylem-bark ratio and the inception of phellogen. Treatment with gibberellic acid or naphthaleneacetic acid delayed the onset of phellogen activity to an older internode as compared with control plants. Since formation of phellogen was also retarded in long days and under conditions of high temperature, both of which stimulated extension growth, it is possible

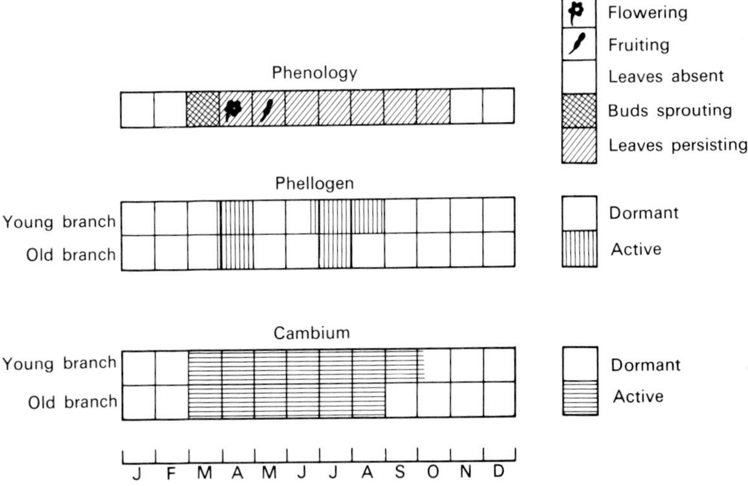

Fig. 12.8 Annual periods of activity of the phellogen and vascular cambium of young and old branches of a tree of *Robinia pseudacacia* growing outdoors in Tel-Aviv. (From Waisel, Liphschitz and Arzee,[688] Fig. 1, p. 333.)

to speculate that phellogen formation may be inhibited by hormones produced by an actively growing shoot apex or other regions of the young shoot. Whatever the validity of such a speculation, this is possibly an interesting field for future experimental work.

Environmental factors seem to be involved in the induction of phellogen.[461] Several such factors were investigated experimentally using green branches of *Eucalyptus camaldulensis* in which phellogen had not yet been formed. Plastic sleeves were placed round these branches, and gases or water were injected into the sleeves. Neither light nor external physical pressure had any effect. High humidity led to early phellogen formation, and oxygen showed some effect.[421] In roots, the cytokinin benzyladenine stimulated the formation of a periderm-like tissue.[508]

Wound periderm

Many of the investigations of factors controlling phellogen formation have been made with wounded tissues, often with potato tubers. Wounding frequently stimulates mitotic activity,[418] and the healing of the wound is accompanied by suberinization and the formation of a periderm. Wound periderm is beneficial in that it inhibits the entry of pathogens. The number of layers of wound cork formed can be affected by environmental factors, such as the type of soil in which potato tubers were grown.[471] High humidity and increased temperatures promoted wound healing.[720]

Priestley and Woffenden[527] observed that wound healing of potato tubers was characterized by a certain sequence of events. A deposit of suberin or cutin formed in the presence of air, occluding the wounded surface, and this was followed by accumulation of substances ('sap') at the blocked surface, and then the formation of a phellogen. These workers showed that if access of air to the tissue was prevented, there was no suberin formation. For example, covering the cut surface with paraffin wax led to the formation of a phellogen without prior suberinization. These observations are held to mean that a substance accumulates at the cut surface and is blocked there, leading to subsequent phellogen formation.[528] Many years later, however, we are little closer to identifying the nature of any substance involved. It may perhaps be significant that wounding potato tubers leads to a rapid increase in endogenous gibberellin-like substances.[539] The sequence of visible events in wound healing is somewhat similar in leaves.[678]

More recent experiments have also led to the experimental separation of suberin deposition and phellogen formation. Placing potato tissue in Tris buffer suppressed mitotic activity leading to phellogen formation, but synthesis of suberin continued normally.[369] Under other conditions, which included taking care not to bruise the tissue, rinsing it to remove injured tissue, and maintaining 100% humidity, suberin synthesis was inhibited but cell division continued.[403]

Despite quite extensive recent studies of both the vascular cambium and the phellogen, we are still far from a complete understanding of what stimulates their activity, or controls the differentiation of their products, especially with reference to the periderm. Further work is clearly required in this field.

Appendix

CLASS EXPERIMENT

The following experiment has been adapted from published work,* in consultation with the original authors, for class use. Detailed instructions are given.

The role of auxin in the regeneration of phloem and xylem in *Coleus*

Rooted cuttings of *Coleus* sp. about 8–10 weeks old are satisfactory material; avoid plants with abundant anthocyanin. Determine which is the fifth pair of expanded (i.e. clearly visible) leaves below the tip of the shoot. A wound will be made in one flat surface of the internode below this pair of leaves, designated the No. 5 internode.

Each student or pair of students requires three plants as comparable as possible and should undertake the following treatments:

(1) Make a small incision in the No. 5 internode, but otherwise leave the plant intact. A small scalpel should be inserted in the middle of one of the flat sides of the internode, in order to sever the fairly small central vascular bundle (larger bundles are present at the angles of the stem). The wound should be approximately 2–3 mm deep and 1–2 mm wide. (*Warning*: do not make the wound too large.) To avoid this a bent Armolet (a sterile disposable lance available from medical instrument shops) can profitably be used.* Ideally, the treatment should be

* These details are taken from papers by LaMotte and Jacobs[399, 400] and Thompson and Jacobs;[658] I am indebted to Dr. W. P. Jacobs and Dr. N. P. Thompson for additional information, and to Mr. A. C. Shaw for suggesting the use of an Armolet.

performed in a darkened room with the stem illuminated from behind, in order to reveal the silhouette of the vascular bundle to be severed, but this is not essential.

(2) Make a wound in the No. 5 internode, as above, remove all leaves and buds from the stem below the wound, and from node No. 5 above the wound, and decapitate the plant some mm above node No. 5. You are left with a bare stem, wounded in the uppermost remaining internode. Treat the cut stump of stem with plain lanolin paste.†

(3) Proceed as for (2), but treat the cut stump of stem with lanolin paste containing 1% indoleacetic acid (IAA).†

After treatment allow the plants to grow (preferably in a greenhouse or growth chamber) for seven days. Then remove the wounded internode from the plant by a horizontal cut at its base and an oblique cut at its top, so that the apical end can later be identified. Remove the side of the internode opposite the wound with a razor blade, and fix the internode in Craf III. Leave for 24 h (or longer); rinse in water. Then place the tissue in 85% lactic acid for 12–24 h, to harden (this can also be left for longer, if necessary). Pour the lactic acid with the internode into a petri dish. Place the internode with the wounded side down and the side previously exposed with the razor blade uppermost, and observe it under a dissecting microscope. Peel back the right (R) and left (L) sides with a blunt dissecting needle, holding the internode with a spatula or similar instrument. Now hold the peeled-back L side with the spatula and roll the central cylinder of the stem, comprising xylem and pith, towards the R side with the dissecting needle. The wounded strip (on the lowermost side) should thus become cleanly separated from the inner tissues. (N.B. The phloem should remain attached to this strip. If separating does not occur in this way, start rolling the central cylinder again from the opposite side.) The R and L sides should now be cut off, leaving only the wounded strip of tissue. (*Warning*: do not cut off too much of the sides.)

For phloem

Transfer the strips with the phloem attached to 0·1% aniline blue in 85% lactic acid for 6–12 h. (If it is not convenient to proceed with this at this stage, the phloem strips can be left in 85% lactic acid.) After

† To make the lanolin paste, weigh out between 50 and 60 g anhydrous lanolin in a beaker. Place in an oven at about 60°C overnight; it will now be melted, and a deep yellow colour. Add water to make 100 g, stirring vigorously. During this process the hydrated lanolin will turn almost white. Dissolve 1·0 g IAA in a little absolute alcohol and add this to the hydrated lanolin, stirring very vigorously to ensure thorough mixing. Make up the control plain lanolin paste without IAA, but with an equivalent quantity of absolute alcohol. Store in a refrigerator, removing some hours before use.

staining in aniline blue, transfer to 60% alcohol containing 0·5% HCl and peel off the epidermis, scar tissue and cortex with fine forceps under a dissecting microscope. (This is difficult, and it may prove expedient to leave some of this tissue still attached. If left too long in the acid alcohol, the tissue will destain more or less completely.) Stain the remaining thin strips again with aniline blue–lactic acid, and pass them through two changes (each for about 10 min, especially if the pieces of tissue are still fairly thick) of acidified 60% alcohol, two changes of absolute alcohol and two of xylol. Mount in canada balsam. Apply weights to the cover slips and keep the slides warm for several days. Alternatively, the lacmoid clearing method of Aloni and Sachs[11] could be used.

For xylem

After the strip with the phloem, cortex and epidermis has been peeled from the remainder of the stem, rinse the remaining cylinder of xylem and pith in water for about an hour. Then place the cylinder for 1–2 hr (less will do) in concentrated ammonium hydroxide which has been saturated with basic fuchsin. (To make the fuchsin–ammonia solution add a saturated aqueous solution of basic fuchsin to ammonia until a yellow or brownish ('straw') colour is obtained. It is better to do this under a hood.) Now pour off the ammonia solution and rinse the cylinders several times with water; allow them to remain in water until the vascular bundles appear red. It may be beneficial to leave the cylinders in water overnight, since they become too red if subsequent dehydration is begun too soon after the bundles are red.

Place the cylinders in a petri dish containing water under a dissecting microscope with the wounded side of the stem down. With a sharp scalpel, cut the xylem from the pith on the R and L sides of the cylinder. Now the xylem tissue should be free from the pith on sides adjacent to the wounded side but still attached on the wounded side. Gently cut off the pith from the xylem on the wounded side, gradually, with care. When the pith has been removed, cut away part of the xylem on the R and L sides, leaving part of each corner bundle with the wounded side (do not cut away too much of the xylem from the sides). Scrape away remaining pith from the wounded side and place the xylem in 50% alcohol. Pass the tissue through an alcohol-xylol series, leaving it about 10 min in each solution, and mount it pith side down on a slide. Mount in canada balsam. Again apply weights to the cover slips and keep the slides warm for several days.

Further Reading

CARLQUIST, S. (1961). *Comparative Plant Anatomy*. Holt, Rinehart and Winston, New York.

CUTLER, D. F. (1969). *Anatomy of the Monocotyledons*. IV. *Juncales*. Clarendon Press, Oxford.

CUTTER, E. G. (1971). *Plant Anatomy: Experiment and Interpretation*. Part 2. *Organs*. Edward Arnold, London.

EAMES, A. J. and MACDANIELS, L. H. (1947). *An Introduction to Plant Anatomy*. 2nd edition. McGraw-Hill, New York and London.

ESAU, K. (1965). *Plant Anatomy*. 2nd edition. Wiley, New York.

ESAU, K. (1969). The Phloem. *Handb. PflanzenAnat.*, **5**(2). Gebrüder Borntraeger, Berlin.

ESAU, K. (1977). *Anatomy of Seed Plants*. 2nd edition. Wiley, New York.

FAHN, A. (1974). *Plant Anatomy*. 2nd edition. Pergamon Press, Oxford.

FOSTER, A. S. (1949). *Practical Plant Anatomy*. 2nd edition. Van Nostrand, New York.

GRAHAM, C. F. and WAREING, P. F. (1976). *The Developmental Biology of Plants and Animals*. Blackwell Scientific Publications, Oxford.

GUNNING, B. E. S. and ROBARDS, A. W. (eds.) (1976). *Intercellular Communication in Plants: Studies on Plasmodesmata*. Springer-Verlag, Berlin.

GUNNING, B. E. S. and STEER, M. W. (1975). *Ultrastructure and the Biology of Plant Cells*. Edward Arnold, London.

LOTT, J. N. A. (1976). *A Scanning Electron Microscope Study of Green Plants*. C. V. Mosby Company, Saint Louis.

MARTIN, J. T. and JUNIPER, B. E. (1970). *The Cuticles of Plants*. Edward Arnold, London.

METCALFE, C. R. (1960). *Anatomy of the Monocotyledons*. I. *Gramineae*. Clarendon Press, Oxford.

METCALFE, C. R. (1971). *Anatomy of the Monocotyledons*. V. *Cyperaceae*. Clarendon Press, Oxford.

METCALFE, C. R. and CHALK, L. (1950). *Anatomy of the Dicotyledons*. Vols. I and II. Clarendon Press, Oxford.

PHILIPSON, W. R., WARD, J. M. and BUTTERFIELD, B. G. (1971). *The Vascular Cambium: its Development and Activity*. Chapman and Hall, London.

PRESTON, R. D. (1974). *The Physical Biology of Plant Cell Walls*. Chapman and Hall, London.

ROBARDS, A. W. (ed.) (1974). *Dynamic Aspects of Plant Ultrastructure*. McGraw-Hill, New York.

ROBERTS, L. W. (1976). *Cytodifferentiation in Plants. Xylogenesis as a Model System*. Cambridge University Press, London.

TOMLINSON, P. B. (1961). *Anatomy of the Monocotyledons*. II. *Palmae*. Clarendon Press, Oxford.

TOMLINSON, P. B. (1969). *Anatomy of the Monocotyledons*. III. *Commelinales—Zingiberales*. Clarendon Press, Oxford.

TROUGHTON, J. and DONALDSON, L. A. (1972). *Probing Plant Structure*. Chapman and Hall, London.

TROUGHTON, J. H. and SAMPSON, F. B. (1973). *Plants. A Scanning Electron Microscope Survey*. John Wiley & Sons Australasia Pty Ltd., Sydney.

WARDLAW, C. W. (1968). *Morphogenesis in Plants*. Methuen, London.

References

1. ABO EL-NIL, M. M. and HILDEBRANDT, A. C. (1976). Cell wall regeneration and colony formation from isolated single geranium protoplasts in microculture. *Can. J. Bot.*, **54**, 1530–1535.
2. ABREU, S. L., ROTHWELL, N. V. and LEWIS, R. F. (1973). An autoradiographic analysis of the root epidermis of the switch grass (*Panicum virgatum*). *Am. J. Bot.*, **60**, 496–504.
3. AHMAD, K. J. (1975). Cuticular studies in some species of *Lepidagathis* and *Barleria*. *Bot. Gaz.*, **136**, 129–135.
4. AJELLO, L. (1941). Cytology and cellular interrelations of cystolith formation in *Ficus elastica*. *Am. J. Bot.*, **28**, 589–594.
5. ALBERSHEIM, P. (1974). Structure and growth of walls of cells in culture. In *Tissue Culture and Plant Science*, STREET, H. E., (ed.) 379–404. Academic Press, London and New York.
6. ALBERSHEIM, P. and ANDERSON-PROUTY, A. J. (1975). Carbohydrates, proteins, cell surfaces, and the biochemistry of pathogenesis. *A. Rev. Pl. Physiol.*, **26**, 31–52.
7. ALDABA, V. C. (1927). The structure and development of the cell wall in plants. I. Bast fibers of *Boehmeria* and *Linum*. *Am. J. Bot.*, **14**, 16–24.
8. ALLDRIDGE, N. A. (1964). Anomalous vessel elements in wilty-dwarf tomato. *Bot. Gaz.*, **125**, 138–142.
9. ALONI, R. (1976a). Polarity of induction and pattern of primary phloem fiber differentiation in *Coleus*. *Am. J. Bot.*, **63**, 877–889.
10. ALONI, R. (1976b). Regeneration of phloem fibres round a wound: a new experimental system for studying the physiology of fibre differentiation. *Ann. Bot.* **40**, 395–397.
11. ALONI, R. and SACHS, T. (1973). The three-dimensional structure of primary phloem systems. *Planta*, **113**, 345–353.
12. AL-TALIB, K. H. and TORREY, J. G. (1959). The aseptic culture of isolated buds of *Pseudotsuga taxifolia*. *Pl. Physiol., Lancaster*, **34**, 630–637.
13. AL-TALIB, K. H. and TORREY, J. G. (1961). Sclereid distribution in the leaves of *Pseudotsuga* under natural and experimental conditions. *Am. J. Bot.*, **48**, 71–79.

14. AMELUNXEN, F., MORGENROTH, K. and PICKSAK, T. (1967). Untersuchungen an der Epidermis mit dem Stereoscan-Elektron-Mikroskop. *Z. Pflanzenphysiol.*, **57**, 79–95.
15. ANDERSON, R. and CRONSHAW, J. (1970). Sieve-plate pores in tobacco and bean. *Planta*, **91**, 173–180.
16. ARIYANAYAGAM, D. V. and STEBBINS, G. L. (1962). Developmental studies of cell differentiation in the epidermis of monocotyledons. III. Interaction of environmental and genetic factors on stomatal differentiation in three genotypes of barley. *Devl Biol.*, **4**, 117–133.
17. ARMSTRONG, W. (1972). A re-examination of the functional significance of aerenchyma. *Physiologia Pl.*, **27**, 173–177.
18. ARNOTT, H. J. and PAUTARD, F. G. E. (1965a). Mineralization in plants. (Abstr.) *Am. J. Bot.*, **52**, 613.
19. ARNOTT, H. J. and PAUTARD, F. G. E. (1965b). Development of raphide idioblasts in *Lemna*. (Abstr.) *Am. J. Bot.*, **52**, 618–619.
20. ARREGUÍN, B. (1958). Rubber and latex. *Handb. PflPhysiol.*, **10**, 223–248.
21. ARTSCHWAGER, E. (1943a). Contribution to the morphology and anatomy of guayule (*Parthenium argentatum*). *Tech. Bull. U.S. Dep. Agric.*, **842**. (33 pp.).
22. ARTSCHWAGER, E. (1943b). Contribution to the morphology and anatomy of the Russian Dandelion (*Taraxacum koksaghyz*). *Tech. Bull. U.S. Dep. Agric.*, **843**. (24 pp.).
23. ARTSCHWAGER, E. (1946). Contribution to the morphology and anatomy of cryptostegia (*Cryptostegia grandiflora*). *Tech. Bull. U.S. Dep. Agric.*, **915**. (40 pp.).
24. ARZEE, T. (1953). Morphology and ontogeny of foliar sclereids in *Olea europaea*. I. Distribution and structure. II. Ontogeny. *Am. J. Bot.*, **40**, 680–687 and 745–752.
25. ARZEE, T., LIPHSCHITZ, N. and WAISEL, Y. (1968). The origin and development of the phellogen in *Robinia pseudacacia* L. *New Phytol.*, **67**, 87–93.
26. ARZEE, T., WAISEL, Y. and LIPHSCHITZ, N. (1970). Periderm development and phellogen activity in the shoots of *Acacia raddiana* Savi. *New Phytol.*, **69**, 395–398.
27. ASH, A. L. (1948). Hemp—production and utilization. *Econ. Bot.*, **2**, 158–169.
28. ATAL, C. K. (1961). Effect of gibberellin on the fibers of hemp. *Econ. Bot.*, **15**, 133–139.
29. AVANZI, S., MAGGINI, F. and INNOCENTI, A. M. (1973). Amplification of ribosomal cistrons during the maturation of metaxylem in the root of *Allium cepa*. *Protoplasma*, **76**, 197–210.
30. AVERS, C. J. (1958). Histochemical localization of enzyme activity in the root epidermis of *Phleum pratense*. *Am. J. Bot.*, **45**, 609–613.
31. AVERS, C. J. (1961). Histochemical localization of enzyme activities in root meristem cells. *Am. J. Bot.*, **48**, 137–143.
32. AVERS, C. J. (1963). Fine structure studies of *Phleum* root meristem cells. II. Mitotic asymmetry and cellular differentiation. *Am. J. Bot.*, **50**, 140–148.
33. AVERS, C. J. and GRIMM, R. B. (1959). Comparative enzyme differentiation in grass roots. I. Acid phosphatase. *Am. J. Bot.*, **46**, 190–193.
34. AVILA, G., ALJARO, M. E., ARAYA, S., MONTENEGRO, G. and KUMMEROW, J. (1975). The seasonal cambium activity of Chilean and Californian shrubs. *Am. J. Bot.*, **62**, 473–478.

35. BAILEY, I. W. (1957). The potentialities and limitations of wood anatomy in the study of the phylogeny and classification of angiosperms. *J. Arnold Arbor.*, **38**, 243–254.

36. BAILEY, I. W. (1964). *Contributions to Plant Anatomy*. Chronica Botanica, Waltham, Mass.

37. BAKER, H. G. (1965). *Plants and Civilization*. Wadsworth, Belmont, California.

38. BAKER, H. G. and BAKER, I. (1973). Amino-acids in nectar and their evolutionary significance. *Nature, Lond.*, **241**, 543–545.

39. BANNAN, M. W (1964). Tracheid size and anticlinal divisions in the cambium of lodgepole pine. *Can. J. Bot.*, **42**, 1105–1118.

40. BANNAN, M. W. (1967). Anticlinal divisions and cell length in conifer cambium. *For. Prod. J.*, **17**, 63–69.

41. BANNAN, M. W. (1968a). Anticlinal divisions and the organization of conifer cambium. *Bot. Gaz.*, **129**, 107–113.

42. BANNAN, M. W. (1968b). Polarity in the survival and elongation of fusiform initials in conifer cambium. *Can. J. Bot.*, **46**, 1005–1008.

43. BEASLEY, C. A. (1973). Hormonal regulation of growth in unfertilized cotton ovules. *Science, N.Y.*, **179**, 1003–1005.

44. BEASLEY, C. A. (1975). Developmental morphology of cotton flowers and seed as seen with the scanning electron microscope. *Am. J. Bot.*, **62**, 584–592.

45. BEASLEY, C. A. and TING, I. P. (1973). The effects of plant growth substances on *in vitro* fiber development from fertilized cotton ovules. *Am. J. Bot.*, **60**, 130–139.

46. BEASLEY, C. A. and TING, I. P. (1974). Effects of plant growth substances on *in vitro* fiber development from unfertilized cotton ovules. *Am. J. Bot.*, **61**, 188–194.

47. BEASLEY, C. A., TING, I. P., LINKINS, A. E., BIRNBAUM, E. H. and DELMER, D. P. (1974). Cotton ovule culture: a review of progress and a preview of potential. In *Tissue Culture and Plant Science*, STREET, H. E. (ed.), 169–192. Academic Press, London.

48. BEER, M. and SETTERFIELD, G. (1958). Fine structure in thickened primary walls of collenchyma cells of celery petioles. *Am. J. Bot.*, **45**, 571–580.

49. BELFORD, D. S. and PRESTON, R. D. (1961). The structure and growth of root hairs. *J. exp. Bot.*, **12**, 157–168.

50. BELL, P. R. (1965). The structure and origin of mitochondria. *Sci. Prog., Oxf.*, **53**, 33–44.

51. BELL, P. R. (1970). Are plastids autonomous? *Symp. Soc. exp. Biol.*, **24**, 109–127.

52. BENAYOUN, J., ALONI, R. and SACHS, T. (1975). Regeneration around wounds and the control of vascular differentiation. *Ann. Bot.*, **39**, 447–454.

53. BIEGLMAYER, C., RUIS, H. and GRAF, J. (1974). Cytochemical localization of catalase activity in glyoxysomes from castor bean endosperm. *Pl. Physiol., Lancaster*, **53**, 276–278.

54. BIERHORST, D. W. and ZAMORA, P. M. (1965). Primary xylem elements and element associations of angiosperms. *Am. J. Bot.*, **52**, 657–710.

55. BIGOT, C. (1976). Bourgeonnement *in vitro* à partir d'épiderme séparé de feuille de *Bryophyllum Daigremontianum* (Crassulacées). *Can. J. Bot.*, **54**, 852–867.

56. BISALPUTRA, T. and ESAU, K. (1964). Polarized light study of phloem differentiation in embryo of *Chenopodium album*. *Bot. Gaz.*, **125**, 1–7.

57. BLOCH, R. (1944). Developmental potency, differentiation and pattern in meristems of *Monstera deliciosa*. *Am. J. Bot.*, **31**, 71–77.
58. BLOCH, R. (1946). Differentiation and pattern in *Monstera deliciosa*. The idioblastic development of the trichosclereids in the air root. *Am. J. Bot.*, **33**, 544–551.
59. BLOCH, R. (1948). The development of the secretory cells of *Ricinus* and the problem of cellular differentiation. *Growth*, **12**, 271–284.
60. BLOCH, R. (1965). Histological foundations of differentiation and development in plants. *Handb. PflPhysiol.*, **15**, 1, 146–188.
61. BONNEMAIN, J.-L. (1970). Histogénèse du phloème interne et du phloème inclus des Solanacées. *Rev. gén. Bot.*, **77**, 5–51.
62. BONNER, J. (1965). *The Molecular Biology of Development*. Cambridge University Press, London.
63. BONNETT, H. T., Jr. and NEWCOMB, E. H. (1965). Polyribosomes and cisternal accumulations in root cells of radish. *J. Cell. Biol.*, **27**, 423–432.
64. BORNMAN, C. H. and BOTHA, C. E. J. (1973). The role of aphids in phloem research. *Endeavour*, **32**, 129–133.
65. BOUCK, G. B. (1963a). Stratification and subsequent behavior of plant cell organelles. *J. Cell Biol.*, **18**, 441–457.
66. BOUCK, G. B. (1963b). An examination of the effects of ultracentrifugation on the organelles in living root tip cells. *Am. J. Bot.*, **50**, 1046–1054.
67. BOUCK, G. B. and CRONSHAW, J. (1965). The fine structure of differentiating sieve tube elements. *J. Cell Biol.*, **25**, 79–95.
68. BOWEN, M. R. and WAREING, P. F. (1969). The interchange of ^{14}C-kinetin and ^{14}C-gibberellic acid between the bark and xylem of willow. *Planta*, **89**, 108–125.
69. BREWER, J. W., COLLYARD, K. J. and LOTT, C. E., Jr. (1974). Analysis of sugars in dwarf mistletoe nectar. *Can. J. Bot.*, **52**, 2533–2538.
70. BRISSON, J. D. and PETERSON, R. L. (1975). SEM of fractured plant material embedded in glycol methacrylate. *Microscop. Soc. Canada*, **2**, 64–65.
71. BROWN, C. L. (1971). Secondary growth. In *Trees: Structure and Function*, ZIMMERMANN, M. H. and BROWN, C. L. (eds.), 67–123. Springer-Verlag, New York.
72. BROWN, W. V. and JOHNSON, Sr. C. (1962). The fine structure of the grass guard cell. *Am. J. Bot.*, **49**, 110–115.
73. BUETOW, D. E. (1976). Phylogenetic origin of the chloroplast. *J. Protozool.*, **23**, 41–47.
74. BÜNNING, E. (1952). Morphogenesis in plants. *Surv. biol. Prog.*, **2**, 105–140.
75. BÜNNING, E. (1957). Polarität und inaquale Teilung des pflanzlichen Protoplasten. *Protoplasmatologia*, **8**, 1–86.
76. BÜNNING, E. (1965). Die Entstehung von Mustern in der Entwicklung von Pflanzen. *Handb. PflPhysiol.*, **15**, 1, 383–408.
77. BÜNNING, E. and BIEGERT, F. (1953). Die Bildung der Spaltöffnungsinitialen bei *Allium Cepa*. *Z. Bot.*, **41**, 17–39.
78. BÜNNING, E. and SAGROMSKY, H. (1948). Die Bildung des Spaltöffnungsmusters in der Blattepidermis. *Z. Naturf.*, **3**b, 203–216.
79. BURGESS, J. (1976). Scanning electron microscopy of cell wall formation around isolated plant protoplasts. *Planta*, **131**, 173–178.
80. BUTTERFIELD, B. G. and MEYLAN, B. A. (1972). Scalariform perforation plate development in *Laurelia novae-zelandiae* A Cunn.: a scanning electron microscope study. *Aust. J. Bot.*, **20**, 253–259.

81. BUTTROSE, M. (1963). Ultrastructure of the developing aleurone cells of wheat grain. *Aust. J. biol. Sci.*, **16**, 768–774.
82. BUVAT, R. (1963). Sur la présence d'acide ribonucléique dans les 'corpuscules muqueux' des cellules criblées de *Cucurbita pepo*. *C. r. hebd. Séanc. Acad. Sci., Paris*, **257**, 733–735.
83. CAMUS, G. (1943). Sur le greffage de bourgeons a Endive sur des fragments de tissus cultivés in vitro. *C. r. Séanc. Soc. Biol.*, **137**, 184–185.
84. CANNY, M. J. (1973a). Translocation and distance. I. The growth of the fruit of the sausage tree, *Kigelia pinnata*. *New Phytol.*, **72**, 1269–1280.
85. CANNY, M. J. (1973b). *Phloem Translocation*. Cambridge Univ. Press.
86. CARCELLER, M., DAVEY, M. R., FOWLER, M. W. and STREET, H. E. (1971). The influence of sucrose, 2,4-D, and kinetin on the growth, fine structure, and lignin content of cultured sycamore cells. *Protoplasma*, **73**, 367–385.
87. CARLQUIST, S. (1961). *Comparative Plant Anatomy*. Holt, Rinehart and Winston, New York.
88. CARLQUIST, S. (1972). Plant anatomy's 300th Anniversary. In *Research Trends in Plant Anatomy*, GHOUSE, A. K. M. and YUNUS, M. (eds.), 1–6. Tata McGraw-Hill, Bombay and New Delhi.
89. CARMI, A., SACHS, T. and FAHN, A. (1972). The relation of ray spacing to cambial growth. *New Phytol.*, **71**, 349–353.
90. CARR, D. J. (1976a). Plasmodesmata in growth and development. In *Intercellular Communication in Plants: Studies on Plasmodesmata*, GUNNING, B. E. S. and ROBARDS, A. W. (eds.), 243–289. Springer-Verlag, Berlin.
91. CARR, D. J. (1976b). Historical perspectives on plasmodesmata. In *Intercellular Communication in Plants: Studies on Plasmodesmata*, GUNNING, B. E. S. and ROBARDS, A. W. (eds.), 291–295. Springer-Verlag, Berlin.
92. CARR, D. J. and CARR, S. G. M. (1970). Oil glands and ducts in *Eucalyptus* l'Hérit. II. Development and structure of oil glands in the embryo. *Aust. J. Bot.*, **18**, 191–212.
93. CARR, S. G. M. and CARR, D. J. (1969). Oil glands and ducts in *Eucalyptus* l'Hérit. I. The phloem and the pith. *Aust. J. Bot.*, **17**, 471–513.
94. CASS, D. D. (1968). Observations on the ultrastructure of the non-articulated laticifers of *Jatropha podagrica* (Euphorbiaceae). *Experientia*, **24**, 961–962.
95. CATESSON, A. M. (1974). Cambial cells. In *Dynamic Aspects of Plant Ultrastructure*, ROBARDS, A. W. (ed.), 358–390. McGraw-Hill, London.
96. CHAFE, S. C. (1970). The fine structure of the collenchyma cell wall. *Planta*, **90**, 12–21.
97. CHAFE, S. C. and WARDROP, A. B. (1970). Microfibril orientation in plant cell walls. *Planta*, **92**, 13–24.
98. CHANDLER, J. A. (1973). Recent developments in analytical electron microscopy. *J. Microsc.*, **98**, 359–378.
99. CHEADLE, V. I. (1937). Secondary growth by means of a thickening ring in certain monocotyledons. *Bot. Gaz.*, **98**, 535–555.
100. CHEADLE, V. I. (1956). Research on xylem and phloem—progress in fifty years. *Am. J. Bot.*, **43**, 719–731.
101. CHEADLE, V. I. (1963). Vessels in Iridaceae. *Phytomorphology*, **13**, 245–248.
102. CHEADLE, V. I. and ESAU, K. (1958). Secondary phloem of Calycanthaceae. *Univ. Calif. Publs Bot.*, **29**, 397–510.

103. CHEADLE, V. I. and TUCKER, J. (1961). Vessels and phylogeny of Monocotyledoneae. In *Recent Advances in Botany*, 161–165. Toronto University Press.

104. CHEADLE, V. I. and WHITFORD, N. B. (1941). Observations on the phloem in Monocotyledoneae. I. The occurrence and phylogenetic specialization in structure of the sieve tubes in the metaphloem. *Am. J. Bot.*, **28**, 623–627.

105. CHILD, C. M. (1917). Experimental alteration of the axial gradient in the alga, *Griffithsia bornetiana*. *Biol. Bull.*, **32**, 213–233.

106. CHLYAH, H. (1974). Formation and propagation of cell division-centers in the epidermal layer of internodal segments of *Torenia fournieri* grown in vitro. Simultaneous surface observations of all the epidermal cells. *Can. J. Bot.*, **52**, 867–872.

107. CHLYAH, H., TRAN THANH VAN, M. and DEMARLY, G. (1975). Distribution pattern of cell division centers on the epidermis of stem segments of *Torenia fournieri* during *de novo* bud formation. *Pl. Physiol., Lancaster*, **56**, 28–33.

108. CLELAND, R. (1971). Cell wall extension. *A. Rev. Pl. Physiol.*, **22**, 197–222.

109. CLUTTER, M. E. (1960). Hormonal induction of vascular tissue in tobacco pith in vitro. *Science, N. Y.*, **132**, 548–549.

110. CLUTTER, M. E. (1963). Effects of IAA and kinetin on cell differentiation in cabbage pith cultures. *Pl. Physiol., Lancaster*, **38** (Suppl.), xii.

111. COBLEY, L. S. (1956). *An Introduction to the Botany of Tropical Crops.* Longmans, Green, London.

112. COCKING, E. C. (1972). Plant cell protoplasts—isolation and development. *A. Rev. Pl. Physiol.*, **23**, 29–50.

113. COCKING, E. C. (1974). Ultrastructure of cultured plant cells. In *Dynamic Aspects of Plant Ultrastructure*, ROBARDS, A. W. (ed.), 310–330. McGraw-Hill, London.

114. COHEN, S. S. (1973). Mitochondria and chloroplasts revisited. *Am. Scient.*, **61**, 437–445.

115. COMMONER, B. and ZUCKER, M. L. (1953). Cellular differentiation: an experimental approach. In *Growth and Differentiation in Plants*, LOOMIS, W. E. (ed.), 339–392. Iowa State College Press, Ames, Iowa.

116. CONWAY, V. M. (1937). Studies on the autecology of *Cladium mariscus* R. Br. III. The aeration of the subterranean parts of the plant. *New Phytol.*, **36**, 64–96.

117. COOK, G. M. W. (1975). *The Golgi apparatus. Oxford Biol. Readers*, **77**. Oxford University Press.

118. COOKE, G. B. (1948). Cork and cork products. *Econ. Bot.*, **2**, 393–402.

119. CORMACK, R. G. H. (1937). The development of root hairs by *Elodea canadensis*. *New Phytol.*, **36**, 19–25.

120. CORMACK, R. G. H. (1949). The development of root hairs in angiosperms. *Bot. Rev.*, **15**, 583–612.

121. CORMACK, R. G. H. (1962). Development of root hairs in angiosperms. II. *Bot. Rev.*, **28**, 446–464.

122. COWAN, J. M. (1950). *The Rhododendron Leaf.* Oliver and Boyd, Edinburgh.

123. CRAFTS, A. S. and CRISP, C. E. (1971). *Phloem Transport in Plants.* W. H. Freeman & Co., San Francisco.

124. CRAFTS, A. S. and CURRIER, H. B. (1963). On sieve tube function. *Protoplasma*, **57**, 188–202.

125. CRAIG, I. W. and GUNNING, B. E. S. (1976). Organelle development, 270–301, in Ref. 271.

126. CRAN, D. G. and POSSINGHAM, J. V. (1974). Plastid thylakoid formation. *Ann. Bot.*, **38**, 843–847.

127. CRICK, F. (1970). Diffusion in embryogenesis. *Nature, Lond.*, **225**, 420–422.

128. CRICK, F. H. C. (1971). The scale of pattern formation. *Symp. Soc. exp. Biol.*, **25**, 429–438.

129. CRONSHAW, J. (1967). Tracheid differentiation in tobacco pith cultures. *Planta*, **72**, 78–90.

130. CRONSHAW, J. (1974). Phloem differentiation and development. In *Dynamic Aspects of Plant Ultrastructure*, ROBARDS, A. E. (ed.), 391–413. McGraw-Hill, London.

131. CRONSHAW, J. and ANDERSON, R. (1971). Phloem differentiation in tobacco pith culture. *J. Ultrastruct. Res.*, **34**, 244–259.

132. CRONSHAW, J. and BOUCK, G. B. (1965). The fine structure of differentiating xylem elements. *J. Cell Biol.*, **24**, 415–431.

133. CRONSHAW, J. and ESAU, K. (1967). Tubular and fibrillar components of mature and differentiating sieve elements. *J. Cell. Biol.*, **34**, 801–815.

134. CRÜGER, H. (1855). Zur Entwicklungsgeschichte der Zellenwand. *Bot. Ztg*, **13**, 601–613 and 617–629.

135. CUTTER, E. G. (1967). Differentiation of trichoblasts in roots of *Hydrocharis*. (Abstr.) *Am. J. Bot.*, **54**, 632.

136. CUTTER, E. G. (1971). *Plant Anatomy : Experiment and Interpretation.* Part 2: *Organs.* Edward Arnold, London.

137. CUTTER, E. G. and FELDMAN, L. J. (1970a). Trichoblasts in *Hydrocharis*. I. Origin, differentiation, dimensions and growth. *Am. J. Bot.*, **57**, 190–201.

138. CUTTER, E. G. and FELDMAN, L. J. (1970b). Trichoblasts in *Hydrocharis*. II. Nucleic acids, proteins and a consideration of cell growth in relation to endopolyploidy. *Am. J. Bot.*, 57, 202–211.

139. CUTTER, E. G. and HUNG, C.-Y. (1972). Symmetric and asymmetric mitosis and cytokinesis in the root tip of *Hydrocharis morsus-ranae* L. *J. Cell Sci.*, **11**, 723–737.

140. CZAJA, A. T. (1930). Zellphysiologische Untersuchungen an *Cladophora glomerata*. Isolierung, Regeneration und Polarität. *Protoplasma*, **11**, 601–627.

141. CZERNIK, C. A. and AVERS, C. J. (1964). Phosphatase activity and cellular differentiation in *Phleum* root meristem. *Am. J. Bot.*, **51**, 424–431.

142. DALE, H. M. (1951). Carbon dioxide and root hair development in *Anacharis (Elodea)*. *Science, N. Y.*, **114**, 438–439.

143. DALESSANDRO, G. (1973). Hormonal control of xylogenesis in pith parenchyma explants of *Lactuca*. *Ann. Bot.*, **37**, 375–382.

144. DALESSANDRO, G. and ROBERTS, L. W. (1971). Induction of xylogenesis in pith parenchyma explants of *Latuca*. *Am. J. Bot.*, **58**, 378–385.

145. DAVEY, M. R. and STREET, H. E. (1971). Studies on the growth in culture of plant cells. IX. Additional features of the fine structure of *Acer pseudoplatanus* L. cells cultured in suspension. *J. exp. Bot.*, **22**, 90–95.

146. DAWES, C. J. and BOWLER, E. (1959). Light and electron microscope studies of the cell wall structure of the root hairs of *Raphanus sativus*. *Am. J. Bot.*, **46**, 561–565.

147. DAYANANDAN, P. and KAUFMAN, P. B. (1973). Stomata in *Equisetum*. *Can. J. Bot.*, **51**, 1555–1564.

148. DAYANANDAN, P. and KAUFMAN, P. B. (1975). Stomatal movements associated with potassium fluxes. *Am. J. Bot.*, **62**, 221–231.

149. DAYANANDAN, P. and KAUFMAN, P. B. (1976). Trichomes of *Cannabis sativa* L. (Cannabaceae). *Am. J. Bot.*, **63**, 578–591.

150. DE MAGGIO, A. E. (1966). Phloem differentiation: induced stimulation by gibberellic acid. *Science, N. Y.*, **152**, 370–372.

151. DEMAGGIO, A., WETMORE, R. and MOREL, G. (1963). Induction de tissu vasculaire dans le prothalle de Fougère. *C. r. Séanc. Acad. Sci., Paris*, **256**, 5196–5199.

152. DE MARIA, M. E., THAINE, R. and SARISALO, H. I. M. (1975). Fine structure of sieve tubes prepared mainly for observation in the electron microscope by a cryogenic method. *J. exp. Bot.*, **26**, 145–160.

153. DERR, W. F. and EVERT, R. F. (1967). The cambium and seasonal development of the phloem in *Robinia pseudoacacia*. *Am. J. Bot.*, **54**, 147–153.

154. DESHPANDE, B. P. (1974a). Development of the sieve plate in *Saxifraga sarmentosa* L. *Ann. Bot.*, **38**, 151–158.

155. DESHPANDE, B. P. (1974b). On the occurrence of spiny vesicles in the phloem of *Salix*. *Ann. Bot.*, **38**, 865–868.

156. DESHPANDE, B. P. (1975). Differentiation of the sieve plate of *Cucurbita*: a further view. *Ann. Bot.*, **39**, 1015–1022.

157. DESHPANDE, B. P. (1976a). Observations on the fine structure of plant cell walls. I. Use of permanganate staining. *Ann. Bot.*, **40**, 433–437.

158. DESHPANDE, B. P. (1976b). Observations on the fine structure of plant cell walls. II. The microfibrillar framework of the parenchymatous cell wall in *Cucurbita*. *Ann. Bot.*, **40**, 439–442.

159. DESHPANDE, B. P. (1976c). Observations on the fine structure of plant cell walls. III. The sieve tube wall in *Cucurbita*. *Ann. Bot.*, **40**, 443–446.

160. DHINDSA, R. S., BEASLEY, C. A. and TING, I. P. (1975). Osmoregulation in cotton fiber. Accumulation of potassium and malate during growth. *Pl. Physiol., Lancaster*, **56**, 394–398.

161. DICKENSON, P. B. and FAIRBAIRN, J. W. (1975). The ultrastructure of the alkaloidal vesicles of *Papaver somniferum* latex. *Ann. Bot.*, **39**, 707–712.

162. DIGBY, J. and WAREING, P. F. (1966). The relationship between endogenous hormone levels in the plant and seasonal aspects of cambial activity. *Ann. Bot.*, **30**, 607–622.

163. DOLEY, D. and LEYTON, L. (1968). Effects of growth regulating substances and water potential on the development of secondary xylem in *Fraxinus*. *New Phytol.*, **67**, 579–594.

164. DOLEY, D. and LEYTON, L. (1970). Effects of growth regulating substances and water potential on the development of wound callus in *Fraxinus*. *New Phytol.*, **69**, 87–102.

165. DORMER, K. J. (1961). The crystals in the ovaries of certain Compositae. *Ann. Bot., N. S.*, **25**, 241–254.

166. DORMER, K. J. (1962). The taxonomic significance of crystal forms in *Centaurea*. *New Phytol.*, **61**, 32–35.

167. DRIESCH, H. (1908). *The Science and Philosophy of the Organism*. A. & C. Black, London.

168. DUCHAIGNE, A. (1955). Les divers types de collenchymes chez les Dicotylédones; leur ontogénie et leur lignification. *Annls Sci. nat., Bot.*, sér. 11, **16**, 455–479.

169. DUCKETT, J. G., TOTH, R. and SONI, S. L. (1975). An ultrastructural study of the *Azolla, Anabaena azollae* relationship. *New Phytol.*, **75**, 111–118.

170. DUFFIELD, E. C. S., WAALAND, S. D. and CLELAND, R. (1972). Morphogenesis in the red alga, *Griffithsia pacifica*: regeneration from single cells. *Planta*, **105**, 185–195.

171. EARLE, E. D. (1968). Induction of xylem elements in isolated *Coleus* pith. *Am. J. Bot.*, **55**, 302–305.

172. EDWARDS, G. E., LEE, S. S., CHEN, T. M. and BLACK, C. C. (1970) Carboxylation reactions and photosynthesis of carbon compounds in isolated mesophyll and bundle sheath cells of *Digitaria sanguinalis* (L.) Scop. *Biochem. biophys. Res. Commun.*, **39**, 389–395.

173. EGLINTON, G. and HAMILTON, R. J. (1967). Leaf epicuticular waxes. *Science, N. Y.*, **156**, 1322–1335.

174. EISNER, T., JOHNESSEE, J. S., CARREL, J., HENDRY, L. B. and MEINWALD, J. (1974). Defensive use by an insect of a plant resin. *Science, N. Y.*, **184**, 996–999.

175. ELIAS, T. S., ROZICH, W. R. and NEWCOMBE, L. (1975). The foliar and floral nectaries of *Turnera ulmifolia* L. *Am. J. Bot.*, **62**, 570–576.

176. ENGLEMAN, E. M. (1965a). Sieve element of *Impatiens sultanii*. 1. Wound reaction. *Ann. Bot., N.S.*, **29**, 83–101.

177. ENGLEMAN, E. M. (1965b). Sieve element of *Impatiens sultanii*. 2. Developmental aspects. *Ann. Bot., N.S.*, **29**, 103–118.

178. ERICKSON, R. O. (1961). Probability of division of cells in the epidermis of the *Phleum* root. *Am. J. Bot.*, **48**, 268–274.

179. ERVIN, E. L. and EVERT, R. F. (1970). Observations on sieve elements in three perennial monocotyledons. *Am. J. Bot.*, **57**, 218–224.

180. ESAU, K. (1936a). Ontogeny and structure of collenchyma and of vascular tissues in celery petioles. *Hilgardia*, **10**, 431–476.

181. ESAU, K. (1936b). Vessel development in celery. *Hilgardia*, **10**, 479–484.

182. ESAU, K. (1943). Vascular differentiation in the vegetative shoot of *Linum*. III. The origin of the bast fibers. *Am. J. Bot.*, **30**, 579–586.

183. ESAU, K. (1948). Phloem structure in the grapevine, and its seasonal changes. *Hilgardia*, **18**, 217–296.

184. ESAU, K. (1961). *Plants, Viruses, and Insects*. Harvard University Press, Cambridge, Mass.

185. ESAU, K. (1964a). Aspects of ultrastructure of phloem. In *The Formation of Wood in Forest Trees*, ZIMMERMANN, M. H. (ed.), 51–63. Academic Press, New York.

186. ESAU, K. (1964b). Structure and development of the bark in dicotyledons. In *The Formation of Wood in Forest Trees*, ZIMMERMANN, M. H. (ed.), 37–50. Academic Press, New York.

187. ESAU, K. (1965a). *Plant Anatomy*. 2nd edition. Wiley, New York.

188. ESAU, K. (1965b). *Vascular Differentiation in Plants*. Holt, Rinehart and Winston, New York.

189. ESAU, K. (1969). The Phloem. *Handb. PflanzenAnat.* **5**(2), Gebrüder Borntraeger, Berlin.

190. ESAU, K. (1971a). The sieve element and its immediate environment: thoughts on research of the past fifty years. *J. Indian bot. Soc.*, **50A**, 115–129.

191. ESAU, K. (1971b). Development of P-protein in sieve elements of *Mimosa pudica*. *Protoplasma*, **73**, 225–238.

192. ESAU, K. (1972a). Changes in the nucleus and the endoplasmic reticulum during differentiation of a sieve element in *Mimosa pudica* L. *Ann. Bot.*, **36**, 703–710.

193. ESAU, K. (1972b). Cytology of sieve elements in minor veins of sugar beet leaves. *New Phytol.*, **71**, 161–168.

194. ESAU, K. (1973). Comparative structure of companion cells and phloem parenchyma cells in *Mimosa pudica* L. *Ann. Bot.*, **37**, 625–632.

195. ESAU, K. (1974). Ultrastructure of secretory cells in the phloem of *Mimosa pudica* L. *Ann. Bot.*, **38**, 159–164.

196. ESAU, K. and CHARVAT, I. D. (1975). An ultrastructural study of acid phosphatase localization in cells of *Phaseolus vulgaris* phloem by the use of the azo dye method. *Tissue & Cell*, **7**, 619–630.

197. ESAU, K. and CHEADLE, V. I. (1958). Wall thickening in sieve elements. *Proc. natn. Acad. Sci. U.S.A.*, **44**, 546–553.

198. ESAU, K. and CHEADLE, V. I. (1959). Size of pores and their content in sieve elements of dicotyledons. *Proc. natn. Acad. Sci. U.S.A.*, **45**, 156–162.

199. ESAU, K. and CHEADLE, V. I. (1961). An evaluation of studies on ultrastructure of sieve plates. *Proc. natn. Acad. Sci. U.S.A.*, **47**, 1716–1726.

200. ESAU, K. and CHEADLE, V. I. (1962). Mitochondria in the phloem of *Cucurbita*. *Bot. Gaz.*, **124**, 79–85.

201. ESAU, K. and CHEADLE, V. I. (1965). Cytologic studies on phloem. *Univ. Calif. Publs Bot.*, **36**, 253–344.

202. ESAU, K., CHEADLE, V. I. and GILL, R. H. (1966a). Cytology of differentiating tracheary elements. I. Organelles and membrane systems. *Am. J. Bot.*, **53**, 756–764.

203. ESAU, K., CHEADLE, V. I. and GILL, R. H. (1966b). Cytology of differentiating tracheary elements. II. Structures associated with cell surfaces. *Am. J. Bot.*, **53**, 765–771.

204. ESAU, K., CHEADLE, V. I. and RISLEY, E. B. (1962). Development of sieve-plate pores. *Bot. Gaz.*, **123**, 233–243.

205. ESAU, K., CHEADLE, V. I. and RISLEY, E. B. (1963). A view of ultrastructure of *Cucurbita* xylem. *Bot. Gaz.*, **124**, 311–316.

206. ESAU, K., CURRIER, H. B. and CHEADLE, V. I. (1957). Physiology of phloem. *A. Rev. Pl. Physiol.*, **8**, 349–374.

207. ESAU, K. and GILL, R. H. (1971). Aggregation of endoplasmic reticulum and its relation to the nucleus in a differentiating sieve element. *J. Ultrastruct. Res.*, **34**, 144–158.

208. ESAU, K. and KOSAKAI, H. (1975). Laticifers in *Nelumbo nucifera* Gaertn.: distribution and structure. *Ann. Bot.*, **39**, 713–719.

209. ESCHRICH, W. (1953). Beiträge zur Kenntniss der Wundsiebröhrenentwicklung bei *Impatiens Holsti*. *Planta*, **43**, 37–74.

210. EVANS, L. S. and VAN'T HOF, J. (1975). Is polyploidy necessary for tissue differentiation in higher plants? *Am. J. Bot.*, **62**, 1060–1064.

211. EVENARI, M. and GUTTERMAN, Y. (1973). Some notes on *Salvadora persica* L. in Sinai and its use as a toothbrush. *Flora*, **162**, 118–125.

212. EVERT, R. F., BORNMAN, C. H., BUTLER, V. and GILLILAND, M. G. (1973). Structure and development of the sieve-cell protoplast in leaf veins of *Welwitschia*. *Protoplasma*, **76**, 1–21.

213. EVERT, R. F., DAVIS, J. D., TUCKER, C. M. and ALFIERI, F. J. (1970). On the occurrence of nuclei in mature sieve elements. *Planta*, **95**, 281–296.

214. EVERT, R. F. and EICHHORN, S. E. (1974). Sieve-element ultrastructure in *Platycerium bifurcatum* and some other Polypodiaceous ferns: the nucleus. *Planta*, **119**, 301–318.

215. EVERT, R. F., ESCHRICH, W. and EICHHORN, S. E. (1973). P-protein distribution in mature sieve elements of *Cucurbita maxima*. *Planta*, **109**, 193–210.

216. EVERT, R. F., ESCHRICH, W., EICHHORN, S. E. and LIMBACH, S. T. (1973). Observations on penetration of barley leaves by the aphid *Rhopalosiphum maidis* (Fitch). *Protoplasma*, **77**, 95–110.

217. EVERT, R. F. and KOZLOWSKI, T. T. (1967). Effect of isolation of bark on cambial activity and development of xylem and phloem in trembling aspen. *Am. J. Bot.*, **54**, 1045–1055.

218. EYMÉ, J. and SUIRE, C. (1967). Au sujet de l'infrastructure des cellules de la région placentaire de *Mnium cuspidatum* Hedw. (Mousse bryale acrocarpe). *C. r. hebd. Séanc. Acad. Sci., Paris, sér. D*, Paris, **265**, 1788–1791.

219. FAHN, A. (1952). On the structure of floral nectaries. *Bot. Gaz.*, **113**, 464–470.

220. FAHN, A. (1974). *Plant Anatomy*. 2nd edition. Pergamon Press, Oxford.

221. FAHN, A. and ARNON, N. (1963). The living wood fibres of *Tamarix aphylla* and the changes occurring in them in transition from sapwood to heartwood. *New Phytol.*, **62**, 99–104.

222. FAHN, A., BEN-SASSON, R. and SACHS, T. (1972). The relation between the procambium and the cambium. In *Research Trends in Plant Anatomy*, GHOUSE, A. K. M. and YUNUS, M. (eds.), 161–170. Tata McGraw-Hill, New Delhi.

223. FAHN, A. and EVERT, R. F. (1974). Ultrastructure of the secretory ducts of *Rhus glabra* L. *Am. J. Bot.*, **61**, 1–14.

224. FAHN, A. and LESHEM, B. (1963). Wood fibres with living protoplasts. *New Phytol.*, **62**, 91–98.

225. FAHN, A. and RACHMILEVITZ, T. (1970). Ultrastructure and nectar secretion in *Lonicera japonica*. In *New Research in Plant Anatomy*, ROBSON, N. K. B., CUTLER, D. F. and GREGORY, M. (eds.) Suppl. 1 to *Bot. J. Linn. Soc.*, **63**, 51–56.

226. FAHN, A. and RACHMILEVITZ, T. (1975). An autoradiographical study of nectar secretion in *Lonicera japonica*. Thunb. *Ann. Bot.*, **39**, 975–976.

227. FAHN, A., SHOMER, I. and BEN-GERA, I. (1974). Occurrence and structure of epicuticular wax on the juice vesicles of citrus fruits. *Ann. Bot.*, **38**, 869–872.

228. FAHN, A., WAISEL, Y. and BENJAMINI, L. (1968). Cambial activity in *Acacia raddiana* Savi. *Ann. Bot.*, **32**, 677–686.

229. FAHN, A. and ZAMSKI, E. (1970). The influence of pressure, wind, wounding and growth substances on the rate of resin duct formation in *Pinus halepensis* wood. *Israel J. Bot.*, **19**, 429–446.

230. FAIRBAIRN, J. W. and KAPOOR, L. D. (1960). The laticiferous vessels of *Papaver somniferum* L. *Planta med.*, **8**, 49–61.

231. FAIRBAIRN, J. W., PALMER, J. M. and PATERSON, A. (1968). The alkaloids of *Papaver somniferum* L. VIII. Organelle activity of the isolated latex. *Phytochemistry*, **7**, 2117–2121.

232. FAIRBAIRN, J. W. and WASSEL, G. (1964a). The alkaloids of *Papaver somniferum* L. I. Evidence for a rapid turnover of the major alkaloids. *Phytochemistry*, **3**, 253–258.

233. FAIRBAIRN, J. W. and WASSEL, G. (1964b). The alkaloids of *Papaver somniferum* L. II. ^{14}C isotopic studies of the rapid changes in the major alkaloids. *Phytochemistry*, **3**, 577–582.

234. FAIRBAIRN, J. W. and WASSEL, G. (1964c). The alkaloids of *Papaver somniferum* L. III. Biosynthesis in the isolated latex. *Phytochemistry*, **3**, 583–585.

235. FIGIER, J. (1969). Incorporation de glycine-^3H chez les glandes pétiolaires de *Mercurialis annua* L. Étude radioautographique en microscopie électronique. *Planta*, **87**, 275–289.

236. FISCHER, A. (1884). *Untersuchungen über das Siebröhren-System der Cucurbitaceen.* Gebrüder Borntraeger, Berlin.

237. FISCHER, R. A. (1968). Stomatal opening: role of potassium uptake by guard cells. *Science, N.Y.*, **160**, 784–785.

238. FISHER, J. B. (1975). Eccentric secondary growth in *Cordyline* and other Agavaceae (Monocotyledonae) and its correlation with auxin distribution. *Am. J. Bot.*, **62**, 292–302.

239. FISHER, J. B., BURG, S. P. and KANG, B. G. (1974). Relationship of auxin transport to branch dimorphism in *Cordyline*, a woody monocotyledon. *Physiologia. Pl.*, **31**, 284–287.

240. FOARD, D. E. (1958). An experimental study of sclereid development in the leaf of *Camellia japonica. Pl. Physiol., Lancaster*, **33** (Suppl.), xli.

241. FOARD, D. E. (1959). Pattern and control of sclereid formation in the leaf of *Camellia japonica. Nature, Lond.*, **184**, 1663–1664.

242. FOARD, D. E. (1970). Differentiation in plant cells. In *Cell Differentiation*, SCHJEIDE, O. A. and DE VELLIS, J. (eds.), 575–602. Van Nostrand Reinhold Co., New York.

243. FOARD, D. E. and HABER, A. H. (1961). Anatomic studies of gamma-irradiated wheat growing without cell division. *Am. J. Bot.*, **48**, 438–446.

244. FOARD, D. E., HABER, A. H. and FISHMAN, T. N. (1965). Initiation of lateral root primordia without completion of mitosis and without cytokinesis in uniseriate pericycle. *Am. J. Bot.*, **52**, 580–590.

245. FORD, P. J. (1976). Control of gene expression during differentiation and development, 302–345 in Ref. 271.

246. FOSKET, D. E. (1970). The time course of xylem differentiation and its relation to deoxyribonucleic acid synthesis in cultured *Coleus* stem segments. *Pl. Physiol., Lancaster*, **46**, 64–68.

247. FOSKET, D. E. and ROBERTS, L. W. (1964). Induction of wound-vessel differentiation in isolated *Coleus* stem segments in vitro. *Am. J. Bot.*, **51**, 19–25.

248. FOSKET, D. E. and TORREY, J. G. (1969). Hormonal control of cell proliferation and xylem differentiation in cultured tissues of *Glycine max* var. Biloxi. *Pl. Physiol., Lancaster*, **44**, 871–880.

249. FOSTER, A. S. (1947). Structure and ontogeny of the terminal sclereids in the leaf of *Mouriria Huberi* Cogn. *Am. J. Bot.*, **34**, 501–514.

250. FOSTER, A. S. (1949). *Practical Plant Anatomy.* 2nd edition. Van Nostrand, New York.

251. FOSTER, A. S. (1955). Structure and ontogeny of terminal sclereids in *Boronia serrulata. Am. J. Bot.*, **42**, 551–560.

252. FRANK, E. (1969). Zur Bildung des Kristallidioblastenmusters bei *Canavalia ensiformis* DC. II. Zur Zellteilung in der Epidermis. *Z. Pflanzenphysiol.*, **60**, 403–413.

253. FRANK, E. and JENSEN, W. A. (1970). On the formation of the pattern of crystal idioblasts in *Canavalia ensiformis* DC. IV. The fine structure of the crystal cells. *Planta*, **95**, 202–217.

254. FRANKE, W. (1961). Ectodesmata and foliar absorption. *Am. J. Bot.*, **48**, 683–691.

255. FRENCH, J. C. and PAOLILLO, D. J., Jr. (1975). The effect of the calyptra on the plane of guard cell mother cell division in *Funaria* and *Physcomitrium* capsules. *Ann. Bot.*, **39**, 233–236.

256. FREY-WYSSLING, A. and MÜHLETHALER, K. (1965). *Ultrastructural Plant Cytology*. Elsevier, Amsterdam, London, New York.

257. FRYNS-CLAESSENS, E. and VAN COTTHEM, W. (1973). A new classification of the ontogenetic types of stomata. *Bot. Rev.*, **39**, 71–138.

258. FUKUDA, Y. (1967). Anatomical study of the internal phloem in the stems of dicotyledons, with special reference to its histogenesis. *J. Fac. Sci. Univ. Tokyo*, Sect. III, Bot., **9**, 313–375.

259. GALSTON, A. W. and DAVIES, P. J. (1969). Hormonal regulation in higher plants. *Science, N.Y.*, **163**, 1288–1297.

260. GAMBORG, O. L., CONSTABEL, F., KAO, K. N. and OHYAMA, K. (1975). Plant protoplasts in genetic modifications and production of intergeneric hybrids. In *Modification of the Information Content of Plant Cells* (Proc. 2nd John Innes Symp.), MARKHAM, R., DAVIES, D. R., HOPWOOD, D. A., HORNE, R. W. (eds.), 181–196. North-Holland Publishing Co., Amsterdam.

261. GAUDET, J. (1960). Ontogeny of foliar sclereids in *Nymphaea odorata*. *Am. J. Bot.*, **47**, 525–532.

262. GAY, A. P. and HURD, R. G. (1975). The influence of light on stomatal density in the tomato. *New Phytol.*, **75**, 37–46.

263. GEE, H. (1972). Localisation and uptake of ^{14}C-IAA in relation to xylem regeneration in *Coleus* internodes. *Planta*, **108**, 1–9.

264. GEESTERANUS, R. A. M. (1941). On the development of the stellate form of the pith cells of *Juncus* species. *Proc. K. ned. Akad. Wet.*, **44**, 489–501 and 648–653.

265. GIBSON, R. W. (1971). Glandular hairs providing resistance to aphids in certain wild potato species. *Ann. appl. Biol.*, **68**, 113–119.

266. GIBSON, R. W. (1974). Aphid-trapping glandular hairs on hybrids of *Solanum tuberosum* and *S. berthaultii*. *Potato Res.*, **17**, 152–154.

267. GILBERT, L. (1971). Butterfly-plant coevolution: Has *Passiflora adenopoda* won the selectional race with Heliconiine butterflies. *Science, N.Y.*, **172**, 585–586.

268. GILDER, J. and CRONSHAW, J. (1973a). Adenosine triphosphatase in the phloem of *Cucurbita*. *Planta*, **110**, 189–204.

269. GILDER, J. and CRONSHAW, J. (1973b). The distribution of adenosine triphosphatase activity in differentiating and mature phloem cells of *Nicotiana tabacum* and its relationship to phloem transport. *J. Ultrastruct. Res.*, **44**, 388–404.

270. GOODWIN, R. H. (1942). On the development of xylary elements in the first internode of *Avena* in dark and light. *Am. J. Bot.*, **29**, 818–828.

271. GRAHAM, C. F. and WAREING, P. F., eds. (1976). *The Developmental Biology of Plants and Animals*. Blackwell Scientific Publications, Oxford.

272. GRAHAM, J. S. D., JENNINGS, A. C., MORTON, R. K., PALK, B. A. and RAISON, J. K. (1962). Protein bodies and protein synthesis in developing wheat endosperm. *Nature, Lond.*, **196**, 967–969.

273. GRANICK, S. (1961). The chloroplasts: inheritance, structure, and function. In *The Cell*, Vol. II, BRACHET, J. and MIRSKY, A. E. (eds.), 489–619. Academic Press, New York and London.

274. GREW, N. (1682). *The Anatomy of Plants*. Johnson Reprint Corp., New York and London, 1965.

275. GROBSTEIN, C. (1966). What we do not know about differentiation. *Am. Zoologist*, **6**, 89–95.

276. GUNNING, B. E. S. (1976a). Introduction to plasmodesmata. In *Intercellular Communication in Plants: Studies on Plasmodesmata*, GUNNING, B. E. S. and ROBARDS, A. W. (eds.), 1–13. Springer-Verlag, Berlin.

277. GUNNING, B. E. S. (1976b). The role of plasmodesmata in short distance transport to and from the phloem. In *Intercellular Communication in Plants: Studies on Plasmodesmata*, GUNNING, B. E. S. and ROBARDS, A. W. (eds.), 203–227. Springer-Verlag, Berlin.

278. GUNNING, B. E. S. and HUGHES, J. E. (1976). Quantitative assessment of symplastic transport of pre-nectar into the trichomes of *Abutilon* nectaries. *Aust. J. Pl. Physiol.*, **3**, 619–637.

279. GUNNING, B. E. S. and PATE, J. S. (1969a). 'Transfer cells'. Plant cells with wall ingrowths, specialized in relation to short distance transport of solutes—their occurrence, structure, and development. *Protoplasma*, **68**, 107–133.

280. GUNNING, B. E. S. and PATE, J. S. (1969b). Cells with wall ingrowths (transfer cells) in the placenta of ferns. *Planta*, **87**, 271–274.

281. GUNNING, B. E. S. and PATE, J. S. (1974). Transfer cells. In *Dynamic Aspects of Plant Ultrastructure*, ROBARDS, A. W. (ed.), 441–480. McGraw-Hill, London.

282. GUNNING, B. E. S., PATE, J. S. and BRIARTY, L. G. (1968). Specialized 'transfer cells' in minor veins of leaves and their possible significance in phloem translocation. *J. Cell Biol.*, **37**, C7–C12.

283. GUNNING, B. E. S., PATE, J. S. and GREEN, L. W. (1970). Transfer cells in the vascular system of stems: taxonomy, association with nodes, and structure. *Protoplasma*, **71**, 147–171.

284. GUNNING, B. E. S. and ROBARDS, A. W. (eds.) (1976a). *Intercellular Communication in Plants: Studies on Plasmodesmata*. Springer-Verlag, Berlin.

285. GUNNING, B. E. S. and ROBARDS, A. W. (1976b). Plasmodesmata: current knowledge and outstanding problems. In *Intercellular Communication in Plants: Studies on Plasmodesmata*, GUNNING, B. E. S. and ROBARDS, A. W. (eds.), 297–311. Springer-Verlag, Berlin.

286. GUNNING, B. E. S. and STEER, M. W. (1975). *Ultrastructure and the Biology of Plant Cells*. Edward Arnold, London.

287. GURDON, J. B. (1970). The autonomy of nuclear activity in multicellular organisms. *Symp. Soc. exp. Biol.*, **24**, 369–378.

288. GURDON, J. B. (1973). *Gene expression during cell differentiation. Oxford Biol. Readers*, **25**. Oxford University Press.

289. GURDON, J. B. (1974). *The Control of Gene Expression in Animal Development*. Oxford University Press.

290. GURDON, J. B. (1976). The pluripotentiality of cell nuclei, 55–72 in Ref. 271.

291. HABER, A. H. (1962). Nonessentiality of concurrent cell divisions for degree of polarization of leaf growth. I. Studies with radiation-induced mitotic inhibition. *Am. J. Bot.*, **49**, 583–589.

292. HABER, A. H. and FOARD, D. E. (1963). Nonessentiality of concurrent cell divisions for degree of polarization of leaf growth. II. Evidence from untreated plants and from chemically induced changes of the degree of polarization. *Am. J. Bot.*, **50**, 937–944.

293. HABERLANDT, G. (1914). *Physiological Plant Anatomy*. Macmillan, London.

293a. HALL, D. M. (1967). Wax microchannels in the epidermis of white clover. *Science, N.Y.*, **158**, 505–506.

294. HALL, M. A. (1976). Hormones and differentiation in plants, 216–231 in Ref. 271.

295. HALL, S. M. and MEDLOW, G. C. (1974). Identification of IAA in phloem and root pressure saps of *Ricinus communis* L. by mass spectrometry. *Planta*, **119**, 257–261.

296. HALLAM, N. D. (1970). Growth and regeneration of waxes on the leaves of *Eucalyptus*. *Planta*, **93**, 257–268.

297. HALLAM, N. D. and JUNIPER, B. E. (1971). The anatomy of the leaf surface. In *Ecology of Leaf Surface Micro-organisms*, PREECE, T. F. and DICKINSON, C. H. (eds.), 3–37. Academic Press, London.

298. HALLICK, R. B., LIPPER, C., RICHARDS, O. C. and RUTTER, W. J. (1976). Isolation of a transcriptionally active chromosome from chloroplasts of *Euglena gracilis*. *Biochemistry*, **15**, 3039–3045.

299. HALPERIN, W. (1966). Alternative morphogenetic events in cell suspensions. *Am. J. Bot.*, **53**, 443–453.

300. HALPERIN, W. (1969). Morphogenesis in cell cultures. *A. Rev. Pl. Physiol.*, **20**, 395–418.

301. HALPERIN, W. and JENSEN, W. A. (1967). Ultrastructural changes during growth and embryogenesis in carrot cell cultures. *J. Ultrastruct. Res.*, **18**, 428–443.

302. HÄMMERLING, J. (1963). Nucleo-cytoplasmic interactions in *Acetabularia* and other cells. *A. Rev. Pl. Physiol.*, **14**, 65–92.

303. HAMMOND, C. T. and MAHLBERG, P. G. (1973). Morphology of glandular hairs of *Cannabis sativa* from scanning electron microscopy. *Am. J. Bot.*, **60**, 524–528.

304. HANSON, A. D. and EDELMAN, J. (1970). Phloem in carrot calluses. *Planta*, **93**, 171–174.

305. HARRIS, K. F. and BRADLEY, R. H. E. (1973). Importance of leaf hairs in the transmission of tobacco mosaic virus by aphids. *Virology*, **52**, 295–300.

306. HARRIS, T. M. (1956). The fossil plant cuticle. *Endeavour*, **15**, 210–214.

307. HATCH, M. D. and SLACK, C. R. (1966). Photosynthesis by sugar-cane leaves. A new carboxylation reaction and the pathway of sugar formation. *Biochem. J.*, **101**, 103–111.

308. HAYWARD, D. M. and PARRY, D. W. (1975). Scanning electron microscopy of silica deposition in the leaves of barley (*Hordeum sativum* L.). *Ann. Bot.*, **39**, 1003–1009.

309. HAYWARD, P. (1974). Waxy structures in the lenticels of potato tubers and their possible effects on gas exchange. *Planta*, **120**, 273–277.

310. HEATH, O. V. S. (1975). *Stomata*. *Oxford Biol. Readers*, **37**. Oxford University Press.

311. HEINRICH, G. (1970). Elektronenmikroskopische Untersuchung der Milchröhren von *Ficus elastica*. *Protoplasma*, **70**, 317–323.

312. HELBAEK, H. (1972). Plant anatomy in culture—historical research. In *Research Trends in Plant Anatomy*, GHOUSE, A. K. M. and YUNUS, M. (eds.), 19–32. Tata McGraw-Hill, Bombay and New Delhi.

313. HEPLER, P. K. and FOSKET, D. E. (1971). The role of microtubules in vessel member differentiation in *Coleus*. *Protoplasma*, **72**, 213–236.

314. HEPLER, P. K. and NEWCOMB, E. H. (1963). The fine structure of young tracheary xylem elements arising by redifferentiation of parenchyma in wounded *Coleus* stem. *J. exp. Bot.*, **14**, 496–503.

315. HEPLER, P. K. and NEWCOMB, E. H. (1964). Microtubules and fibrils in the cytoplasm of *Coleus* cells undergoing secondary wall deposition. *J. Cell Biol.*, **20**, 529–533.

316. HEPLER, P. K. and PALEVITZ, B. A. (1974). Microtubules and microfilaments. *A. Rev. Pl. Physiol.*, **25**, 309–362.

317. HEPLER, P. K., RICE, R. M. and TERRANOVA, W. A. (1972). Cytochemical localization of peroxidase activity in wound vessel members of *Coleus*. *Can. J. Bot.*, **50**, 977–983.

318. HEPTON, C. E. L. and PRESTON, R. D. (1960). Electron microscopic observations of the structure of sieve-connexions in the phloem of angiosperms and gymnosperms. *J. exp. Bot.*, **11**, 381–394.

319. HESLOP-HARRISON, J. (1963). Structure and morphogenesis of lamellar systems in grana-containing chloroplasts. I. Membrane structure and lamellar architecture. *Planta*, **60**, 243–260.

320. HESLOP-HARRISON, J. (1967). Differentiation. *A. Rev. Pl. Physiol.*, **18**, 325–348.

321. HESLOP-HARRISON, J. (1975). Incompatibility and the pollen-stigma interaction. *A. Rev. Pl. Physiol.*, **26**, 403–425.

322. HESLOP-HARRISON, J., KNOX, R. B., HESLOP-HARRISON, Y. and MATTSSON, O. (1975). Pollen-wall proteins: emission and role in incompatibility responses. In *The Biology of the Male Gamete*, DUCKETT, J. G. and RACEY, P. A. (eds.), *Biol. J. Linn. Soc.*, **7**, *suppl.* **1**, 189–202.

323. HESLOP-HARRISON, Y. (1970). Scanning electron microscopy of fresh leaves of *Pinguicula*. *Science, N.Y.*, **167**, 172–174.

324. HESLOP-HARRISON, Y. (1975). Enzyme release in carnivorous plants. In *Lysosomes in Biology and Pathology*, DINGLE, J. T. and DEAN, R. T. (eds.), 525–578. North Holland Publishing Co., Amsterdam.

325. HESLOP-HARRISON, Y. (1976). Enzyme secretion and digest uptake in carnivorous plants. In *Perspectives in Experimental Biology*, SUNDERLAND, N. (ed.), Vol. **2**, Botany, 463–476. Pergamon Press, Oxford.

326. HESLOP-HARRISON, Y. and KNOX, R. B. (1971). A cytochemical study of the leaf-gland enzymes of insectivorous plants of the genus *Pinguicula*. *Planta*, **96**, 183–211.

327. HESS, T. and SACHS, T. (1972). The influence of a mature leaf on xylem differentiation. *New Phytol.*, **71**, 903–914.

328. HINCHMAN, R. R. and GORDON, S. A. (1974). Amyloplast size and number in gravity-compensated oat seedlings. *Pl. Physiol., Lancaster*, **53**, 398–401.

329. HOFF, J. E. (1971). Potato protein crystals. *Hortscience*, **6**, 190.

330. HORNER, H. T., Jr. and LERSTEN, N. R. (1968). Development, structure and function of secretory trichomes in *Psychotria bacteriophila* (Rubiaceae). *Am. J. Bot.*, **55**, 1089–1099.

331. HOWE, K. J. and STEWARD, F. C. (1962). Anatomy and development of *Mentha piperita* L. *Mem. Cornell Univ. agric. Exp. Stn*, **379**, 11–40.

332. HUANG, C. S. and MAGGENTI, A. R. (1969). Wall modifications in developing giant cells of *Vicia faba* and *Cucumis sativus* induced by root knot nematode, *Meloidogyne javanica*. *Phytopathology*, **59**, 931–937.

333. HULBARY, R. L. (1944). The influence of air spaces on the three-dimensional shapes of cells in *Elodea* stems, and a comparison with pith cells of *Ailanthus*. *Am. J. Bot.*, **31**, 561–580.

334. HUMBLE, G. D., FISCHER, R. A. and HSIAO, T. C. (1970). Potassium role found essential in stomatal functioning for plant life. *Calif. Agric.*, **4**, no. 4, 10.

335. INAMDAR, J. A. (1968a). Epidermal structure and ontogeny of stomata in some Verbenaceae. *Ann. Bot.*, **33**, 55–66.

336. INAMDAR, J. A. (1968b). Ontogeny of stomata in some Oleaceae. *Proc. Indian Acad. Sci.*, **67**, 157–164.

337. INAMDAR, J. A. (1968c). Development of stomata in vegetative and floral organs in some Caryophyllaceae. *Aust. J. Bot.*, **16**, 445–449.

338. INAMDAR, J. A., GOPAL, B. V. and CHOHAM, A. J. (1968). Development of normal and abnormal stomata in some Araliaceae. *Ann. Bot.*, **33**, 67–73.

339. INGLE, J. and TIMMIS, J. N. (1975). A role for differential replication of DNA in development. In *Modification of the Information Content of Plant Cells* (Proc. 2nd John Innes Symp.), MARKHAM, R., DAVIES, D. R., HOPWOOD, D. A., HORNE, R. W. (eds.), 37–52. North-Holland Publishing Co., Amsterdam.

340. JACOBS, W. P. (1952). The role of auxin in differentiation of xylem around a wound. *Am. J. Bot.*, **39**, 301–309.

341. JACOBS, W. P. (1954). Acropetal auxin transport and xylem regeneration—a quantitative study. *Am. Nat.*, **88**, 327–337.

342. JACOBS, W. P. (1956). Internal factors controlling cell differentiation in the flowering plants. *Am. Nat.*, **90**, 163–169.

343. JACOBS, W. P. (1970). Regeneration and differentiation of sieve tube elements. *Int. Rev. Cytol.*, **28**, 239–273.

344. JACOBS, W. P. and MORROW, I. B. (1957). A quantitative study of xylem development in the vegetative shoot apex of *Coleus*. *Am. J. Bot.*, **44**, 823–842.

345. JACOBS, W. P. and MORROW, I. B. (1958). Quantitative relations between stages of leaf development and differentiation of sieve tubes. *Science, N.Y.*, **128**, 1084–1085.

346. JACOBS, W. P. and MORROW, I. B. (1967). A quantitative study of sieve-tube differentiation in vegetative shoot apices of *Coleus*. *Am. J. Bot.*, **54**, 425–431.

347. JACOBSON, S. L. (1974). Effect of ionic environment on the response of the sensory hair of Venus's-flytrap. *Can. J. Bot.*, **52**, 1293–1302.

348. JACOLI, G. G. (1974). Translocation of mycoplasma-like bodies through sieve pores in plant tissue cultures infected with aster yellows. *Can. J. Bot.*, **52**, 2085–2088.

349. JAFFE, L. F. (1968). Localization in the developing *Fucus* egg and the general role of localizing currents. *Adv. Morphogen.*, **7**, 295–328.

350. JAGELS, R. (1973). Studies of a marine grass, *Thalassia testudinum*. I. Ultrastructure of the osmoregulatory leaf cells. *Am. J. Bot.*, **60**, 1003–1009.

351. JEFFREE, C. E., BAKER, E. A. and HOLLOWAY, P. J. (1975). Ultrastructure and recrystallization of plant epicuticular waxes. *New Phytol.*, **75**, 539–549.

352. JEFFS, R. A. and NORTHCOTE, D. H. (1967). The influence of indol-3yl-acetic acid and sugar on the pattern of induced differentiation in plant tissue culture. *J. Cell Sci.*, **2**, 77–88.

353. JENNINGS, J. D. (1957). Danger Cave. *Anthrop. Pap. Univ. Utah*, **27**, 1–328.

354. JENSEN, W. A. (1961). Relation of primary cell wall formation to cell development in plants. In *Synthesis of Molecular and Cellular Structure*, RUDNICK, D. (ed.) (Symp. Dev. Growth, **19**), 89–110.

355. JENSEN, W. A. (1964a). *The Plant Cell.* Wadsworth, Belmont, California.

356. JENSEN, W. A. (1964b). Cell development during plant embryogenesis. *Brookhaven Symp. Biol.*, **16**, 179–202.

357. JOHNSON, B. (1956). The influence on aphids of the glandular hairs on tomato plants. *Plant Path.*, **5**, 131–132.

358. JOHNSON, L. E. B., WILCOXSON, R. D. and FROSHEISER, F. I. (1975). Transfer cells in tissues of the reproductive system of alfalfa. *Can. J. Bot.*, **53**, 952–956.

359. JONES, M. G. K. (1976). The origin and development of plasmodesmata. In *Intercellular Communication in Plants: Studies on Plasmodesmata*, GUNNING, B. E. S. and ROBARDS, A. W. (eds.), 81–105. Springer-Verlag, Berlin.

360. JONES, M. G. K. and DROPKIN, V. H. (1976). Scanning electron microscopy of nematode-induced giant transfer cells. *Cytobios*, **15**, 149–161.

361. JONES, M. G. K. and NORTHCOTE, D. H. (1972a). Nematode-induced syncytium—a multinucleate transfer cell. *J. Cell Sci.*, **10**, 789–809.

362. JONES, M. G. K. and NORTHCOTE, D. H. (1972b). Multinucleate transfer cells induced in *Coleus* roots by the root-knot nematode, *Meloidogyne arenaria. Protoplasma*, **75**, 381–395.

363. JONES, M. G. K., NOVACKY, A. and DROPKIN, V. H. (1974). 'Action potentials' in nematode-induced plant transfer cells. *Protoplasma*, **80**, 401–405.

364. JOST, L. (1893). Über Beziehungen zwischen der Blattentwickelung und der Gefässbildung in den Pflanze. *Bot. Ztg*, **51**, 89–138.

365. JUNIPER, B. E. (1959). Growth, development, and effect of the environment on the ultra-structure of plant surfaces. *J. Linn. Soc. (Bot.)*, **56**, 413–419.

366. JUNIPER, B. E. and GILCHRIST, A. J. (1976). Absorption and transport of calcium in the stalked glands of *Drosera capensis* L. In *Perspectives in Experimental Biology*, SUNDERLAND, N. (ed.), Vol. **2**, Botany, 477–486. Pergamon Press, Oxford.

367. JUNIPER, B. E. and ROBERTS, R. M. (1966). Polysaccharide synthesis and the fine structure of root cells. *Jl R. microsc. Soc.*, **85**, 63–72.

368. KAGAWA, T. and BEEVERS, H. (1975). The development of microbodies (glyoxysomes and leaf peroxisomes) in cotyledons of germinating watermelon seedlings. *Pl. Physiol., Lancaster*, **55**, 258–264.

369. KAHL, G., ROSENSTOCK, G. and LANGE, H. (1969). Die Trennung von Zellteilung und Suberinsynthese in dereprimiertem pflanzlichem Speichergewebe durch tris-(hydroxymethyl-) Aminomethan. *Planta*, **87**, 365–371.

370. KARLING, J. S. (1916). The laticiferous system of *Achras zapota* L. I. A preliminary account of the origin, structure, and distribution of the latex vessels in the apical meristem. *Am. J. Bot.*, **16**, 803–824.

371. KAUFMAN, P. B., BIGELOW, W. C., SCHMID, R. and GHOSHEH, N. S. (1971). Electron microprobe analysis of silica in epidermal cells of *Equisetum*. *Am. J. Bot.*, **58**, 309–316.

372. KAUFMAN, P. B. and CASSELL, S. J. (1963). Striking features in the development of internodal epidermis in the oat plant (*Avena sativa*). *Mich. Bot.*, **2**, 115–121.

373. KAUFMAN, P. B., PETERING, L. B. and SMITH, J. G. (1970). Ultrastructural development of cork-silica cell pairs in *Avena* internodal epidermis. *Bot. Gaz.*, **131**, 173–185.

374. KAUFMAN, P. B., PETERING, L. B. and SONI, S. L. (1970). Ultrastructural studies on cellular differentiation in internodal epidermis of *Avena sativa*. *Phytomorphology*, **20**, 281–309.

375. KAUFMAN, P. B., PETERING, L. B., YOCUM, C. S. and BAIC, D. (1970). Ultrastructural studies on stomata development in internodes of *Avena sativa*. *Am. J. Bot.*, **57**, 33–49.

376. KAUL, R. B. (1971). Diaphragms and aerenchyma in *Scirpus validus*. *Am. J. Bot.*, **58**, 808–816.

377. KAUL, R. B. (1972). Adaptive leaf architecture in emergent and floating *Sparganium*. *Am. J. Bot.*, **59**, 270–278.

378. KAUL, R. B. (1973). Development of foliar diaphragms in *Sparganium eurycarpum*. *Am. J. Bot.*, **60**, 944–949.

379. KIDWAI, P. and ROBARDS, A. W. (1969). On the ultrastructure of resting cambium of *Fagus sylvatica* L. *Planta*, **89**, 361–368.

380. KING, N. J. and BAYLEY, S. T. (1965). A preliminary analysis of the proteins of the primary walls of some plant cells. *J. exp. Bot.*, **16**, 294–303.

381. KIRCHANSKI, S. J. (1975). The ultrastructural development of the dimorphic plastids of *Zea mays* L. *Am. J. Bot.*, **62**, 695–705.

382. KIRK, J. T. O. and TILNEY-BASSETT, R. A. E. (1967). *The Plastids*. Freeman, San Francisco and London.

383. KIRSCHNER, H., SACHS, T. and FAHN, A. (1971). Secondary xylem reorientation as a special case of vascular tissue differentiation. *Israel J. Bot.*, **20**, 184–198.

384. KOHLENBACH, H. W. and SCHMIDT, B. (1975). Cytodifferenzierung in Form einer direkten Umwandlung isolierter Mesophyllzellen zu Tracheiden. *Z. PflPhysiol.*, **75**, 369–374.

385. KOLLMANN, R. (1964). On the fine structure of the sieve element protoplast. *Phytomorphology*, **14**, 247–264.

386. KORN, R. W. (1972). Arrangement of stomata on the leaves of *Pelargonium zonale* and *Sedum stahlii*. *Ann. Bot.*, **36**, 325–333.

387. KORN, R. W. (1974). The three-dimensional shape of plant cells and its relationship to pattern of tissue growth. *New Phytol.*, **73**, 927–935.

388. KORN, R. W. and FREDRICK, G. W. (1973). Development of D-type stomata in the leaves of *Ilex crenata* var. *convexa*. *Ann. Bot.*, **37**, 647–656.

389. KORN, R. W. and SPALDING, R. M. (1973). The geometry of plant epidermal cells. *New Phytol.*, **72**, 1357–1365.

390. KORTSCHAK, H. P., HARTT, C. E. and BURR, G. O. (1965). Carbon dioxide fixation in sugarcane leaves. *Pl. Physiol., Lancaster*, **40**, 209–213.

391. KROPFITSCH, M. (1951). Apfelgas—Wirkung auf Stomatazahl. *Protoplasma*, **40**, 256–265.

392. KUNDU, B. C. (1942). The anatomy of two Indian fibre plants, *Cannabis* and *Corchorus*, with special reference to the fibre distribution and development. *J. Indian bot. Soc.*, **21**, 93–128.

393. KUNDU, B. C. and SEN, S. (1961). Origin and development of fibres in ramie (*Boehmeria nivea* Gaud.). *Proc. natn. Inst. Sci. India*, B (Suppl.), **26**, 190–198.

394. LAETSCH, W. M. (1968). Chloroplast specialization in dicotyledons possessing the C_4-dicarboxylic acid pathway of photosynthetic CO_2 fixation. *Am. J. Bot.*, **55**, 875–883.

395. LAETSCH, W. M. (1969). Relationship between chloroplast structure and photosynthetic carbon-fixation pathways. *Sci. Prog., Oxf.*, **57**, 323–351.
396. LAETSCH, W. M. (1974). The C_4 syndrome: a structural analysis. *A. Rev. Pl. Physiol.*, **25**, 27–52.
397. LAETSCH, W. M. and KORTSCHAK, H. P. (1972). Chloroplast structure and function in tissue cultures of a C_4 plant. *Pl. Physiol., Lancaster*, **49**, 1021–1023.
398. LAETSCH, W. M. and PRICE, I. (1969). Development of the dimorphic chloroplasts of sugar cane. *Am. J. Bot.*, **56**, 77–87.
399. LAMOTTE, C. E. and JACOBS, W. P. (1962). Quantitative estimation of phloem regeneration in *Coleus* internodes. *Stain Technol.*, **37**, 63–73.
400. LAMOTTE, C. E. and JACOBS, W. P. (1963). A role of auxin in phloem regeneration in *Coleus* internodes. *Devl Biol.*, **8**, 80–98.
401. LAMPORT, D. T. A. (1965). The protein component of primary cell walls. *Adv. Bot. Res.*, **2**, 151–218.
402. LANG, A. (1974). Inductive phenomena in plant development. In *Basic Mechanisms in Plant Morphogenesis. Brookhaven Symp. Biol.*, **25**, 129–144.
403. LANGE, H., ROSENSTOCK, G. and KAHL, G. (1970) Induktionsbedingungen der Suberinsynthese und Zellproliferation bei Parenchymfragmenten der Kartoffelknolle. *Planta*, **90**, 109–118.
404. LÄUCHLI, A., KRAMER, D., PITMAN, M. G. and LÜTTGE, U. (1974). Ultrastructure of xylem parenchyma cells of barley roots in relation to ion transport to the xylem. *Planta*, **119**, 85–99.
405. LAWRENCE, P. A. and HAYWARD, P. (1971). The development of a simple pattern: spaced hairs in *Oncopeltus fasciatus. J. Cell Sci.*, **8**, 513–524.
406. LAWTON, J. R. S. and CANNY, M. J. (1970). The proportion of sieve elements in the phloem of some tropical trees. *Planta*, **95**, 351–354.
407. LEDBETTER, M. C. and KRIKORIAN, A. D. (1975). Trichomes of *Cannabis sativa* as viewed with scanning electron microscope. *Phytomorphology*, **25**, 166–176.
408. LEDBETTER, M. C. and PORTER, K. R. (1963). A 'microtubule' in plant cell fine structure. *J. Cell Biol.*, **19**, 239–250.
409. LEECH, J. H., MOLLENHAUER, H. H. and WHALEY, W. G. (1963). Ultrastructural changes in the root apex. *Symp. Soc. exp. Biol.*, **17**, 74–84.
410. LEPP, N. W. and PEEL, A. J. (1969). Some effects of IAA and kinetin upon the movement of sugars in the phloem of willow. *Planta*, **90**, 230–235.
411. LEROUX, R. (1954). Recherches sur les modifications anatomiques de trois espèces d'osiers (*Salix viminalis* L., *Salix purpurea* L., *Salix fragilis* L.) provoquées par l'acide naphtalène-acétique. *C. r. Séanc. Soc. Biol.*, **148**, 284–286.
412. LETVENUK, L. J. and PETERSON, R. L. (1976). Occurrence of transfer cells in vascular parenchyma of *Hieracium florentinum* roots. *Can. J. Bot.*, **54**, 1458–1471.
413. LEVERING, C. A. and THOMSON, W. W. (1971). The ultrastructure of the salt gland of *Spartina foliosa. Planta*, **97**, 183–196.
414. LEVIN, D. A. (1973). The role of trichomes in plant defense. *Q. Rev. Biol.*, **48**, 3–15.
415. LEWIS, D. (1970). Conclusions: organelles as membrane complexes. *Symp. Soc. exp. Biol.*, **24**, 497–501.

416. LIESE, W. (1965). The fine structure of bordered pits in softwoods. In *Cellular Ultrastructure of Woody Plants*, CÔTÉ, W. A., Jr. (ed.), 271–290. Syracuse University Press.

417. LINSBAUER, K. (1930). *Die Epidermis. Handb. Pflanzenanat.* Band IV. Abteilung 1. Teil 2: *Histologie.* Gebrüder Borntraeger, Berlin.

418. LIPETZ, J. (1970). Wound healing in higher plants. *Int. Rev. Cytol.*, **27**, 1–28.

419. LIPHSCHITZ, N., ADIVA-SHOMER-ILAN, ESHEL, A. and WAISEL, Y. (1974). Salt glands on leaves of Rhodes grass (*Chloris gayana* Kth). *Ann. Bot.*, **38**, 459 462.

420. LIPHSCHITZ, N. and WAISEL, Y. (1970a). Environmental effects on wood production and cambial activity in *Ziziphus spina-christi* (L.) Willd. *Israel J. Bot.*, **19**, 592–598.

421. LIPHSCHITZ, N. and WAISEL, Y. (1970b). Phellogen initiation in the stems of *Eucalyptus camaldulensis* Dehnh. *Aust. J. Bot.*, **18**, 185–189.

422. LIST, A., Jr. (1963). Some observations on DNA content and cell and nuclear volume growth in the developing xylem cells of certain higher plants. *Am. J. Bot.*, **50**, 320–329.

423. LLOYD, F. E. (1942) *Carnivorous Plants.* Chronica Botanica, Waltham, Mass.

423a. LOTT, J. N. A. (1976). *A Scanning Electron Microscope Study of Green Plants.* C. V. Mosby Company, St. Louis.

424. LOTT, J. N. A., LARSEN, P. L. and DARLEY, J. J. (1971). Protein bodies from the cotyledons of *Cucurbita maxima. Can. J. Bot.*, **49**, 1777–1782.

425. LOWARY, P. A. and AVERS, C. J. (1965). Nucleolar variation during differentiation of *Phleum* root epidermis. *Am. J. Bot.*, **52**, 199–203.

426. LÜTTGE, U. (1971). Structure and function of plant glands. *A. Rev. Pl. Physiol.*, **22**, 23–44.

427. LYON, N. C. and MUELLER, W. C. (1974). A freeze-etch study of plant cell walls for ectodesmata. *Can. J. Bot.*, **52**, 2033–2036.

428. MACDOUGAL, D. T. (1926). Growth and permeability of century-old cells. *Am. Nat.*, **60**, 393–415.

429. MACNEISH, R. S. (1964). Ancient Mesoamerican civilization. *Science, N.Y.*, **143**, 531–537.

430. MACROBBIE, E. A. C. (1971). Phloem translocation. Facts and mechanisms: a comparative survey. *Biol. Rev.*, **46**, 429–481.

431. MAHLBERG, P. G. (1959a). Karyokinesis in the non-articulated laticifers of *Nerium oleander* L. *Phytomorphology*, **9**, 110–118.

432. MAHLBERG, P. G. (1959b). Development of the non-articulated laticifer in proliferated embryos of *Euphorbia marginata* Pursh. *Phytomorphology*, **9**, 156–162.

433. MAHLBERG, P. G. (1961). Embryogeny and histogenesis in *Nerium oleander*. II. Origin and development of the non-articulated laticifer. *Am. J. Bot.*, **48**, 90–99.

434. MAHLBERG, P. G. (1963). Development of non-articulated laticifer in seedling axis of *Nerium oleander. Bot. Gaz.*, **124**, 224–231.

435. MAHLBERG, P. G. (1973). Scanning electron microscopy of starch grains from latex of *Euphorbia terracina* and *E. tirucalli. Planta*, **110**, 77–80.

436. MAHLBERG, P. G. (1975). Evolution of the laticifer in *Euphorbia* as interpreted from starch grain morphology. *Am. J. Bot.*, **62**, 577–583.

437. MAHLBERG, P. G. and SABHARWAL, P. S. (1967). Mitosis in the non-articulated laticifer of *Euphorbia marginata. Am. J. Bot.*, **54**, 465–472.

438. MAHLBERG, P. G. and SABHARWAL, P. S. (1968). Origin and early development of nonarticulated laticifers in embryos of *Euphorbia marginata*. *Am. J. Bot.*, **55**, 375–381.
439. MARGULIS, L. (1970). *Origin of Eukaryotic Cells*. Yale University Press, New Haven.
440. MARTIN, D. J. (1955). Features on plant cuticle. An aid to the analysis of the natural diet of grazing animals, with especial reference to Scottish hill sheep. *Trans. Proc. bot. Soc. Edinb.*, **36**, 278–288.
441. MARTIN, J. T. and JUNIPER, B. E. (1970). *The Cuticles of Plants*. Edward Arnold, London.
442. MARTIN, P. G. (1961). Evidence for the continuity of nucleolar material in mitosis. *Nature, Lond.*, **190**, 1078–1079.
443. MARTIN, R. J. L. and STOTT, G. L. (1957). The physical factors involved in the drying of sultana grapes. *Aust. J. agric. Res.*, **8**, 444–459.
444. MATILE, Ph., JANS, B. and RICKENBACHER, R. (1970). Vacuoles of *Chelidonium* latex: lysosomal property and accumulation of alcaloids. *Biochem. Physiol. Pflanzen*, **161**, 447–458.
445. MELARAGNO, J. E. and WALSH, M. A. (1976). Ultrastructural features of developing sieve elements in *Lemna minor* L.—the protoplast. *Am. J. Bot.*, **63**, 1145–1157.
446. MELCHERS, G., KELLER, W. and LABIB, G. (1975). Somatic hybridisation. In *Modification of the Information Content of Plant Cells* (Proc. 2nd John Innes Symp.), MARKHAM, R., DAVIES, D. R., HOPWOOD, D. A., HORNE, R. W. (eds), 161–168. North-Holland Publishing Co., Amsterdam.
447. MERCER, F. V. and RATHGEBER, N. (1962). Nectar secretion and cell membranes. In *Electron Microscopy*. Breese, S. S., Jr. (ed.), 5th Int. Congr. for Electron Microscopy, Philadelphia. Vol. **2**, WW11–12. Academic Press, New York and London.
448. METCALFE, C. R. (1961). The anatomical approach to systematics. General introduction with special reference to recent work on monocotyledons. In *Recent Advances in Botany*, 146–150. Toronto University Press.
449. METCALFE, C. R. (1967). Distribution of latex in the plant kingdom. *Econ. Bot.*, **21**, 115–127.
450. METCALFE, C. R. (1972). Botanical communication with special reference to plant anatomy. In *Research Trends in Plant Anatomy*, GHOUSE, A. K. M. and YUNUS, M. (eds.), 7–17. Tata McGraw-Hill, Bombay and New Delhi.
451. METCALFE, C. R. and CHALK, L. (1950). *Anatomy of the Dicotyledons*. Vols. I and II. Clarendon Press, Oxford.
452. MEYER, J. (1959). Le caractère précocement idioblastique des initiales stomatiques du pétiole de *Populus pyramidalis* Rozier. *Protoplasma*, **51**, 313–319.
453. MEYLAN, B. A. and BUTTERFIELD, B. G. (1972a). *Three-dimensional Structure of Wood*. Reed Education, Hong Kong.
454. MEYLAN, B. A. and BUTTERFIELD, B. G. (1972b). Perforation plate development in *Knightia excelsa* R. Br.: a scanning electron microscope study. *Aust. J. Bot.*, **20**, 79–86.
455. MEYLAN, B. A. and BUTTERFIELD, B. G. (1973). Unusual perforation plates: observations using scanning electron microscopy. *Micron*, **4**, 47–59.
456. MIA, A. J. and PATHAK, S. M. (1965). Histochemical studies of sclereid induction in the shoot of *Rauwolfia* species. *J. exp. Bot.*, **16**, 177–181.

457. MINOCHA, S. C. and HALPERIN, W. (1976). Enzymatic changes and lignification in relation to tracheid differentiation in cultured tuber tissue of Jerusalem artichoke (*Helianthus tuberosus*). *Can. J. Bot.*, **54**, 79–89.

458. MITCHELL, J. W. and WORLEY, J. F. (1964). Intracellular transport apparatus of phloem fibers. *Science, N.Y.*, **145**, 409–410.

459. MOLLENHAUER, H. H. and MORRÉ, D. J. (1966). Golgi apparatus and plant secretion. *A. Rev. Pl. Physiol.*, **17**, 27–46.

460. MOOR, H. (1959). Platin-Kohle-Abdrouk-Technik angewandt auf den Feinbau der Milchröhren. *J. Ultrastruct. Res.*, **2**, 393–422.

461. MOREY, P. R. (1973). *How Trees Grow*. Studies in Biol., **39**. Edward Arnold, London.

462. MOREY, P. R. and DAHL, B. E. (1975). Histological and morphological effects of auxin transport inhibitors on honey mesquite. *Bot. Gaz.*, **136**, 274–280.

463. MORRÉ, D. J., JONES, D. D. and MOLLENHAUER, H. H. (1967). Golgi apparatus mediated polysaccharide secretion by outer root cap cells of *Zea mays*. *Planta*, **74**, 286–301.

464. MORTLOCK, C. (1952). The structure and development of the hydathodes of *Ranunculus fluitans* Lam. *New Phytol.*, **51**, 129–138.

465. MÜHLETHALER, K. (1961). Plant cell walls. In *The Cell*, Vol. II, BRACHET, J. and MIRSKY, A. E. (eds.), 85–134. Academic Press, New York and London.

466. MÜHLETHALER, K. (1965). Growth theories and the development of the cell wall. In *Cellular Ultrastructure of Woody Plants*, CÔTÉ, W. A., Jr. (ed.), 51–60. Syracuse University Press.

467. MÜHLETHALER, K. (1967). Ultrastructure and formation of plant cell walls. *A. Rev. Pl. Physiol.*, **18**, 1–24.

468. MURMANIS, L. and EVERT, R. F. (1966). Some aspects of sieve cell ultrastructure in *Pinus strobus*. *Am. J. Bot.*, **53**, 1065–1078.

469. NAGATA, T. and TAKEBE, I. (1970). Cell wall regeneration and cell division in isolated tobacco mesophyll protoplasts. *Planta*, **92**, 301–308.

470. NEWCOMB, E. H. and BONNETT, H. T., Jr. (1965). Cytoplasmic microtubule and wall microfibril orientation in root hairs of radish. *J. Cell Biol.*, **27**, 575–589.

471. NIELSEN, N. K. (1968). An investigation of the regenerative power of periderm in potato tubers after wounding. *Acta Agric. scand.*, **18**, 113–120.

472. NOEL, A. R. A. (1974). Aspects of cell wall structure and the development of the velamen in *Ansellia gigantea* Reichb. f. *Ann. Bot.*, **38**, 495–504.

473. NORTHCOTE, D. H. (1972). Chemistry of the plant cell wall. *A. Rev. Pl. Physiol.*, **23**, 113–132.

474. NORTHCOTE, D. H. and PICKETT-HEAPS, J. D. (1966). A function of the Golgi apparatus in polysaccharide synthesis and transport in the root-cap cells of wheat. *Biochem. J.*, **98**, 159–167.

475. NOUGARÈDE, A. (1963). Premières observations sur l'infrastructure et sur l'évolution des cellules des jeunes ébauches foliaires embryonnaires du *Tropaeolum majus* L.: cytologie de la deshydratation de maturation. *C. r. hebd. Séanc. Acad. Sci., Paris*, **257**, 1335–1338.

476. O'BRIEN, T. P. (1972). The cytology of cell-wall formation in some eukaryotic cells. *Bot. Rev.*, **38**, 87–118.

477. O'BRIEN, T. P. (1974). Primary vascular tissues. In *Dynamic Aspects of Plant Ultrastructure*, ROBARDS, A. W. (ed.), 414–440. McGraw-Hill, London.

478. O'BRIEN, T. P. and CARR, D. J. (1970). A suberized layer in the cell walls of the bundle sheath of grasses. *Aust. J. biol. Sci.*, **23**, 275–287.

479. O'BRIEN, T. P., ZEE, S. and SWIFT, J. G. (1970). The occurrence of transfer cells in the vascular tissues of the coleoptilar node of wheat. *Aust. J. biol. Sci.*, **23**, 709–712.

480. ÖPIK, H. (1974). Mitochondria. In *Dynamic Aspects of Plant Ultrastructure*, ROBARDS, A. W. (ed.), 52–83. McGraw-Hill, London.

481. PALEG, L. G. (1965). Physiological effects of gibberellins. *A. Rev. Pl. Physiol.*, **16**, 291–322.

482. PALEVITZ, B. A. and HEPLER, P. K. (1974a). The control of the plane of division during stomatal differentiation in *Allium*. I. Spindle reorientation. *Chromosoma*, **46**, 297–326.

483. PALEVITZ, B. A. and HEPLER, P. K. (1974b). The control of the plane of division during stomatal differentiation in *Allium*. II. Drug studies. *Chromosoma*, **46**, 327–341.

484. PALEVITZ, B. A. and HEPLER, P. K. (1976). Cellulose microfibril orientation and cell shaping in developing guard cells of *Allium*: the role of microtubules and ion accumulation. *Planta*, **132**, 71–93.

485. PALIWAL, G. S. (1969). Stomatal ontogeny and phylogeny. I. Monocotyledons. *Acta bot. neerl.*, **18**, 654–668.

486. PALLAS, J. E., Jr. and MOLLENHAUER, H. H. (1972). Electron microscopic evidence for plasmodesmata in dicotyledonous guard cells. *Science, N.Y.*, **175**, 1275–1276.

487. PALMER, P. G. (1976). Grass cuticles: a new paleoecological tool for East African lake sediments. *Can. J. Bot.*, **54**, 1725–1734.

488. PANT, D. D. (1965). On the ontogeny of stomata and other homologous structures. *Plant Sci. Ser. (Allahabad)*, **1**, 1–24.

489. PANT, D. D. and KIDWAI, P. F. (1967). Development of stomata in some Cruciferae. *Ann. Bot., N. S.*, **31**, 513–521.

490. PANT, D. D. and KIDWAI, P. F. (1968). Structure and ontogeny of stomata in some Caryophyllaceae. *J. Linn. Soc. (Bot.)*, **60**, 309–314.

491. PANT, D. D. and MEHRA, B. (1964). Ontogeny of stomata in some Ranunculaceae. *Flora, Jena*, **155**, 179–188.

492. PAOLILLO, D. J., Jr. (1970). The three-dimensional arrangement of intergranal lamellae in chloroplasts. *J. Cell Sci.*, **6**, 243–255.

493. PARTANEN, C. R. (1965). On the chromosomal basis for cellular differentiation. *Am. J. Bot.*, **52**, 204–209.

494. PARTHASARATHY, M. V. (1968). Observations on metaphloem in the vegetative parts of palms. *Am. J. Bot.*, **55**, 1140–1168.

495. PARTHASARATHY, M. V. (1974a). Ultrastructure of phloem in palms. I. Immature sieve elements and parenchymatic elements. *Protoplasma*, **79**, 59–91.

496. PARTHASARATHY, M. V. (1974b). Ultrastructure of phloem in palms. II. Structural changes, and fate of the organelles in differentiating sieve elements. *Protoplasma*, **79**, 93–125.

497. PARTHASARATHY, M. V. (1974c). Ultrastructure of phloem in palms. III. Mature phloem. *Protoplasma*, **79**, 265–315.

498. PARTHASARATHY, M. V. (1975). Sieve-element structure. In *Transport in Plants. I. Phloem Transport*. ZIMMERMANN, M. H. and MILBURN, J. A. (eds.), 3–38. Springer-Verlag, Berlin.

499. PARTHASARATHY, M. V. and MÜHLETHALER, K. (1969). Ultrastructure of protein tubules in differentiating sieve elements. *Cytobiologie*, **1**, 17–36.

500. PATE, J. S. (1975). Exchange of solutes between phloem and xylem and circulation in the whole plant. *Encyclopedia of Plant Physiology, new series*, **1**, 451–473. (*Transport in Plants I. Phloem Transport*, ZIMMERMANN, M. H. and KILBURN, J. A. (eds.), Springer-Verlag, Berlin).

501. PATE, J. S. and GUNNING, B. E. S. (1969). Vascular transfer cells in angiosperm leaves. A taxonomic and morphological survey. *Protoplasma*, **68**, 135–156.

502. PATE, J. S. and GUNNING, B. E. S. (1972). Transfer cells. *A. Rev. Pl. Physiol.*, **23**, 173–196.

503. PATE, J. S., GUNNING, B. E. S., and BRIARTY, L. G. (1969). Ultrastructure and functioning of the transport system of the leguminous root nodule. *Planta*, **85**, 11–34.

504. PATE, J. S., GUNNING, B. E. S., and MILLIKEN, F. F. (1970). Function of transfer cells in the nodal regions of stems, particularly in relation to the nutrition of young seedlings. *Protoplasma*, **71**, 313–334.

505. PAZOUREK, J. (1970). The effect of light intensity on stomatal frequency in leaves of *Iris hollandica* hort., var. Wedgwood. *Biol. Plant.*, **12**, 208–215.

506. PEDERSEN, M. W., LE FEVRE, C. W. and WIEBE, H. H. (1958). Absorption of C^{14}-labeled sucrose by alfalfa nectaries. *Science, N.Y.*, **127**, 758–759.

507. PENNY, M. G. and BOWLING, D. J. F. (1974). A study of potassium gradients in the epidermis of intact leaves of *Commelina communis* L. in relation to stomatal opening. *Planta*, **119**, 17–25.

508. PETERSON, R. L. (1971). Induction of a 'periderm-like' tissue in excised roots of the fern *Ophioglossum petiolatum* Hook. *Ann. Bot.*, **35**, 165–167.

509. PETERSON, R. L., FIRMINGER, M. S. and DOBRINDT, L. A. (1975). Nature of the guard cell wall in leaf stomata of three *Ophioglossum* species. *Can. J. Bot.*, **53**, 1698–1711.

510. PETERSON, R. L., PETERSON, C. A. and ROBARDS, A. W. (1976). Unpublished observations on onion epidermal cells.

511. PETERSON, R. L. and YEUNG, E. C. (1975). Ontogeny of phloem transfer cells in *Hieracium floribundum*. *Can. J. Bot.*, **53**, 2745–2758.

512. PHILIPSON, W. R., WARD, J. M. and BUTTERFIELD, B. G. (1971). *The Vascular Cambium: its Development and Activity*. Chapman and Hall, London.

513. PHILLIPS, R. and TORREY, J. G. (1973). DNA synthesis, cell division and specific cytodifferentiation in cultured pea root cortical explants. *Devl Biol.*, **31**, 336–347.

514. PHILLIPS, R. and TORREY, J. G. (1974). DNA levels in differentiating tracheary elements. *Devl Biol.* **39**, 322–325.

515. PICKETT-HEAPS, J. D. (1969a). Preprophase microtubules and stomatal differentiation; some effects of centrifugation on symmetrical and asymmetrical cell division. *J. Ultrastruct. Res.*, **27**, 24–44.

516. PICKETT-HEAPS, J. D. (1969b). Preprophase microtubules and stomatal differentiation in *Commelina cyanea*. *Aust. J. biol. Sci.*, **22**, 375–391.

517. PICKETT-HEAPS, J. D. (1974). Plant microtubules. In *Dynamic Aspects of Plant Ultrastructure*, ROBARDS, A. W. (ed.), 219–255. McGraw-Hill, London.

518. PICKETT-HEAPS, J. D. and NORTHCOTE, D. H. (1966). Cell division in the formation of the stomatal complex of the young leaves of wheat. *J. Cell Sci.*, **1**, 121–128.

519. PIZZOLATO, T. D. and HEIMSCH, C. (1975a). Ontogeny of the protophloem fibers and secondary xylem fibers within the stem of *Coleus*. I. A light microscope study. *Can. J. Bot.*, **53**, 1658–1671.

520. PIZZOLATO, T. D. and HEIMSCH, C. (1975b). Ontogeny of the protophloem fibers and secondary xylem fibers within the stem of *Coleus*. II. An electron microscope study. *Can. J. Bot.*, **53**, 1672–1697.

521. POPHAM, R. A. (1958). Some causes underlying cellular differentiation. *Ohio J. Sci.*, **58**, 347–353.

522. PORTER, K. R. (1961). The ground substance; observations from electron microscopy. In *The Cell*, Vol. II, BRACHET, J. and MIRSKY, A. E. (eds.), 621–675. Academic Press, New York and London.

523. POSTLETHWAIT, S. N. and NELSON, O. E., Jr. (1957). A chronically wilted mutant of maize. *Am. J. Bot.*, **44**, 628–633.

524. POWER, J. B. and COCKING, E. C. (1971). Fusion of plant protoplasts. *Sci. Prog., Oxf.*, **59**, 181–198.

525. PRESTON, R. D. (1974a). Plant cell walls. In *Dynamic Aspects of Plant Ultrastructure*, ROBARDS, A. W. (ed.), 256–309. McGraw-Hill, London.

526. PRESTON, R. D. (1974b). *The Physical Biology of Plant Cell Walls*. Chapman & Hall, London.

527. PRIESTLEY, J. H. and WOFFENDEN, L. M. (1922). Physiological studies in plant anatomy. V. Causal factors in cork formation. *New Phytol.*, **21**, 252–268.

528. PRIESTLEY, J. H. and WOFFENDEN, L. M. (1923). The healing of wounds in potato tubers and their propagation by cut sets. *Ann. appl. Biol.*, **10**, 96–115.

529. PUISEUX-DAO, S. (1970). Acetabularia *and Cell Biology*. Logos Press Ltd., London.

530. RACHMILEVITZ, T. and FAHN, A. (1973). Ultrastructure of nectaries of *Vinca rosea* L., *Vinca major* L. and *Citrus sinensis* Osback cv. *Valencia* and its relation to the mechanism of nectar secretion. *Ann. Bot.*, **37**, 1–9.

531. RACHMILEVITZ, T. and FAHN, A. (1975). The floral nectary of *Tropaeolum majus* L. The nature of the secretory cells and the manner of nectar secretion. *Ann. Bot.*, **39**, 721–728.

532. RAGETLI, H. W. J., WEINTRAUB, M. and LO, E. (1972). Characteristics of *Drosera* tentacles. I. Anatomical and cytological detail. *Can. J. Bot.*, **50**, 159–168.

533. RAINBOW, A. and WHITE, D. J. B. (1972). Preliminary observations on the ultrastructure of maturing cork-cells from tubers of *Solanum tuberosum* L. *New Phytol.*, **71**, 899–902.

534. RAMSEY, J. C. and BERLIN, J. D. (1976). Ultrastructure of early stages of cotton fiber differentiation. *Bot. Gaz.*, **137**, 11–19.

535. RAO, A. N. (1969). Effect of injury on the foliar sclereid development in *Fagraea fragrans*. *Experientia*, **25**, 884–885.

536. RAO, A. N. and SINGARAYAR, M. (1968). Controlled differentiation of foliar sclereids in *Fagraea fragrans*. *Experientia*, **24**, 298–299.

537. RAO, A. N. and VAZ, S. J. (1969). Morphogenesis of foliar sclereids. I. Ontogeny and distribution of foliar sclereids in *Fagraea fragrans*. *Phytomorphology*, **19**, 159–169.

538. RAO, A. N. and VAZ, S. J. (1970). Morphogenesis of foliar sclereids. II. Effects of experimental wounds on leaf sclereid development and distribution in *Fagraea fragrans*. *J. Singapore natn. Acad. Sci.*, **1**, 1–7.

539. RAPPAPORT, L. and SACHS, M. (1967). Wound-induced gibberellins. *Nature, Lond.*, **214**, 1149–1150.

540. REAMS, W. M., Jr. (1953). The occurrence and ontogeny of hydathodes in *Hygrophila polysperma* T. Anders. *New Phytol.*, **52**, 8–13.

541. RIER, J. P. and BESLOW, D. T. (1967). Sucrose concentration and the differentiation of xylem in callus. *Bot. Gaz.*, **128**, 73–77.

542. ROBARDS, A. W. (1969). The effect of gravity on the formation of wood. *Sci. Prog., Oxf.*, **57**, 513–532.

543. ROBARDS, A. W. (1975). Plasmodesmata. *A. Rev. Pl. Physiol.*, **26**, 13–29.

544. ROBARDS, A. W. (1976). Plasmodesmata in higher plants. In *Intercellular Communication in Plants: Studies on Plasmodesmata*, GUNNING, B. E. S. and ROBARDS, A. W. (eds.), 15–57. Springer-Verlag, Berlin.

545. ROBERTS, K. and NORTHCOTE, D. H. (1972). Hydroxyproline: observations on its chemical and autoradiographical localization in plant cell wall protein. *Planta*, 107, 43–51.

546. ROBERTS, L. W. (1969). The initiation of xylem differentiation. *Bot. Rev.*, **35**, 201–250.

547. ROBERTS, L. W. (1976). *Cytodifferentiation in Plants. Xylogenesis as a Model System.* Cambridge University Press, London.

548. ROBERTS, L. W. and FOSKET, D. E. (1962). Further experiments on wound-vessel formation in stem wounds of *Coleus*. *Bot. Gaz.*, **123**, 247–254.

549. ROBINSON, K. R. and JAFFE, L. F. (1975). Polarizing fucoid eggs drive a calcium current through themselves. *Science, N.Y.*, **187**, 70–72.

550. ROELOFSEN, P. A. (1959). *The Plant Cell-Wall. Handb. Pflanzenanat.* Band III, Teil 4. Abteilung *Cytologie*. Gebrüder Borntraeger, Berlin.

551. ROELOFSEN, P. A. (1965). Ultrastructure of the wall in growing cells and its relation to the direction of the growth. *Adv. Bot. Res.*, **2**, 69–149.

552. ROSENE, H. F. (1954). A comparative study of the rates of water influx into the hairless epidermal surface and the root hairs of onion roots. *Physiologia Pl.*, **7**, 676–686.

553. ROST, T. L. and LERSTEN, N. R. (1970). Transfer aleurone cells in *Setaria lutescens* (Gramineae). *Protoplasma*, **71**, 403–408.

554. ROTHWELL, N. V. (1964). Nucleolar size differences in the grass root epidermis. *Am. J. Bot.*, **51**, 172–179.

555. ROTHWELL, N. V. (1966). Evidence for diverse cell types in the apical region of the root epidermis of *Panicum virgatum*. *Am. J. Bot.*, **53**, 7–11.

556. ROYLE, D. J. and THOMAS, G. G. (1973). Factors affecting zoospore response towards stomata in hop downy mildew (*Pseudoperonospora humuli*) including some comparisons with grapevine downy mildew (*Plasmopara viticola*). *Physiol. Plant Path.*, **3**, 405–417.

557. SABNIS, D. D., HIRSHBERG, G. and JACOBS, W. P. (1969). Radioautographic analysis of the distribution of label from ^{3}H-indoleacetic acid supplied to isolated *Coleus* internodes. *Pl. Physiol., Lancaster*, **44**, 27–36.

558. SACHS, T. (1968a). The role of the root in the induction of xylem differentiation in peas. *Ann. Bot.*, **32**, 391–399.

559. SACHS, T. (1968b). On the determination of the pattern of vascular tissue in peas. *Ann. Bot.*, **32**, 781–790.

560. SACHS, T. (1969). Polarity and the induction of organized vascular tissues. *Ann. Bot.*, **33**, 263–275.

561. SACHS, T. (1970). A control of bud growth by vascular tissue differentiation. *Israel J. Bot.*, **19**, 484–498.

562. SACHS, T. (1972). The induction of fibre differentiation in peas. *Ann. Bot.*, **36**, 189–197.

563. SACHS, T. (1974a). The induction of vessel differentiation by auxin. *Proc. 8th Internat. Conf. Plant Growth Substs*, 900–906.

564. SACHS, T. (1974b). The developmental origin of stomata pattern in *Crinum*. *Bot. Gaz.*, **135**, 314–318.
565. SACHS, T. (1975). The control of the differentiation of vascular networks. *Ann. Bot.*, **39**, 177–204.
566. SAKAI, W. A. (1974). Scanning electron microscopy and energy dispersive x-ray analysis of chalk secreting leaf glands of *Plumbago capensis*. *Am. J. Bot.*, **61**, 94–99.
567. SAKAI, W. S. and HANSON, M. (1974). Mature raphid and raphid idioblast structure in plants of the edible aroid genera *Colocasia*, *Alocasia*, and *Xanthosoma*. *Ann. Bot.*, **38**, 739–748.
568. SAMPSON, J. (1961). A method of replicating dry or moist surfaces for examination by light microscopy. *Nature, Lond.*, **191**, 932–933.
569. SASSEN, M. M. A. (1965). Breakdown of the plant cell wall during the cell-fusion process. *Acta bot. neerl.*, **14**, 165–196.
570. SAX, H. J. (1938). The relation between stomata counts and chromosome number. *J. Arnold Arbor.*, **19**, 437–441.
571. SCHIEFERSTEIN, R. H. and LOOMIS, W. E. (1956). Wax deposits on leaf surfaces. *Pl. Physiol., Lancaster*, **31**, 240–247.
572. SCHMID, R. (1965). The fine structure of pits in hardwoods. In *Cellular Ultrastructure of Woody Plants*, CÔTÉ, W. A., Jr. (ed.), 291–304. Syracuse University Press.
573. SCHNEPF, E. (1960) Zur Feinstruktur der Drüsen von *Drosophyllum lusitanicum*. *Planta*, **54**, 641–674.
574. SCHNEPF, E. (1963a). Zur Cytologie und Physiologie pflanzlicher Drüsen. 1. Über den Fangschleim der Insektivoren. *Flora, Jena*, **153**, 1–22.
575. SCHNEPF, E. (1963b). Zur Cytologie und Physiologie pflanzlicher Drüsen. 2. Über die Wirkung von Sauerstoffentzug und von Atmungsinhibitoren auf die Sekretion des Fangschleimes von *Drosophyllum* und auf die Feinstruktur der Drüsenzellen. *Flora, Jena*, **153**, 23–48.
576. SCHNEPF, E. (1964). Zur Cytologie und Physiologie pflanzlicher Drüsen. IV. Teil: Licht- und elektronen-mikroskopische Untersuchungen an Septalnektarien. *Protoplasma*, **58**, 137–171.
577. SCHNEPF, E. (1965a). Die Morphologie der Sekretion in pflanzlichen Drüsen. *Ber. dt. bot. Ges.*, **78**, 478–483.
578. SCHNEPF, E. (1965b). Physiologie und Morphologie sekretarischer Pflanzenzellen. In *Sekretion und Exkretion* (2 wissenschaftliche Konferenz der Gesellschaft Deutscher Naturforscher und Ärzte, 1964), 72–88 Springer-Verlag, Berlin.
579. SCHNEPF, E. (1969). Sekretion und Exkretion bei Pflanzen. *Protoplasmatologia*, **8**, 1–181.
580. SCHNEPF, E. (1974). Gland cells. In *Dynamic Aspects of Plant Ultrastructure*, ROBARDS, A. E. (ed.), 331–357. McGraw-Hill, London.
581. SCHNEPF, E. and KLASOVÁ, A. (1972). Zur Feinstruktur von Öl- und Flavon-Drüsen. *Ber dt. bot. Ges.*, **85**, 249–258.
582. SCHÖNHERR, J. and BUKOVAC, M. J. (1970). Preferential polar pathways in the cuticle and their relationship to ectodesmata. *Planta*, **92**, 189–201.
583. SCHRÖTER, K. and SIEVERS, A. (1971). Wirkung der Turgorreduktion auf den Golgi-Apparat und die Bildung der Zellwand bei Wurzelhaaren. *Protoplasma*, **72**, 203–211.
584. SCHWAB, D. W., SIMMONS, E. and SCALA, J. (1969). Fine structure changes during function of the digestive gland of Venus's-flytrap. *Am. J. Bot.*, **56**, 88–100.

585. SCOTT, D. H. (1886). On the occurrence of articulated laticiferous vessels in *Hevea*. *J. Linn. Soc. (Bot.)*, **21**, 566–573.

586. SCOTT, F. M., HAMNER, K. S., BAKER, E. and BOWLER, E. (1956). Electron microscope studies of cell wall growth in the onion root. *Am. J. Bot.*, **43**, 313–324.

587. SCULTHORPE, C. D. (1967). *The Biology of Aquatic Vascular Plants.* Edward Arnold, London.

588. SHANKS, R. (1965). Differentiation in leaf epidermis. *Aust. J. Bot.*, **13**, 143–151.

589. SHARMA, G. K. and BUTLER, J. (1975). Environmental pollution: leaf cuticular patterns in *Trifolium pratense* L. *Ann. Bot.*, **39**, 1087–1090.

590. SHELDRAKE, A. R. (1971). Auxin in the cambium and its differentiating derivatives. *J. exp. Bot.*, **22**, 735–740.

591. SHELDRAKE, A. R. (1973a). Auxin transport in secondary tissues. *J. exp. Bot.*, **24**, 87–96.

592. SHELDRAKE, A. R. (1973b). The production of hormones in higher plants. *Biol. Rev.*, **48**, 509–559.

593. SHELDRAKE, A. R. (1973c). Do coleoptile tips produce auxin? *New Phytol.*, **72**, 433–447.

594. SHELDRAKE, A. R. and NORTHCOTE, D. H. (1968). Some constituents of xylem sap and their possible relationship to xylem differentiation. *J. exp. Bot.*, **19**, 681–689.

595. SHIMONY, C. and FAHN, A. (1968). Light- and electron-microscopical studies on the structure of salt glands of *Tamarix aphylla* L. *J. Linn. Soc. (Bot.)*, **60**, 283–288.

596. SHIMONY, C., FAHN, A. and REINHOLD, L. (1973). Ultrastructure and ion gradients in the salt glands of *Avicennia marina* (Forssk.) Vierh. *New Phytol.*, **72**, 27–36.

597. SHININGER, T. (1975). Is DNA synthesis required for the induction of differentiation in quiescent root cortical parenchyma? *Devl Biol.*, **45**, 137–150.

598. SHUEL, R. W. (1964). Nectar secretion in excised flowers. III. The dual effect of indolyl-3-acetic acid. *J. Apicult. Res.*, **3**, 99–111.

599. SIDDIQUI, A. W., JONES, R. L. and SPANNER, D. C. (1974). Translocation in the stolon of *Saxifraga sarmentosa* L. The ultrastructural background. *Ann. Bot.*, **38**, 145–149.

600. SIEBERS, A. M. (1971a). Initiation of radial polarity in the interfascicular cambium of *Ricinus communis* L. *Acta bot. neerl.*, **20**, 211–220.

601. SIEBERS, A. M. (1971b). Differentiation of isolated interfascicular tissue of *Ricinus communis* L. *Acta bot. neerl.*, **20**, 343–355.

602. SIEBERS, A. M. (1972). Vascular bundle differentiation and cambial development in cultured tissue blocks excised from the embryo of *Ricinus communis* L. *Acta bot. neerl.*, **21**, 327–342.

603. SIEBERS, A. M. and LADAGE, C. A. (1973). Factors controlling cambial development in the hypocotyl of *Ricinus communis* L. *Acta bot. neerl.*, **22**, 416–432.

604. SIEGEL, A. (1975). Gene amplification in plants. In *Modification of the Information Content of Plant Cells* (Proc. 2nd John Innes Symp.), MARKHAM, R., DAVIES, D. R., HOPWOOD, D. A., HORNE, R. W. (eds.), 15–26. North-Holland Publishing Co., Amsterdam.

605. SIMON, S. (1908). Experimentelle Untersuchungen über die Entstehung von Gefässverbindungen. *Ber. dt. bot. Ges.*, **26**, 364–396.

606. SINGH, A. P. and SRIVASTAVA, L. M. (1973). The fine structure of pea stomata. *Protoplasma*, **76**, 61–82.
607. SINNOTT, E. W. and BLOCH, R. (1939). Cell polarity and the differentiation of root hairs. *Proc. natn. Acad. Sci. U.S.A.*, **25**, 248–252.
608. SINNOTT, E. W. and BLOCH, R. (1943). Development of the fibrous net in the fruit of various races of *Luffa cylindrica*. *Bot. Gaz.*, **105**, 90–99.
609. SINNOTT, E. W. and BLOCH, R. (1944). Visible expression of cytoplasmic pattern in the differentiation of xylem strands. *Proc. natn. Acad. Sci. U.S.A.*, **30**, 388–392.
610. SINNOTT, E. W. and BLOCH, R. (1945). The cytoplasmic basis of intercellular patterns in vascular differentiation. *Am. J. Bot.*, **32**, 151–156.
611. SINNOTT, E. W. and BLOCH, R. (1946). Comparative differentiation in the air roots of *Monstera deliciosa*. *Am. J. Bot.*, **33**, 587–590.
612. SIRCAR, S. M. and CHAKRAVERTY, R. (1960). The effect of gibberellic acid on jute (*Corchorus capsularis* Linn.). *Sci. Cult.*, **26**, 141–143.
613. SJOLUND, R. D. (1968). *Chloroplast development and cellular differentiation in tissue cultures of* Streptanthus tortuosus *Kell.* (*Cruciferae*). Ph.D. thesis, University of California, Davis.
614. SMITH, C. EARLE, Jr. (1965). Plant fibers and civilization—cotton, a case in point. *Econ. Bot.*, **19**, 71–82.
615. SNOW, R. (1935). Activation of cambial growth by pure hormones. *New Phytol.*, **34**, 347–360.
616. SOLEREDER, H. (1908). *Systematic Anatomy of the Dicotyledons*. Clarendon Press, Oxford.
617. SONI, S. L., KAUFMAN, P. B. and BIGELOW, W. C. (1970). Electron microprobe analysis of the distribution of silicon in leaf epidermal cells of the oat plant. *Phytomorphology*, **20**, 350–363.
618. SONI, S. L., KAUFMAN, P. B. and BIGELOW, W. C. (1972). Electron microprobe analysis of silicon and other elements in developing silica cells of the leaf and internode of *Cyperus alternifolius*. *Ann. Bot.*, **36**, 611–619.
619. SONI, S. L. and PARRY, D. W. (1973). Electron probe microanalysis of silicon deposition in the inflorescence bracts of the rice plant (*Oryza sativa*). *Am. J. Bot.*, **60**, 111–116.
620. SOROKIN, H. P. and THIMANN, K. V. (1964). The histological basis for inhibition of axillary buds in *Pisum sativum* and the effects of auxins and kinetin on xylem development. *Protoplasma*, **59**, 326–350.
621. SOUTHORN, W. A. (1960). Complex particles in *Hevea* latex. *Nature, Lond.*, **188**, 165–166.
622. SPENCER, H. G. (1939). The effect of puncturing individual latex tubes of *Euphorbia wulfenii*. *Ann. Bot., N.S.*, **3**, 227–229.
623. SPURR, S. H. and VAUX, H. J. (1976). Timber: biological and economic potential. *Science, N.Y.*, **191**, 752–756.
624. SRIVASTAVA, L. M. (1966). On the fine structure of the cambium of *Fraxinus americana* L. *J. Cell Biol.*, **31**, 79–93.
625. SRIVASTAVA, L. M. and SINGH, A. P. (1972). Stomatal structure in corn leaves. *J. Ultrastruct. Res.*, **39**, 345–363.
626. STACE, C. A. (1963). *A Guide to Subcellular Botany*. Longmans, Green, London.
627. STACE, C. A. (1965). Cuticular studies as an aid to plant taxonomy. *Bull. Br. Mus. nat. Hist.* (*Bot.*), **4**, 1–78.

628. STAFFORD, H. A. (1948). Studies on the growth and xylary development of *Phleum pratense* seedlings in darkness and in light. *Am. J. Bot.*, **35**, 706–715.
629. STANGE, L. (1965). Plant cell differentiation. *A. Rev. Pl. Physiol.*, **16**, 119–140.
630. STANT, M. Y. (1961). The effect of gibberellic acid on fibre-cell length. *Ann. Bot., N.S.*, **25**, 453–462.
631. STANT, M. Y. (1963). The effect of gibberellic acid on cell width and the cell-wall of some phloem fibres. *Ann. Bot., N.S.*, **27**, 185–196.
632. STANT, M. Y. (1973a). Scanning electron microscopy of silica bodies and other epidermal features in *Gibasis* (*Tradescantia*) leaf. *J. Linn. Soc.* (*Bot.*), **66**, 233–244.
633. STANT, M. Y. (1973b). The role of the scanning electron microscope in plant anatomy. *Kew Bull.*, **28**, 105–115.
634. STEBBINS, G. L. (1965). From gene to character in higher plants. *Am. Scient.*, **53**, 104–126.
635. STEBBINS, G. L. (1974). Evolution of morphogenetic patterns. *Brookhaven Symp. Biol.*, **25**, 227–243.
636. STEBBINS, G. L. and JAIN, S. K. (1960). Developmental studies of cell differentiation in the epidermis of monocotyledons. I. *Allium*, *Rhoeo* and *Commelina*. *Devl Biol.*, **2**, 409–426.
637. STEBBINS, G. L. and KHUSH, G. S. (1961). Variation in the organization of the stomatal complex in the leaf epidermis of monocotyledons and its bearing on their phylogeny. *Am. J. Bot.*, **48**, 51–59.
638. STEBBINS, G. L. and SHAH, S. S. (1960). Developmental studies of cell differentiation in the epidermis of monocotyledons. II. Cytological features of stomatal development in the Gramineae. *Devl Biol.*, **2**, 477–500.
639. STEBBINS, G. L., SHAH, S. S., JAMIN, D. and JURA, P. (1967). Changed orientation of the mitotic spindle of stomatal guard cell divisions in *Hordeum vulgare*. *Am. J. Bot.*, **54**, 71–80.
640. STERLING, C. (1947). Sclereid formation in the shoot of *Pseudotsuga taxifolia*. *Am. J. Bot.*, **34**, 45–52.
641. STERLING, C. (1954). Sclereid development and the texture of Bartlett pears. *Fd Res.*, **19**, 433–443.
642. STEVENS, A. B. P. (1956). The structure and development of the hydathodes of *Caltha palustris* L. *New Phytol.*, **55**, 339–345.
643. STEWARD, F. C. (1963). The control of growth in plant cells. *Scient. Am.*, **209**, 104–113.
644. STEWARD, F. C. (1970). From cultured cell to whole plants: the induction and control of their growth and differentiation. *Proc. R. Soc.*, B, **175**, 1–30.
645. STEWARD, F. C. with MAPES, M. O., KENT, A. E. and HOLSTEN, R. D. (1964). Growth and development of cultured plant cells. *Science, N.Y.*, **143**, 20–27.
646. STEWARD, F. C., MAPES, M. O. and MEARS, K. (1958). Growth and organized development of cultured cells. II. Organization in cultures grown from freely suspended cells. *Am. J. Bot.*, **45**, 704–708.
647. STEWART, J. MCD. (1975). Fiber initiation on the cotton ovule (*Gossypium hirsutum*). *Am. J. Bot.*, **62**, 723–730.
648. STEWART, K. D. and CUTTER, E. G. (1967). Ultrastructure of trichoblasts and root tip cells in *Hydrocharis*. (Abstr.) *Am. J. Bot.*, **54**, 632.

649. STREET, H. E. (1976). Experimental embryogenesis—the totipotency of cultured plant cells, 73–90 in Ref. 271.

650. SUSSEX, I. M. (1974). Do concepts of animal development apply to plant systems? *Brookhaven Symp. Biol.*, **25**, 145–151.

651. SUSSEX, I. M., CLUTTER, M. E. and GOLDSMITH, M. H. M. (1972). Wound recovery by pith cell redifferentiation: structural changes. *Am. J. Bot.*, **59**, 797–804.

652. THAINE, R. (1964). Protoplast structure in sieve tube elements. *New Phytol.*, **63**, 236–243.

653. THAINE, R. (1969). Movement of sugars through plants by cytoplasmic pumping. *Nature, Lond.*, **222**, 873–875.

654. THAINE, R. and DE MARIA, M. E. (1973). Transcellular strands of cytoplasm in sieve tubes of squash. *Nature, Lond.*, **245**, 161–163.

655. THAINE, R., DE MARIA, M. E. and MARJATTA SARISALO, H. I. (1975). Evidence of transcellular strands in transverse cryostat sections of *Cucurbita pepo* sieve tubes. *J. exp. Bot.*, **26**, 91–101.

656. THAIR, B. W. and STEEVES, T. A. (1976). Response of the vascular cambium to reorientation in patch grafts. *Can. J. Bot.*, **54**, 361–373.

657. THOMPSON, N. P. (1967). The time course of sieve tube and xylem cell regeneration and their anatomical orientation in *Coleus* stems. *Am. J. Bot.*, **54**, 588–595.

658. THOMPSON, N. P. and JACOBS, W. P. (1966). Polarity of IAA effect on sieve-tube and xylem regeneration in *Coleus* and tomato stems. *Pl. Physiol., Lancaster*, **41**, 673–682.

659. THOMSON, W. W. (1974). Ultrastructure of mature chloroplasts. In *Dynamic Aspects of Plant Ultrastructure*, ROBARDS, A. W. (ed.), 138–177. McGraw-Hill, London.

660. THOMSON, W. W. and LIU, L. L. (1967). Ultrastructural features of the salt gland of *Tamarix aphylla* L. *Planta*, **73**, 201–220.

661. THORNTON, J. J. and NAKAMURA, G. R. (1972). The identification of marijuana. *J. Forens. Sci. Soc.*, **12**, 461–519.

662. THURESON-KLEIN, A. (1970). Observations on the development and fine structure of the articulated laticifers of *Papaver somniferum*. *Ann. Bot.*, **34**, 751–759.

663. THURSTON, E. L. and LERSTEN, N. R. (1969). The morphology and toxicology of plant stinging hairs. *Bot. Rev.*, **35**, 393–412.

664. THURSTON, E. L. and SEABURY, F. (1975). A scanning electron microscopic study of the utricle trichomes in *Utricularia biflora* Lam. *Bot. Gaz.*, **136**, 87–93.

665. TOLBERT, N. E. (1971). Microbodies—peroxisomes and glyoxysomes. *A. Rev. Pl. Physiol.*, **22**, 45–74.

666. TORREY, J. G. (1971). Cytodifferentiation in plant cell and tissue culture. In *Les Cultures de Tissus de Plantes, Colloques internat. de Centre National de la Recherche Scientifique*, **193**, 177–186.

667. TORREY, J. G. (1975). Tracheary element formation from single isolated cells in culture. *Physiologia Pl.*, **35**, 158–165.

668. TORREY, J. G., FOSKET, D. E. and HEPLER, P. K. (1971). Xylem formation: a paradigm of cytodifferentiation in higher plants. *Am. Sci.*, **59**, 338–352.

669. TRAN THANH VAN, M., CHLYAH, H. and CHLYAH, A. (1974). Regulation of organogenesis in thin layers of epidermal and sub-epidermal cells. In *Tissue Culture and Plant Science*, STREET, H. E. (ed.), 101–139. Academic Press, London.

670. TRAN THANH VAN, M., THI DIEN, N. and CHLYAH, A. (1974). Regulation of organogenesis in small explants of superficial tissue of *Nicotiana tabacum* L. *Planta*, **119**, 149–159.
671. TRIBE, M. and WHITTAKER, P. (1972). *Chloroplasts and Mitochondria. Studies in Biology* **31**. Edward Arnold, London.
672. TROUGHTON, J. and DONALDSON, L. A. (1972). *Probing Plant Structure.* Chapman and Hall, London.
673. TSCHERMAK-WOESS, E. and HASITSCHKA, G. (1953). Über Musterbildung in der Rhizodermis und Exodermis bei einigen Angiospermen und einer Polypodiacee. *Öst. bot. Z.*, **100**, 646–651.
674. TSOUMIS, G. (1965). Light and electron microscopic evidence on the structure of the membrane of bordered pits in the tracheids of conifers. In *Cellular Ultrastructure of Woody Plants*, CÔTÉ, W. A., Jr. (ed.), 305–317. Syracuse University Press.
675. TU, J. C. and HIRUKI, C. (1971). Electron microscopy of cell wall thickening in local lesions of potato virus-M infected red kidney bean. *Phytopathology*, **61**, 862–868.
676. TUCKER, S. C. (1964). The terminal idioblasts in Magnoliaceous leaves. *Am. J. Bot.*, **51**, 1051–1062.
677. TUCKER, S. C. (1974). Dedifferentiated guard cells in magnoliaceous leaves. *Science, N.Y.*, **185**, 445–447.
678. TUCKER, S. C. (1975). Wound regeneration in the lamina of magnoliaceous leaves. *Can. J. Bot.*, **53**, 1352–1364.
679. UNZELMAN, J. M. and HEALEY, P. L. (1972). Development and histochemistry of nuclear crystals in the secretory trichome of *Pharbitis nil. J. Ultrastruct. Res.*, **39**, 301–309.
680. UNZELMAN, J. M. and HEALEY, P. L. (1974). Development, structure, and occurrence of secretory trichomes of *Pharbitis. Protoplasma*, **80**, 285–303.
681. UPHOF, J. C. TH. (1962). *Plant Hairs. Handb. Pflanzenanat.* Band IV, Teil 5. Abteilung: *Histologie.* Gebrüder Borntraeger, Berlin.
682. VAN COTTHEM, W. R. J. (1970). A classification of stomatal types. *Bot. J. Linn. Soc.*, **63**, 235–246.
683. VANDERWOUDE, W. J., LEMBI, C. A. and MORRÉ, D. J. (1972). Auxin (2,4-D) stimulation (*in vivo* and *in vitro*) of polysaccharide synthesis in plasma membrane fragments isolated from onion stems. *Biochem. biophys. Res. Commun.*, **46**, 245–253.
684. VARNER, J. E. and RAM CHANDRA, G. (1964). Hormonal control of enzyme synthesis in barley endosperm. *Proc. natn. Acad. Sci. U.S.A.*, **52**, 100–106.
685. VASIL, I. K. and VASIL, V. (1972). Totipotency and embryogenesis in plant cell and tissue cultures. *In Vitro*, **8**, 117–127.
686. VIGIL, E. L. and RUDDAT, M. (1973). Effect of gibberellic acid and actinomycin D on the formation and distribution of rough endoplasmic reticulum in barley aleurone cells. *Pl. Physiol., Lancaster*, **51**, 549–558.
687. VÖCHTING, H. (1878). *Über Organbildung im Pflanzenreich.* Max Cohen, Bonn.
688. WAISEL, Y., LIPHSCHITZ, N. and ARZEE, T. (1967). Phellogen activity in *Robinia pseudacacia* L. *New Phytol.*, **66**, 331–335.
689. WAISEL, Y., LIPHSCHITZ, N. and FAHN, A. (1970). Cambial activity in *Zygophyllum dumosum* Boiss. *Ann. Bot.*, **34**, 409–414.
690. WALKER, W. S. (1957). The effect of mechanical stimulation on the collenchyma of *Apium graveolens* L. *Proc. Iowa Acad. Sci.*, **64**, 177–186.

691. WALKER, W. S. (1960). The effects of mechanical stimulation and etiolation on the collenchyma of *Datura stramonium*. *Am. J. Bot.*, **47**, 717–724.
692. WANGERMANN, E. (1967). The effect of the leaf on differentiation of primary xylem in the internode of *Coleus blumei* Benth. *New Phytol.*, **66**, 747–754.
693. WARDLAW, C. W. (1972). Plant anatomy in retrospect and prospect. *Adv. Pl. Morph.*, 1972, 1–2.
694. WARDLAW, I. F. (1974). Phloem transport: physical chemical or impossible. *A. Rev. Pl. Physiol.* **25**, 515–539.
695. WARDROP, A. B. and HARADA, H. (1965). The formation and structure of the cell wall in fibres and tracheids. *J. exp. Bot.*, **16**, 356–371.
696. WARDROP, A. B., INGLE, H. D. and DAVIES, G. W. (1963). Nature of vestured pits in angiosperms. *Nature, Lond.*, **197**, 202–203.
697. WAREING, P. F. (1958). Interaction between indoleacetic acid and gibberellic acid in cambial activity. *Nature, Lond.*, **181**, 1744–1745.
698. WAREING, P. F. (1976). Origin of cell heterogeneity in plants, 29–42 in Ref. 271.
699. WAREING, P. F. and GRAHAM, C. F. (1976). Nucleus and cytoplasm, 5–13 in Ref. 271.
700. WAREING, P. F., HANNEY, C. E. A. and DIGBY, J. (1964). The role of endogenous hormones in cambial activity and xylem differentiation. In *The Formation of Wood in Forest Trees*, ZIMMERMANN, M. H. (ed.), 323–344. Academic Press, New York.
701. WEATHERLEY, P. E. and JOHNSON, R. P. C. (1968). The form and function of the sieve tube: a problem in reconciliation. *Int. Rev. Cytol.*, **24**, 149–192.
702. WEIER, T. E. (1961). The ultramicro structure of starch-free chloroplasts of fully expanded leaves of *Nicotiana rustica*. *Am. J. Bot.*, **48**, 615–630.
703. WENHAM, M. W. and CUSICK, F. (1975). The growth of secondary wood fibres. *New Phytol.*, **74**, 247–261.
704. WERGIN, W. P., ELMORE, C. D., HANNY, B. W. and INGBER, B. F. (1975). Ultrastructure of the subglandular cells from the foliar nectaries of cotton in relation to the distribution of plasmodesmata and the symplastic transport of nectar. *Am. J. Bot.*, **62**, 842–849.
705. WERKER, E. and FAHN, A. (1969). Resin ducts of *Pinus halepensis* Mill.—their structure, development and pattern of arrangement. *J. Linn. Soc.* (*Bot.*), **62**, 379–411.
706. WETMORE, R. H., DEMAGGIO, A. E. and RIER, J. P. (1964). Contemporary outlook on the differentiation of vascular tissues. *Phytomorphology*, **14**, 203–217.
707. WETMORE, R. H. and RIER, J. P. (1963). Experimental induction of vascular tissues in callus of angiosperms. *Am. J. Bot.*, **50**, 418–430.
708. WETMORE, R. H. and SOROKIN, S. (1955). On the differentiation of xylem. *J. Arnold Arbor.*, **36**, 305–317.
709. WHALEY, W. G. (1948). Rubber—the primary sources for American production. *Econ. Bot.*, **2**, 198–216.
710. WHALEY, W. G. (1975). *The Golgi Apparatus*. Cell Biol. Monogr. **2**. Springer-Verlag, Wien–New York.
711. WHALEY, W. G. and MOLLENHAUER, H. H. (1963). The Golgi apparatus and cell plate formation—a postulate. *J. Cell Biol.*, **17**, 216–221.
712. WHALEY, W. G., MOLLENHAUER, H. H. and LEECH, J. H. (1960). The ultrastructure of the meristematic cell. *Am. J. Bot.*, **47**, 401–448.

713. WHATLEY, J. M. (1972). The ultrastructure of guard cells of *Phaseolus vulgaris*. *New Phytol.*, **71**, 175–179.
714. WHITAKER, D. M. (1937). Determination of polarity by centrifuging eggs of *Fucus furcatus*. *Biol. Bull. mar. biol. Lab., Woods Hole*, **73**, 249–260.
715. WHITAKER, D. M. (1940). Physical factors of growth. *Growth* (Suppl.), (2nd Symp. Dev. Growth), 75–88.
716. WHITE, R. A. (1963). Tracheary elements of the ferns. II. Morphology of tracheary elements; conclusions. *Am. J. Bot.*, **50**, 514–522.
717. WHITMORE, T. C. (1962). Studies in systematic bark morphology. II. General features of bark construction in Dipterocarpaceae. *New Phytol.*, **61**, 208–220.
718. WHITMORE, T. C. (1963). Studies in systematic bark morphology. IV. The bark of beech, oak and sweet chestnut. *New Phytol.*, **62**, 161–169.
719. WIGGINTON, M. J. (1973). Diffusion of oxygen through lenticels in potato tuber. *Potato Res.*, **16**, 85–87.
720. WIGGINTON, M. J. (1974). Effects of temperature, oxygen tension and relative humidity on the wound-healing process in the potato tuber. *Potato Res.*, **17**, 200–214.
721. WILLIAMS, J. A. (1973). A considerably improved method for preparing plastic epidermal imprints. *Bot. Gaz.*, **134**, 87–91.
722. WILLIAMS, L. (1964). Laticiferous plants of economic importance. V. Resources of guttapercha—*Palaquium* species (Sapotaceae). *Econ. Bot.*, **18**, 5–26.
723. WILLIAMS, W. T. and BARBER, D. A. (1961). The functional significance of aerenchyma in plants. *Symp. Soc. exp. Biol.*, **15**, 132–144.
724. WILSON, K. (1936). The production of root-hairs in relation to the development of the piliferous layer. *Ann. Bot.*, **50**, 121–154.
725. WILSON, K. J., NESSLER, C. L. and MAHLBERG, P. G. (1976). Pectinase in *Asclepias* latex and its possible role in laticifer growth and development. *Am. J. Bot.*, **63**, 1140–1144.
726. WODZICKI, T. J. and WODZICKI, A. B. (1973). Auxin stimulation of cambial activity in *Pinus silvestris*. II. Dependence upon basipetal transport. *Physiologia Pl.*, **29**, 288–292.
727. WOLPERT, L., CLARKE, M. R. B., and HORNBRUCH, A. (1972). Positional signalling along *Hydra*. *Nature New Biol.*, **239**, 101–105.
728. WOODING, F. B. P. (1971). *Phloem. Oxford Biol. Readers*, **15**. Oxford University Press.
729. WOODING, F. B. P. and NORTHCOTE, D. H. (1964). The development of the secondary wall of the xylem in *Acer pseudoplatanus*. *J. Cell Biol.*, **23**, 327–337.
730. WOODING, F. B. P. and NORTHCOTE, D. H. (1965a). The fine structure and development of the companion cell of the phloem of *Acer pseudoplatanus*. *J. Cell Biol.*, **24**, 117–128.
731. WOODING, F. B. P. and NORTHCOTE, D. H. (1965b). Association of the endoplasmic reticulum and the plastids in *Acer* and *Pinus*. *Am. J. Bot.*, **52**, 526–531.
732. WOODING, F. B. P. and NORTHCOTE, D. H. (1965c). The fine structure of the mature resin canal cells of *Pinus pinea*. *J. Ultrastruct. Res.*, **13**, 233–244.
733. WORLEY, J. F. (1973). Evidence in support of 'open' sieve tube pores. *Protoplasma*, **76**, 129–132.
734. YEMM, E. W. and WILLIS, A. J. (1954). Chlorophyll and photosynthesis in stomatal guard cells. *Nature, Lond.*, **173**, 726.

735. YEUNG, E. C. and PETERSON, R. L. (1972). Xylem transfer cells in the rosette plant *Hieracium floribundum*. *Planta*, **107**, 183–188.
736. YEUNG, E. C. and PETERSON, R. L. (1974). Ontogeny of xylem transfer cells in *Hieracium floribundum*. *Protoplasma*, **80**, 155–174.
737. YEUNG, E. C. and PETERSON, R. L. (1975). Fine structure during ontogeny of xylem transfer cells in the rhizome of *Hieracium floribundum*. *Can. J. Bot.*, **53**, 432–438.
738. ZAHUR, M. S. (1959). Comparative study of secondary phloem of 423 species of woody dicotyledons belonging to 85 families. *Mem. Cornell Univ. agric. Exp. Stn*, **358**, 1–160.
739. ZAJAÇZKOWSKI, S. (1973). Auxin stimulation of cambial activity in *Pinus silvestris*. I. The differential cambial response. *Physiologia Pl.*, **29**, 281–287.
740. ZAMSKI, E. and FAHN, A. (1972). Observations on resin secretion from isolated portions of resin ducts of *Pinus halepensis* Mill. *Israel J. Bot.*, **21**, 35–38.
741. ZEE, S.-Y. (1974). Distribution of vascular transfer cells in the culm nodes of bamboo. *Can. J. Bot.*, **52**, 345–347.
742. ZEIGER, E. (1971). Organelle distribution and cell differentiation in the formation of stomatal complexes in barley. *Can. J. Bot.*, **49**, 1623–1625.
743. ZEIGER, E. and HEPLER, P. K. (1976). Production of guard cell protoplasts from onion and tobacco. *Pl. Physiol., Lancaster*, **58**, 492–498.
744. ZELITCH, I. (1973). Plant productivity and the control of photorespiration. *Proc. natn. Acad. Sci. U.S.A.*, **70**, 579–584.
745. ZIEGLER, H. and LÜTTGE, U. (1966). Die Salzdrüsen von *Limonium vulgare*. I. Die Feinstruktur. *Planta*, **70**, 193–206.
746. ZIMMERMANN, M. H. (1960). Transport in the phloem. *A. Rev. Pl. Physiol.*, **11**, 167–190.
747. ZIMMERMANN, M. H. (1963). How sap moves in trees. *Scient. Am.*, **208**, 133–142.
748. ZIMMERMANN, M. H. (1973). The monocotyledons: their evolution and comparative biology. IV. Transport problems in arborescent monocotyledons. *Q. Rev. Biol.*, **48**, 314–321.
749. ZIMMERMANN, M. H. and MILBURN, J. A. (eds.) (1975). *Transport in Plants. I. Phloem Transport*. Springer-Verlag, Berlin.
750. ZUCKER, M. (1963). Experimental morphology of stomata. *Bull. Conn. agric. Exp. Stn*, **664**, 1–17.

Index

Major entries are shown in bold type, those referring to illustrations in italic type.